AF594275

*25th Anniversary Edition*

# OPTICS FOR CLINICIANS

*25th Anniversary Edition*

# OPTICS FOR CLINICIANS

Melvin L. Rubin, M.D., M.S.
Professor and Chairman
Richardson Eminent Scholar
Department of Ophthalmology
University of Florida College of Medicine
Gainesville, Florida

 TRIAD PUBLISHING COMPANY GAINESVILLE, FLORIDA

Copyright © 1993, 1974 by Melvin L. Rubin
First edition © 1971
Library of Congress Catalog Card Number 72-97862
ISBN 0-937404-34-9

All rights reserved. No part of this book
may be reproduced in any manner without
written permission of the publisher.

Printed in the United States of America

Published and distributed by
Triad Publishing Company
Post Office Box 13096
Gainesville, Florida 32604

Dedicated to

the memory of my father

MORRIS RUBIN

and to all students

who like things to "make sense"

## A NOTE FROM THE PUBLISHER

It is no coincidence that the 25th anniversary of *Optics for Clinicians* comes at the same time as that of Triad Publishing Company. *Optics for Clinicians* was our first published book and the one by which all our successes have been measured.

We are proud that in an era when most books tend to stay in print less than a year, *Optics* has remained popular over the decades. We are, of course, pleased with its widespread acceptance and look forward to the next twenty-five years!

# PREFACE, 25TH ANNIVERSARY EDITION

Twenty-five years ago I began an outline of what was to become one of my most satisfying educational ventures: a method for teaching optics to clinicians who were not really very interested in this subject.

My experience teaching eye residents and evaluating the results of the annual OKAP (Ophthalmic Knowledge Assessment Program), which I had created specifically for its educational guidance, convinced me that ophthalmology residents had a conspicuous lack of proficiency in handling even rudimentary optics information. The topic was just not getting across.

Though the young clinicians certainly learned the practical "how to" of refracting and prescribing, the majority had only a weak grasp of what they were doing optically — how images are formed by lenses, how corrective lenses function, and how even the simplest ophthalmic instruments work (for example, how an ophthalmoscope allows the fundus to be visualized).

Fortunately, a profound knowledge of optics and memorized formulas are not necessary for understanding the optical concepts related to clinical practice. But at least some basic information in this area is needed at the fingertips of all clinicians who wish to increase their competence in dealing with clinical refraction patients.

Faced with the challenge of maximizing understanding while minimizing math, I prepared this small volume as a new kind of teaching manual, with just enough fundamentals to allow readers to understand optical concepts as they pertain to refraction, corrective lenses, and ophthalmic instruments. My aim was to promote logical reasoning instead of the blind application of mathematical formulas.

The approach and the informal style of *Optics for Clinicians* have been enthusiastically embraced by residents-in-training as well as by practitioners. Over the years it has been adopted as a basic text by training programs and basic science courses (including the American Academy of Ophthalmology's Basic and Clinical Sciences Course).

Although from time to time I have been asked to update the material, on review I find it as topical and valid today as it was a quarter century ago.

So, while no text can ever be really complete and satisfy all desires and answer all criticisms, *Optics for Clinicians* (at least in my view) comes as close as needed to an elementary, general purpose, practitioner's text of optics that is current, reasonable in length and sufficient in depth to satisfy the needs of the everyday clinical refractionist.

April, 1993 MELVIN L. RUBIN

# INTRODUCTION

Ask yourself what you want to learn by studying this manual. I hope you will reply that you wish to understand practical visual optics — the optics of the eye and seeing, of spectacle correction, of magnification, of instrumentation.

Before you can swim, however, you must get wet, and before you plunge headlong into the clinical aspects, you must first strengthen your skill in handling the fundamentals of geometrical optics. Therefore, the book begins with a section on *Basic Optics.* Learn these early lessons well so that you can think and talk the accepted lingo with ease. Then, after completing this initial section, you should be able to proceed swiftly into the real "meat," *Visual Optics.*

This book is a teaching manual and is designed to be read straight through. You will find that the subject development flows naturally and smoothly from one topic to the next.

After introducing the optical basics in a one-sided dialogue with the reader, I bring in details that are more practical. I use a spiral approach in which I touch base with a subject more than once, each time probing deeper, with continual supplementation and re-emphasis. New points are built on previous ones in logical succession. At appropriate sites in the text, *Clinical Points* are slipped in and explained in conjunction with the corresponding optical principles.

Sample problems are sprinkled throughout. Try to work out each before checking the answer and the reasoning. Getting yourself "into the act" personally will expedite your learning.

So, bear with me now and begin by dunking your big toe boldly into the cold water. I promise, it will warm up quickly.

# CONTENTS

## VISUAL OPTICS

## LIGHT: PHYSICAL ASPECTS



## APPENDICES

# BASIC OPTICS

# VERGENCE AND LENSES

## Vergence

Any object — a colorful vase, a black-and-white photograph, a stick, or a letter E on a Snellen chart — can be considered to be made up of an infinite number of points, each of which contributes something to the overall make-up of that object. To enable us to form an image of that object, we must have some light coming forth from it. The amount doesn't have to be very great; in fact it can be very little, but there has to be some. That light energy can be emitted by it, reflected from it, transmitted through it — but somehow light energy (which may even be invisible to the eye) must be given off. The object is then considered "luminous." From it, a myriad of light rays are thrown off; each is infinitely thin and each projects to a specific *direction.*

Before we can deal with these rays intelligently, we all must agree on certain of their characteristics. These have to do with light *conventions* and are simply our rules for dealing with these "animals." One convention is that the light must travel from left to right. Of course, light travels in *all* directions, but for our analyses here, please just accept this first convention about the light ray.

Light rays will always diverge (spread apart) from *every* point comprising a real object, so let's focus our attention on the rays emitted by *one* point — any point X — on that object.

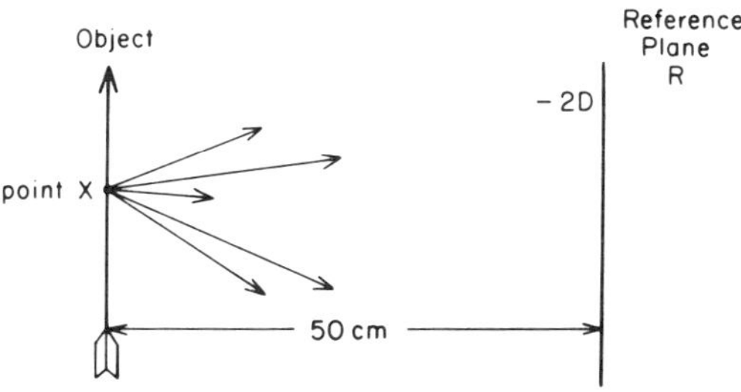

From point X in the above figure five rays out of the many possible ones are drawn and travel to the right. These rays will arrive at a reference plane, say, 50 cm away. When they strike this plane, they are diverging at a certain rate. We define (not explain) that, *at* the reference plane, the *vergence* of those rays from point X is inversely proportional to the distance (in meters) between X and the plane. That is, at reference plane R, the rays from point X have a vergence of $\frac{1}{.5 \text{ meters}} = 2$ Diopters. Again by *our* convention, all rays *diverging* at any reference plane are considered negative (—) so the actual vergence here is written as — 2 Diopters.

*All* rays from *all* real points *di*verge. When you run across rays which are *con*verging, please realize that they must have been created by some other optical system; they cannot occur naturally. These are defined as positive (+) by convention. We will adhere to this sign convention throughout this book. But, please do not consult other texts to confirm this. You may find that they do not agree and you will be confused. Just learn it this way. Again, minus means *di*verging rays and plus *con*verging, whether talking about lenses, mirrors, eyes, or any optical system at all. Simple enough.

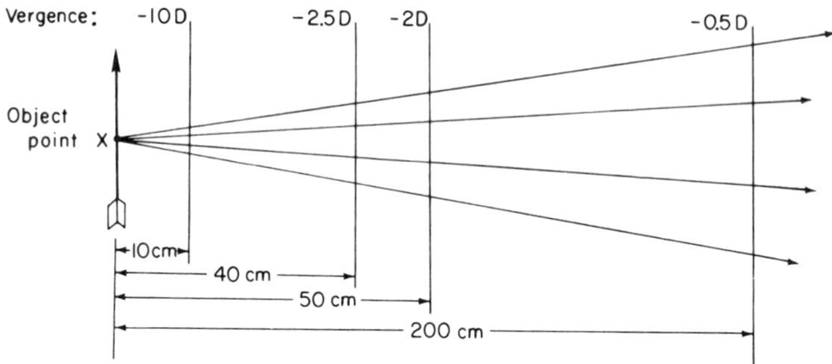


If we move our reference plane closer to point X so that it is only 40 cm away, the vergence at that plane is now — 2.5 Diopters. $\frac{1}{.40 \text{ meters}}$; at 10 cm, it is — 10 D, and at 2 meters, it is — 0.5 D. Do not forget that minus sign.

If we remove the actual plane entirely, we can still talk about the vergence of the rays *at a given distance from X;* this vergence will be the same whether or not there is a real plane there to intercept the rays.

## The Plus Lens

Let us now put a lens — any lens — somewhere to the right of our luminous point X — say, at 66 cm. The vergence of the light from X as measured at the lens plane is — 1.5 D (the same as its vergence at that position if the lens were not there). The lens has certain properties which bestow upon it the ability to *change* the vergence of those light rays falling on it. Its ability to change vergence is also expressed in diopters. A lens is considered plus (+) if it *adds* vergence to the incoming light, and minus (—) if it *subtracts* vergence (or makes the light more *di*vergent.)

If we place a + 3 Diopter lens 50 cm to the right of point X, the lens will add + 3 D of vergence to the incoming rays which already have a vergence of — 2 D at that lens plane. The light rays will leave the lens with a vergence (again, as measured *at the lens plane*) of + 1 Diopter. See the diagram below:

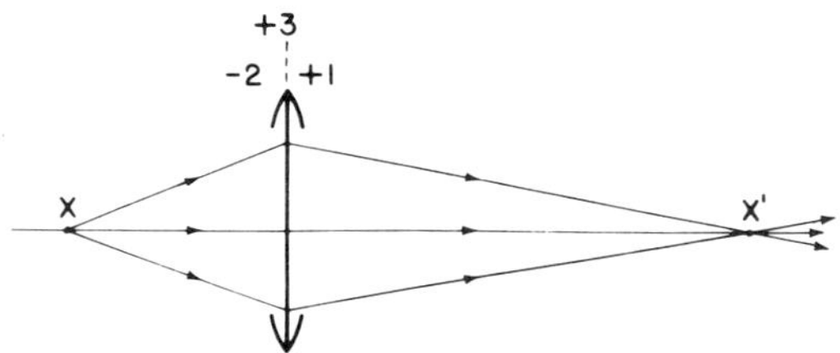


To keep us aware of the fact that our vergence measurements take place at the lens plane, we write a — 2 on our diagram just to the left of the lens, as the incoming rays hit it. This is the vergence of object point X and will be labeled *U* (in diopters). The + 3 Diopters of lens power *P* combines with (is added to) this object vergence, and the resulting *image-point* vergence, $V$, of $+ 1$ Diopter is written just to the right of the lens (again to remind us that the *lens* is the reference plane). What we have just said can be written in math shorthand:

$$U + P = V$$

The object vergence combined with the vergence-change provided by a lens yields the image vergence. This is one of the very few key "formulas" you *must* know; we will continue to rely on it heavily.

Again by convention, *vergences* will be expressed in capital letters: $U$ for the object, $V$ for the image, and small letters $u$ and $v$ corresponding, respectively, to object and image distances:

$$\frac{1}{U} = u,$$

$$\frac{1}{V} = v$$

The previous object-image example is our first encounter (here, in the *image*), with converging or + rays. We see that it took an optical system to create them. The image vergence of "+ 1 D" says that the rays which leave the lens P are *con*verging (the + sign) and do so towards an image point located $\frac{1}{V}$ meters to the right of lens P. Since we know the image vergence in diopters is $+ 1$ D, the actual distance to the image point is $\frac{1}{+1} = + 1$ meter. Thus, the point X′ (the *image* point which corresponds optically to the real object point X) is located 1 meter to the right of lens P. The image point is a *real* one, that is, it can be focused on a screen.

X and X′ are called *conjugate* points — they are related by their being the object and image of one another. Since light rays are completely retraceable, if the real object point were located instead exactly at X′, its image by lens P would always be precisely at X. However, we have agreed (by our convention) to stay away from light rays which move from right to left, but please realize they *can* do so. (A mirror *would* reverse the direction of these rays, but this will be explored later).

Another definition is now in order. The lens *axis:* this line coincides with and represents a ray of light which falls perpendicular to the lens surface *and* whose direction of travel is not disturbed by the lens; the axial ray goes through undeviated.

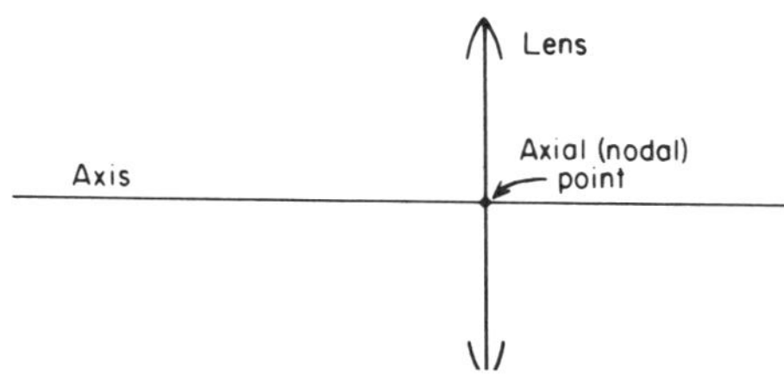


The spot on the lens that this ray strikes is the axial point — a very critical one for all our optical constructions. Actually *any* ray from *any* direction that passes through the axial point will also proceed undeviated; we also call this point the *NODAL POINT.* In our "thin" lens system here, the axial point and nodal point coincide exactly. When we study more complex optical systems, we will investigate such a point further.

Recall, we chose X to represent a *single* point in a much larger object composed of many such points; in the optical example given, we could have chosen *any* X that was equally distant from our lens P. It should, therefore, be clear that all such object points must lie in a plane which is perpendicular to the lens axis and parallel to lens P. These points constitute an *object plane.* (*Other* points on the same object but *not* in that same plane would necessarily yield rays of different vergence at the lens.)

Just as X establishes an object plane, X′ (the image point of X) fixes an *image* plane which is also parallel to lens P. All points in the same object plane as X will be imaged somewhere in the same image plane as X′. The exact, corresponding positions of these points will be determined later.

Referring back to our same + 3 D lens P, let's move object X closer toward it. When X moves from 50 cm to 40 cm from the lens, its vergence at the lens increases from — 2 to — 2.5 Diopters.

$$U + P = V \text{ becomes } -2.5 + 3.0 = +0.5$$

The *image* rays are still converging, but the *distance* from the lens to the new image point X′ has now increased; X′ has moved further to the right, to a position $\frac{1}{.5\text{ D}}$ or 2 meters away.

As X moves even closer — to 33 cm from the lens — something peculiar happens:

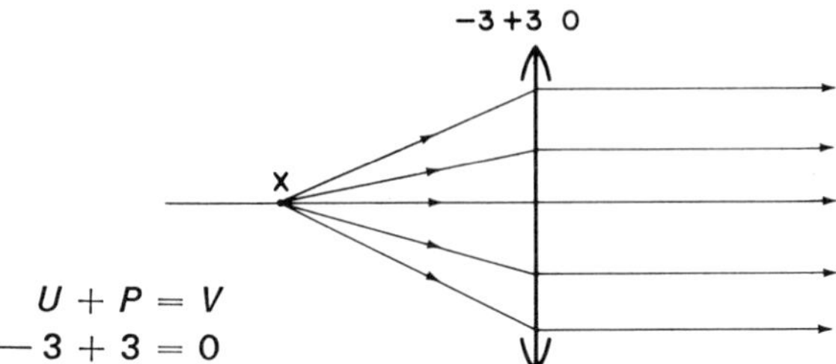


The image point vergence becomes zero; its distance from lens P is $v$.

$$v = \frac{1}{V}$$

Since it is a mathematical commandment that "Thou shalt not divide by zero" and here $V = 0$, now what? We cannot say that there is no image point to correspond to object point X ; but what we *can* say is that as the image vergence *V approaches* zero, $v$ approaches an infinite distance to the right. Thus, image point X′ is located a long distance away — at "right infinity."

Here we have found a special position of object point X relative to lens P — the location of an object point, the image of which is at infinity. That position along the lens axis is called the *PRIMARY FOCAL POINT* (F) of the lens. Light rays which originate there and strike the lens will leave the lens in parallel bundles.

That plane which contains F (parallel to the lens plane) is the *primary focal plane.* Any point Y in this plane emits rays which, of course, diverge — those rays that fall upon lens P will enter the "image space" (*following* the lens' action) as parallel to each other; but, parallel to what direction? Certainly not to the lens axis, since only rays arising specifically at axial point F itself would do that. No; that direction taken by these parallel rays in the image will be indicated by only *one* of the rays leaving object point Y — that particular one that leaves Y aimed directly at the axial (nodal) point of lens P (see the diagram below); that ray then (by our definition of the nodal point) will continue on undeviated and will establish the direction of all the parallel rays in the "image space" which arise at point Y in the primary focal plane.

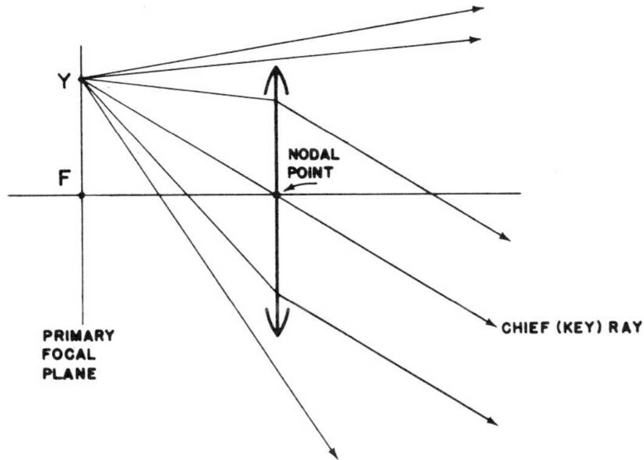


We should now see that the *FOCAL LENGTH f* of the lens (that is, the *axial distance* between F and the lens) is nothing more than that specific object distance *u* which yields an image vergence of zero for this particular lens P. So *f* in meters must equal $\frac{1}{\text{lens power (in Diopters)}}$.

Our next simple "formula" then is

$$f = \frac{1}{P}.$$

A + 14 D lens will have a focal length of $\frac{1}{+14}$ or + 7.1 cm.

If we place a 35 mm slide transparency in the position of the primary focal plane of a projector lens and make the slide very luminous by shining a strong projector light bulb onto it, the lens would project the image of that slide onto a screen. The image would be sharp and clear if the screen were at right infinity. Since that makes for a rather long projection room, the slide can be placed slightly shy of F, that is, a little further from the lens (thus reducing somewhat the vergence of the light rays emanating from each point on the slide); then the projected image would be found at some finite distance in front of the lens — a little more practical than infinity for most projection!

So far, we have been seeing how the image point moves as we brought an axial object point X closer to the lens. Let's backtrack for one moment. When point X on the axis is a long way to the *left* of the lens (at left infinity), the light rays emitted from it are, naturally, divergent. But since the lens is so far away, by the time the rays reach it, they are diverging so minimally that we must consider them arriving in a parallel bundle, that is with an object vergence *U* of zero.

When we look at the lens in the diagram below, we see a number of parallel light rays drawn, but remember, *all of these came from one point,* X, located at infinity on the left and on the lens axis.

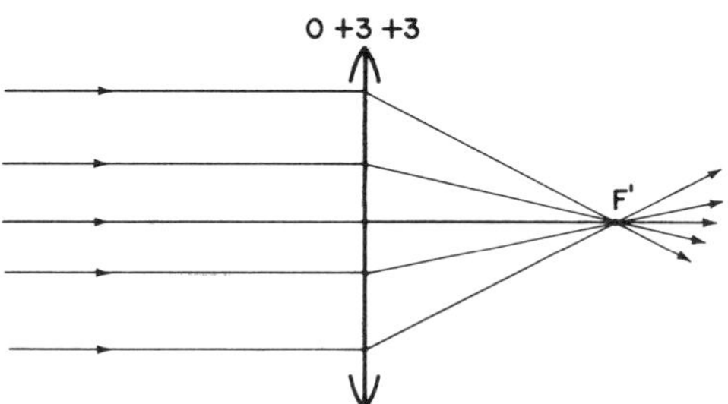


The *image* of axial object point X will be at F' — the *SECONDARY FOCAL POINT* (and another key position). F' is the *image* corresponding (conjugate) to an axial object point situated at infinity.

* * *

As an interjection, there is one particular type of *erroneous* diagram that is often sketched by beginners, but unhappily, it is also occasionally found in elementary physics or physiology texts when simple optics is being "explained" — I quiver each time I see it, or a reasonable facsimile thereof, and now you should too!

The diagram below is correct;

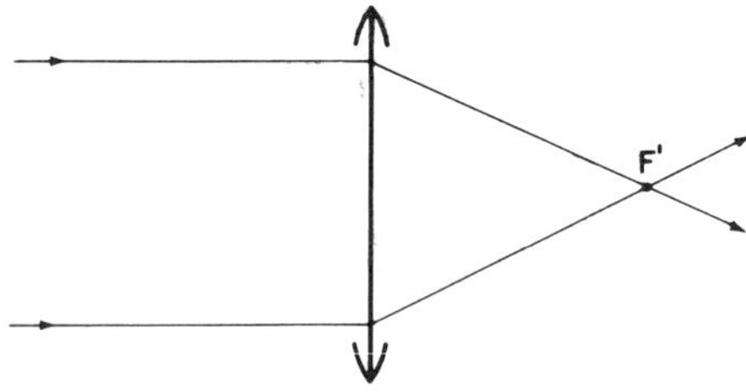

the two rays drawn, of course, signify light emanating from *one* point (the axial one) of a very distant object as they are brought to a point focus at F′.

Now look at this one:

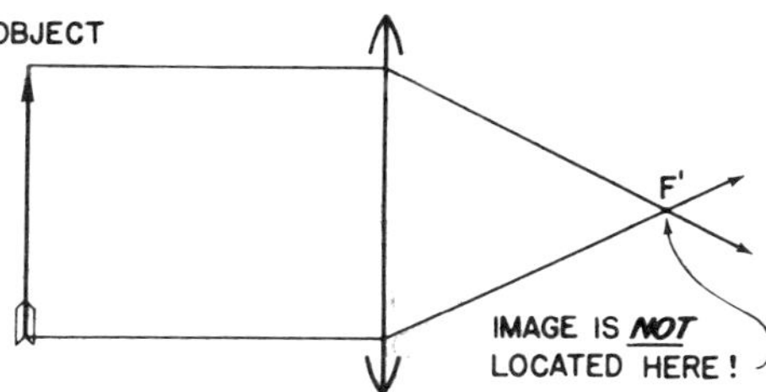


*This* diagram, with one ray *from each end* of an arrow-shaped object, I believe, actually attempts to show the same thing (that is, a final image at the focal point F′), but this one is horribly incorrect. The true image of the arrow object must be further to the right than F′. Diagrams like this and its ilk have done much to disturb students by shattering their confidence in being able to understand something simple. Don't you make this same mistake. 'Nuf said.

* * *

## *"Off-axis" Objects and Images*

When object point X is at infinity but *above* the lens axis, rays arising there also arrive at the lens in parallel bundles, but strike the lens at some angle (inclination) to the axis. How steep an angle? You guessed it. That angle of inclination $\theta$ of all these parallel rays would be given by *the one ray* from point X which was directed at the nodal point of the lens. The image, of course, would be located in the secondary focal plane, since that plane is the home of *all* image points representing object points at infinity. The exact position of the image point in that plane is pinpointed by our undeviated ray through the lens nodal point. (See next figure.)

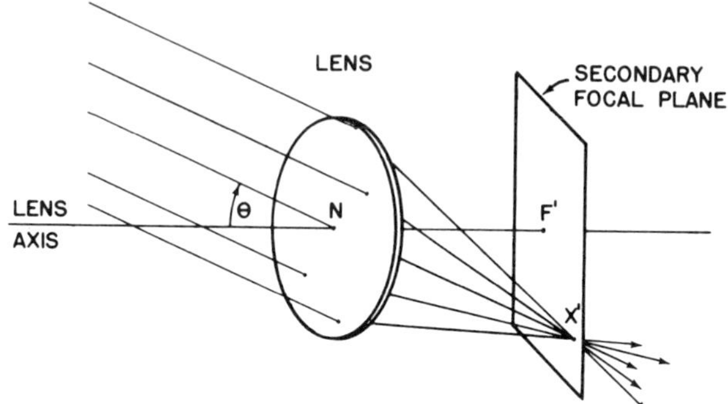


If a sheet of camera film were located at this secondary focal plane and lens P were a simple camera lens, a sharp image of every object point located at optical infinity would be created on the film and captured for posterity, (if the film were properly exposed).

Let us now continue where we left off and move object point X closer to the lens P. Point X will then be closer to the lens than the primary focal point F. Say, X is at a distance of 25 cm from the lens: then,

$$U + P = V$$
$$-4 + 3 = -1$$

The image point vergence at lens P is — 1 D.
Here is our first encounter with a negative vergence for an *image* point; how is this to be interpreted?

By convention, we originally agreed that *minus* meant divergent object *or* image rays. So, here too, we must observe this convention, and the minus sign must mean divergent rays are present in the *"image space"*, that is, *after* leaving the lens. Since these rays are diverging, they *cannot* be brought into focus on a screen; only converging image rays can form a *real* image as mentioned before. We say that (by definition) if *diverging* rays are present in the "image space", the image is *virtual* — it cannot be focused on a screen. Even so, there still *is* an image point X' which is conjugate (corresponds) to object point X; but, the image is located on the *left* of the lens! How far? Well,

$$U + P = V$$
$$-4 + 3 = -1$$
$$\frac{1}{V} = v; \frac{1}{-1\ D} = -1 \text{ meter}$$

Thus, the image point X′ would be 100 cm to the *left* of the lens.

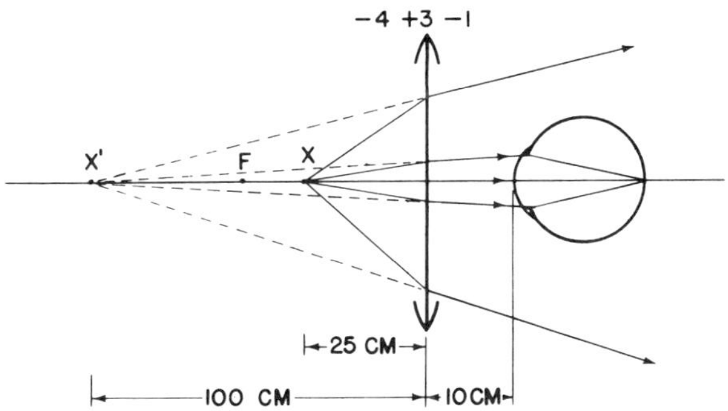


That is, although the *actual* rays (shown in *solid* lines) are divergent as they leave lens P, they simulate rays (dotted lines) which would have come from a real point located 100 cm away. All *real,* honest-to-goodness light rays will, in the diagrams of this book, be indicated as solid lines, while rays from *apparent* (virtual) points will be shown as dotted lines).

For all intents and purposes then, after leaving the lens, the rays will *seem* to arise from X′. Thus an eye looking through lens P would not "see" point X (which is really 25 cm from the lens); it would "see" the image point X′ (as created by the lens) as if it were located 100 cm away from it.

If you wish to know the *vergence* at the *eye* of those light rays that seem to come from X′, you will have to determine the distance between X′ and the eye. Say, the eye is located 10 cm from lens P. Then X′ must be located 110 cm from the eye, and X′ has a vergence of $\frac{1}{-1.1 \text{ meters}}$ or $-.91$ D at the eye. Notice now that not only is X′ an image created by lens P; it is also, simultaneously, an object for the eye and, thus, can be said to lie in the "*object* space" of the eye. (Even though I have already tossed out the terms "object" and "image" space glibly, I would still like to withhold their definitions until a bit later.)

Consider one last example for this + 3 D lens:

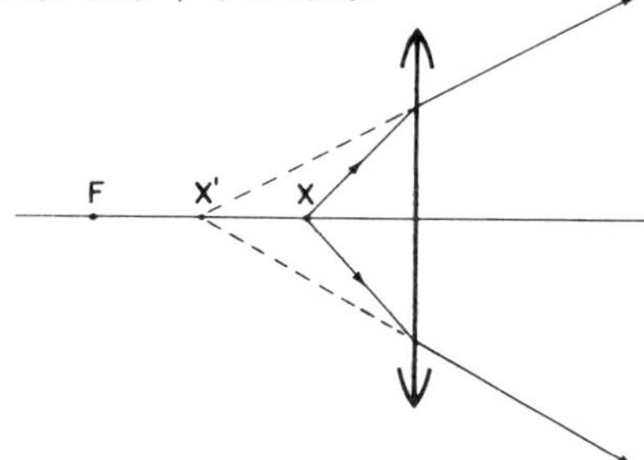


X continues to move closer to P — to a position only 5 cm from it. X presents a vergence $U$ to the lens of $-\frac{1}{.05\text{ m}}$ or $-20$ D.

$$U + P = V$$

$$-20 + 3 = -17$$

The image X' is now located at $\frac{1}{-17\text{ D}}$ or $-5.9$ cm, to the left of the lens. Furthermore, the closer X is to the lens, the smaller the apparent separation between X and X'.

## *Object Movement Vs. Image Movement — Plus*

With the help of a table to recapitulate the examples given so far, let us study what happens to the image movement as the object point moves from left-hand infinity towards a + 3 D lens.

| Object distance *u* in *cm* | $\infty$ | — 50 | — 40 | — 33 (at F) | — 25 | — 5 |
|---|---|---|---|---|---|---|
| Image distance *v* in *cm* | + 33 (at F') | + 100 | + 200 | $\infty$ | — 100 | — 5.9 |

As the object located at left infinity moves to the right (towards the lens), the image point first seen at F′ also moves towards the right, more and more rapidly until the image reaches right infinity precisely when the object arrives at F. With any further approach of the object towards the lens, the image suddenly appears at *left* infinity — (consider that the image initially moved *away* from the lens to right infinity and then "around the earth" and suddenly appeared from the left.)

The image then proceeds to approach the lens from the left, tagging behind the object as the latter moves still closer to the lens. The image finally catches up with the object *at* the lens plane. Thus with the plus lens, as the object moves to the right, the image moves to the right also.

So far we have examined the properties of a + 3 lens. We can generalize that *any* plus lens will have identical properties, aside from variation in its power. Each lens will exert its own particular influence on the incoming object vergence. A + 12 Diopter aphakic spectacle lens would add + 12 Diopters of convergence to *any* object rays arriving at it. A + 0.001 Diopter astronomical telescope objective would add this tiny amount of *con*vergence to incoming object rays.

## The Minus Lens

A *minus* lens has the ability to add *di*vergence to incoming rays. Our same little relationship $U + P = V$ and all our related sign conventions continue to hold true.

If a real object point X is located 25 cm to the left of lens P of — 3 Diopters, the object vergence at the lens would be $\frac{1}{-.25 \text{ meters}}$

or — 4 D. The — 3 D lens adds *its* vergence power of — 3 D to the incoming — 4 D vergence to yield — 7 D of image vergence.

$$U + P = V$$
$$-4 + (-3) = -7$$

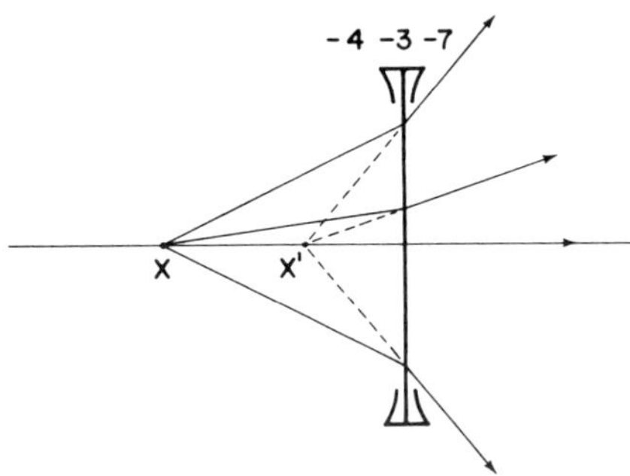


This locates an image point at $-\frac{1}{7\text{ D}}$ or — 14.1 cm, the minus sign again meaning "to the left of lens P".

If the object (in the next figure) were at left infinity, the object vergence *U* would be zero.

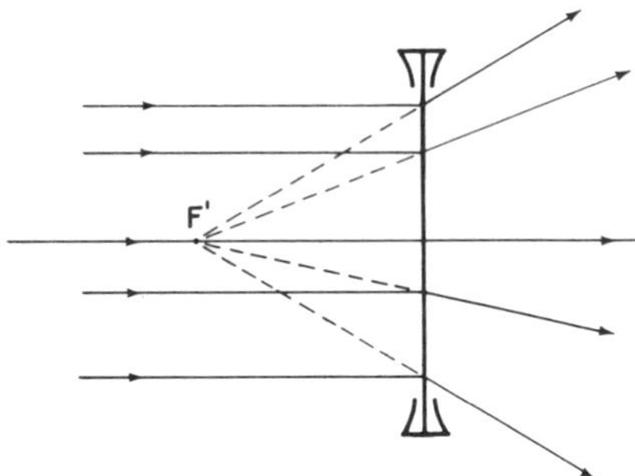


$$U + P = V$$

Therefore, $0 + (-3) = -3$

Thus, the image vergence is — 3 Diopters, and the image point would be found at — 33 cm (33 cm to *left* of lens P). We have already defined that the *image* point conjugate to an axial object point located at infinity is called the secondary focal point (F') of a lens so this minus lens' F' is 33 cm on the *left.*

With any minus lens, F' will always be on the left; but remember also that F' is still an *image,* just as much as is F' of a *plus* lens. The object point at infinity, though it appears to be on the same side of the lens as F', it is not in the same "space". The image point F' exists only in the "image" space, the term signifying that we are dealing with light which has already been influenced by a given lens.

Now, how on earth can we locate the *primary* focal point (F) of a minus lens, since by definition it is the *object,* the image of which is located at right infinity (that is, the image vergence must be zero)? Our trusty relationship $U + P = V$ should help us:

$$U - 3 = 0$$

$U$ must $= +3$, which means that this object point F (the primary focal point of the lens) must be 33 cm to the *right* of the lens. Our light rays (remember this convention?) must travel from left to right; so, for us to obtain an *image* vergence of zero, the object vergence has to be plus (convergent) in the *object space.* Only then can parallel rays be created by the minus lens. (As stated early in this treatise, convergent object rays can only be formed by another optical device; they cannot occur naturally.)

To demonstrate F of a minus lens diagrammatically, we must show object rays converging toward F (the "object" point) *before* they impinge on the lens. The fact that the object rays never really reach F but are only directed towards it should not distress you.

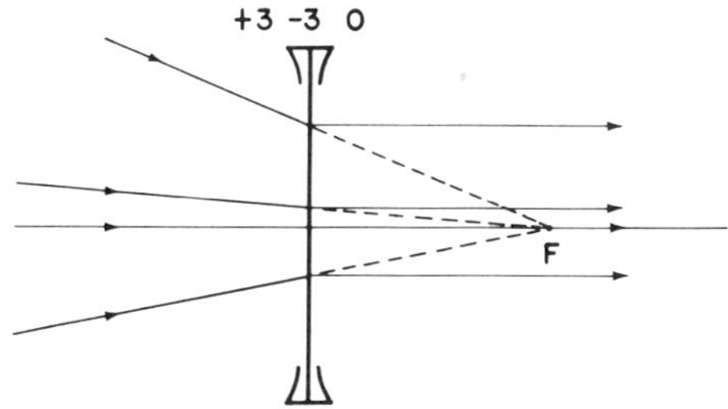


When convergent rays fall on any lens — plus or minus — the object vergence will of course be *plus,* and the object is considered "*virtual*" since it is intangible. Optically, however, a virtual object can be manipulated as readily as a real one, just as can a virtual image. *Virtual* object rays will be shown in the accompanying diagrams as dotted instead of solid lines.

Let us now move the virtual "object" closer to the lens, to somewhere between the lens and F. We will thus have to make the object rays even more convergent than + 3 D. Say the vergence of object point X is + 4 Diopters (see figure below);

$$U + P = V$$

$+ 4 + (-3) = + 1$ D and a *real* image point will be created by a minus lens. If you desire, you can focus this image onto a screen one meter away — to the right.

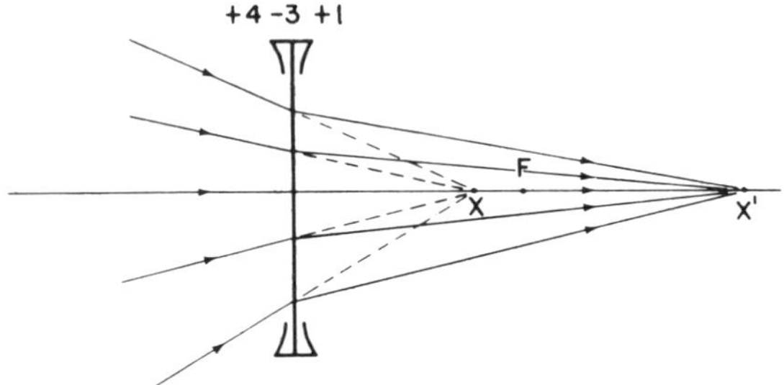


## *Object Movement Vs. Image Movement — Minus*

Again, a table has been constructed relating the object distance *u* and image distance *v* for a minus lens. I have added a few more examples for our — 3 Diopter lens:

| Object distance *u* in *cm* | + 25 | + 33 (at F) | + 100 | ∞ | − 33 | − 25 | − 10 |
|---|---|---|---|---|---|---|---|
| Image distance *v* in *cm* | + 100 | ∞ | − 50 | − 33 (at F′) | − 16.7 | − 14.1 | − 6.9 |

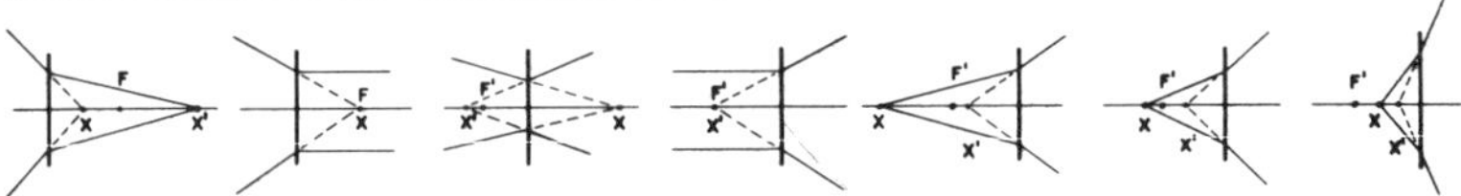

As an object point moves from the right side of the lens towards right infinity (of necessity, these object points are created by *convergent* light beams from some other optical system), the image "moves" ahead of it in the same direction — *through* right infinity, "around the earth" and over to left infinity, and thence toward the lens. With a *real* object point (which moves from *left* infinity towards the lens), the corresponding virtual image point will continue to move ahead of it to the right (still closer to the lens), but also in the same direction as the moving object.

The point of these two tables — one (a few pages ago) for a plus lens, the other for a minus — is to demonstrate that the sign of the lens makes no difference to one clear-cut conclusion: the object and its image *always* will move in the same direction. *When any object moves to the right or left, its image by any type of optical system, plus or minus, will do likewise.* This point is quite important in helping you to understand the movement of images when dealing with the eye and with cases of refractive error.

## Object Space Vs. Image Space

Up to now, I wanted you to get the "feel" of handling these terms yourself, since sometimes when dealing with concepts, intuition is better than definition. Now, however, to set the record straight, I will try to clarify the two conceptual terms, "object space" and "image space", both of which I have already used freely. But by now, you already should have an inkling of what they mean.

Prior to impinging on any lens, all rays which leave an object plane are said to be located in the "object space." After the lens adds its own vergence to those rays, the resultant vergence is said to exist in the "image space." That sounds easy enough. However, the lens which creates those spaces "looks" like it physically separates the "object space" on the left from the "image space" on the right — it does not. Both spaces co-exist simultaneously on *both* sides of the lens, and that is what makes this concept confusing.

Object space exists only for rays and ray directions *before* they are influenced by a lens. Consider that object space is composed of *all* points which could possibly serve as an object for the lens. We have studied about divergent object rays which come from a real object located on the left of the lens; we have also seen how convergent object rays might optically form a virtual object on the right of any lens. The latter object also is in the object space of the lens. So, we can see that object space does completely surround the lens. Remember then, the object space of a lens may contain real or virtual rays, but *before* they have been influenced by the lens.

Image space, on the other hand, exists only for rays *after* they are influenced by a lens or an optical system; this space exists only because the lens has *already* exerted its influence on some object rays. Image space also completely surrounds the lens; if the *image* is on the right of the lens, it is real, if on the left, it is virtual. In either case, the image is still in the "image space."

You should now understand that every lens is completely engulfed — at the same time — by both object and image spaces. Contrary to appearances, these were not created to confuse you. Keep the above descriptions in mind.

So far, we have looked at *axial* object points and image points and diagrammed their conjugate relationship, but we also elaborated (through our discussion of the axial focal points) that the *points* also represent planes. These object and image planes which contain the corresponding axial points are parallel to the lens plane and possess all the vergence properties (at the lens plane) of those points. One locates these planes relative to the lens the same way as one does the axial points ($U + P = V$). Any object point in one plane has its image in its conjugate plane and, at the specific location denoted by that single, undeviated light ray which leaves object point X and passes through the lens nodal point.

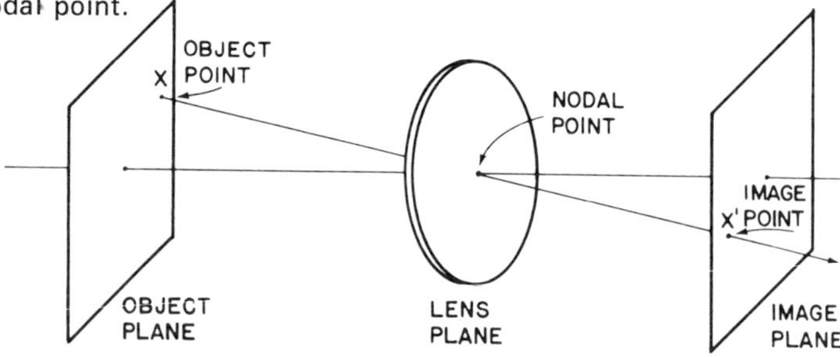

What else can we learn from $U + P = V$?

If we know any two terms, the third can easily be calculated and all sorts of nice little problems become manageable.

*EXAMPLE 1*

What strength lens will image an object located 25 cm to the left of it onto a screen 1 meter distant to the right?

ANS:

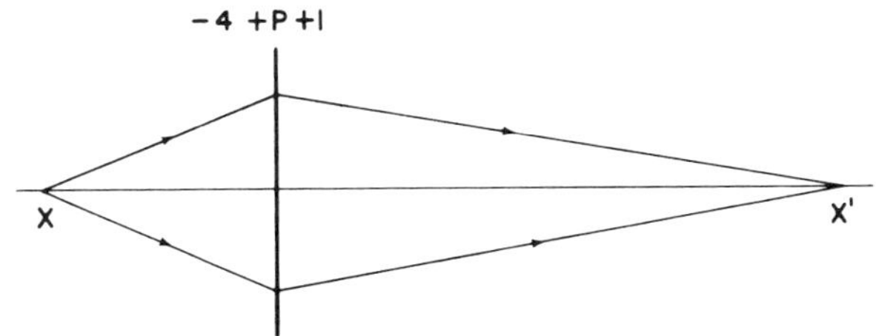


$$U + P = V$$
$$-4 + P = +1$$
$$P = +5 \text{ D}$$

A + 5 D lens fills the bill.

*EXAMPLE 2*

How far away must an object be to have its image sharply defined on film which is located 50 mm behind a + 25 D camera lens?

ANS: $v = 50$ mm $= .05$ meters, and $P = +25$ D.

$$U + P = V$$
$$U + 25 = +\frac{1}{.05} \quad \text{(See next figure.)}$$
$$U = 20 - 25 = -5 \text{ D}$$
$$u = \frac{1}{-5} = -0.2 \text{ m} = -20 \text{ cm}$$

The object X must be 20 cm away (to the left).

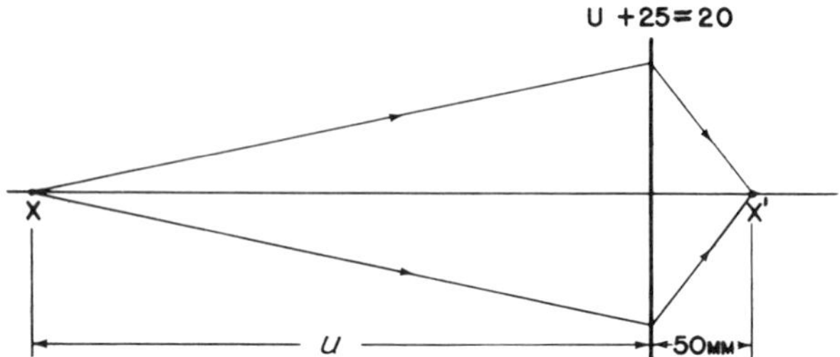


*EXAMPLE 3*

Exactly where, *between* an object and its corresponding image, must a + 4 D lens be placed so as to have the object and image separated by 1 meter?

ANS: X and X′ are 1 meter apart; $u$ is the object distance and $v$ is the image distance. (See diagram below).
In *absolute* distance measurement (without regard to the sign), since $u + v = 1$ meter, $v = (1 - u)$ meters.

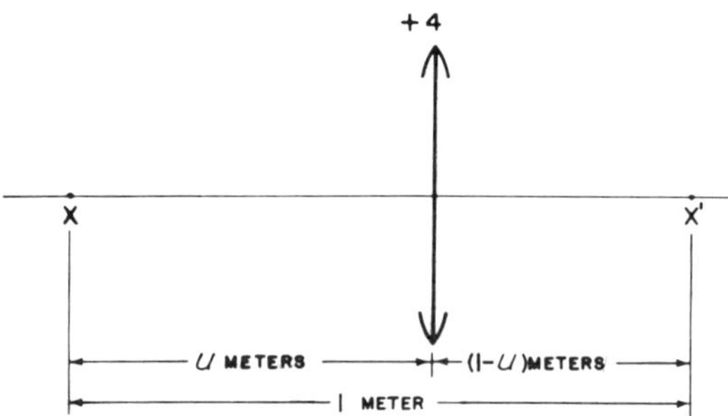


We are given that the lens is *between* the object and the image, so whatever distance $u$ happens to be, the object vergence $U$ must be minus at the lens plane. With comparable reasoning, $V$ must be plus.

So we have $U = -\frac{1}{u}$ and $V = \frac{1}{1 - u}$

Now, substitute these for $U$, $P$, and $V$ in our general formula,

$$U + P = V$$

$$-\frac{1}{u} + 4 = \frac{1}{1 - u}$$

To solve this equation for $u$, we must simplify it by eliminating the denominators. This manipulation requires easy 9th grade algebra.

$$-(1-u)+4(u)(1-u)=u$$

$$-1+u+4u-4u^2=u$$

$$-1+4u-4u^2=0$$

Multiplying by $(-1)$;

$$4u^2-4u+1=0$$

$$(2u-1)^2=0$$

Take the square root;

$$2u-1=0$$

$$2u=1$$

$$u=\tfrac{1}{2}\text{ meter}$$

$$1-u=v=\tfrac{1}{2}\text{ meter}$$

Therefore, the lens would be located 50 cm from both the object and the image, that is, half-way between them.*

*EXAMPLE 4*

A convergent light beam from a projector located on the left creates a clear, 3 cm diameter spot image on a screen on the right. You don't care about the size of the spot, but you want to move that image 25 cm to the right without budging the projector. You have available a $-2$ D lens to help you. Where must you hold this lens to perform this image displacement?

ANS: The distance X to X′ is given as 0.25 meters.

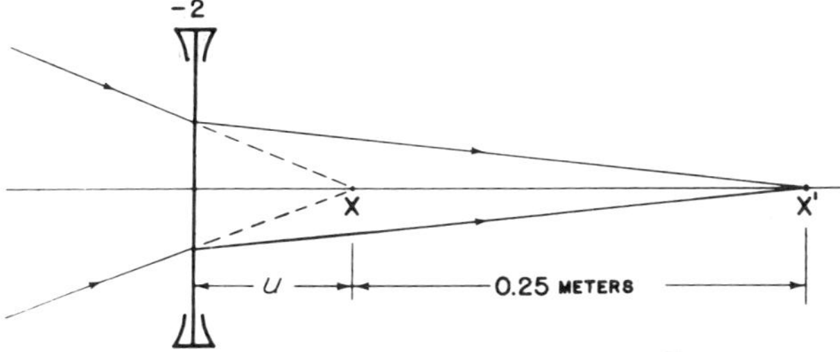


Let $u$ = the distance from our lens to the image at X (which would have been produced by the projector if the lens did not intercept the rays first!) This is the object distance for our $-2$ D lens. The object vergence $U=+\frac{1}{u}$; P $=-2$; the image vergence $V=+\frac{1}{(.25+u)}$.

* If the $+4$ D lens were not specified as being *between* X and X′, two other solutions become possible. Can you find them?

$$U + P = V$$

$$+\frac{1}{u} - 2 = +\frac{1}{.25 + u}$$

$$(.25 + u) - 2(u)\ (.25 + u) = u$$

$$.25 - .5u - 2u^2 = 0$$

Multiply by (— 1): $$2u^2 + .5u - .25 = 0$$

Factor: $$(2u - .5)\ (u + .5) = 0$$

Thus we have *two* answers to the problem:

a) $2u = .5$

$u = .25$ meter

$u = +\ 25$ cm

b) $u = -\ .5$

$u = -\ 50$ cm

How do we interpret these answers?

a) The — 2 lens should be held 25 cm in front of the original screen. This would move the projected image to a position 25 cm further away.

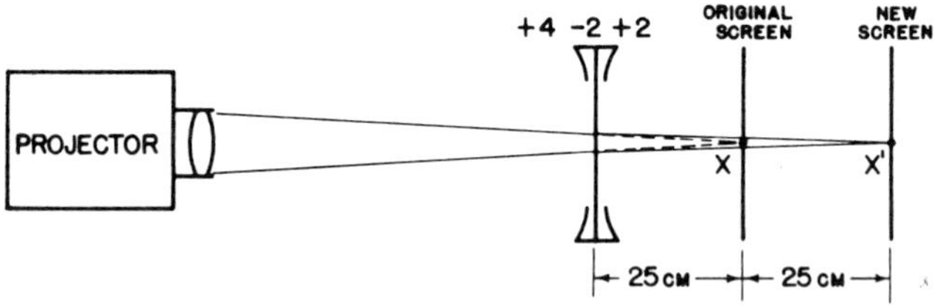


b) But, we see from our algebraic solution that another possibility is suggested (see figure below). An object distance $u$ of — 50 cm tells us to hold the lens in such a way as to present a diverging beam to the lens. This is possible only if we place it in the light path *after* the projection image is formed at X. When there is no screen in this position, the rays will continue on, but then they will be *divergent.* Placed at — 50 cm, the — 2 D lens "sees" an object vergence of — 2 D and creates an image vergence of — 4 D, since

$$U + P = V$$

$$-2 + (-2) = -4$$

An image vergence of — 4 signifies that our lens creates a virtual image located 25 cm to its *left.*

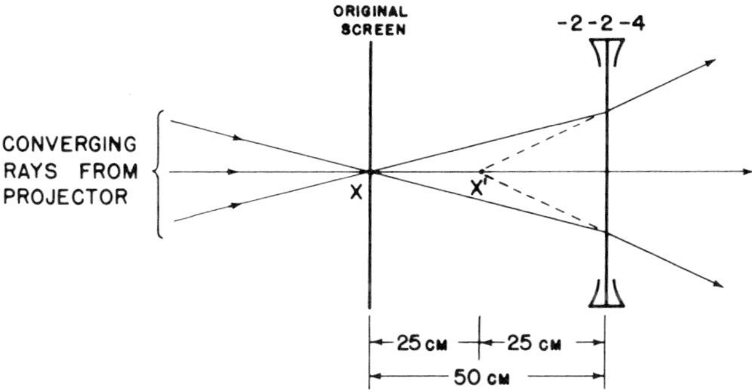


The original object point X would indeed be imaged at X', 25 cm further to its right, as was the condition stated in the original problem, but of course, we had no way of knowing *a priori* that the math solution would include a *virtual* final image. Though this answer fulfills the image criteria, it is impossible to focus this image without additional converging lenses, so this answer is not a practical solution, while answer (a) is correct *and* practical.

*OTHER EXERCISES —*

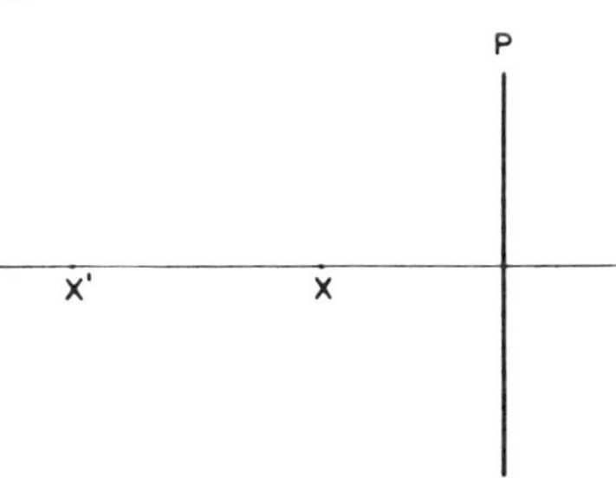


If I asked you to tell me what type of lens (plus or minus) would image axial point X at X' in the above situation, you *should* be able to respond. If you can't, follow along with my reasoning: Since X is on the left of lens P, light rays will diverge toward the lens. After being influenced (refracted) by the lens, all rays from X must be bent as if they came from X'. So, draw *any* ray from X and notice how the lens bends it so that it would appear to come from X'. If it converges the ray, the lens must be plus; If it diverges it, the lens must be minus.

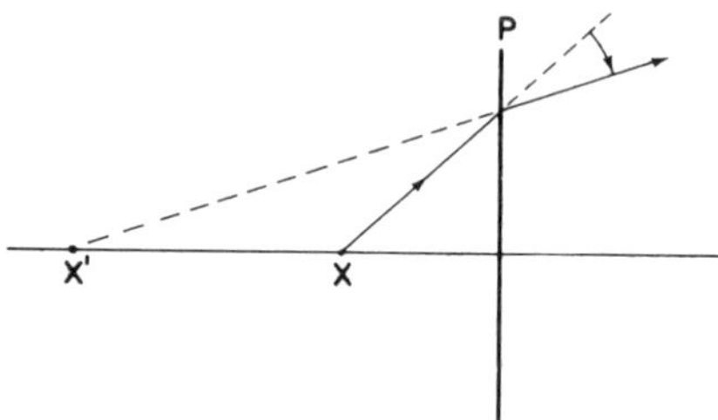


P here must be plus since the light ray is bent downward, towards the lens axis — that is, convergence is added by the lens. The added convergence though was not sufficient to form a *real* image point. X' is thus virtual.

What if X is on the right of lens P?

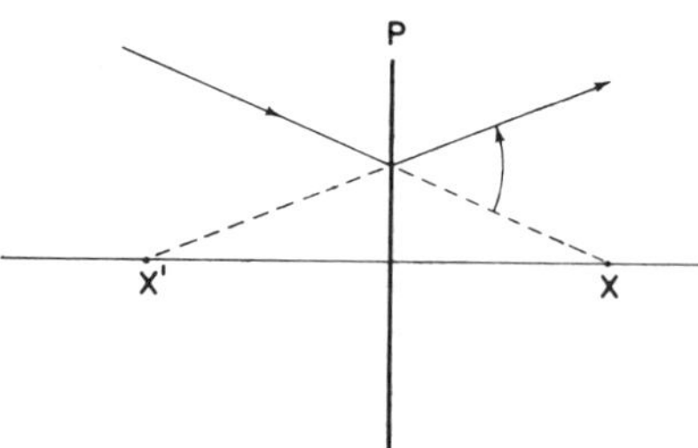


The only difference with X on the right is that this signifies that light is *convergent* (at the lens plane) toward object point X (which must thus be virtual). After a ray drawn toward X hits lens P, it will act as if it were directed from axial point X'. Again note only what happens *at* the lens to that single light ray. Since the light ray is bent upwards, away from the axis, divergence must have been added by it; therefore P here must be a *minus* lens. In this example both X and X' are virtual!

Try to label a few other lenses as plus or minus:

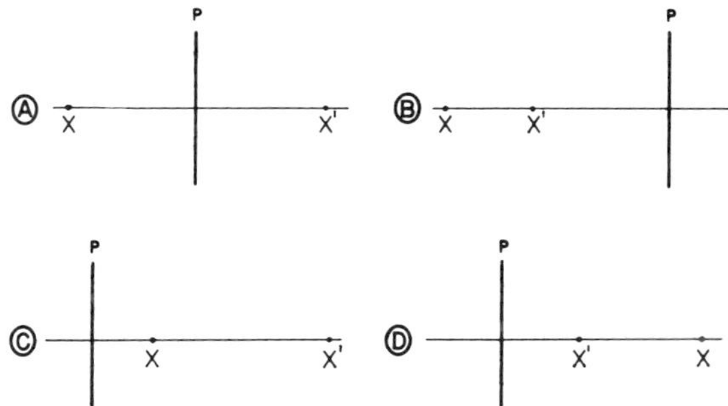


ANSWERS: A. Plus B. Minus C. Minus D. Plus

## MULTIPLE THIN LENS SYSTEMS

When you have to find object-image relationships through more than one lens, you must treat the vergences separately for each lens in succession, always dealing first with the first lens to encounter the incident light. The image position created by the first lens will then be the object position for the second.

But be careful here; we said we are dealing with *positions.* The image *vergence* created by $P_1$ (the first lens in a series) is *not* the same as the object vergence for $P_2$ unless the lenses are in contact; you must *calculate* the appropriate vergences from the known positions. So first, locate the *position* of the image created by $P_1$; then find the distance of this image (now the object) from $P_2$. The reciprocal of this distance ($\frac{1}{\text{distance}}$) is the object vergence presented to $P_2$, and don't forget the *sign* of this object vergence.

*EXAMPLE:* (Consult the figure where each lens is considered independently):

Lens $P_1 = +2$ D
Lens $P_2 = +1$ D
Lens $P_3 = -4$ D

A real object is 1 meter to left of $P_1$.
The distance between $P_1$ and $P_2 = 25$ cm.
The distance between $P_2$ and $P_3 = 23$ cm.

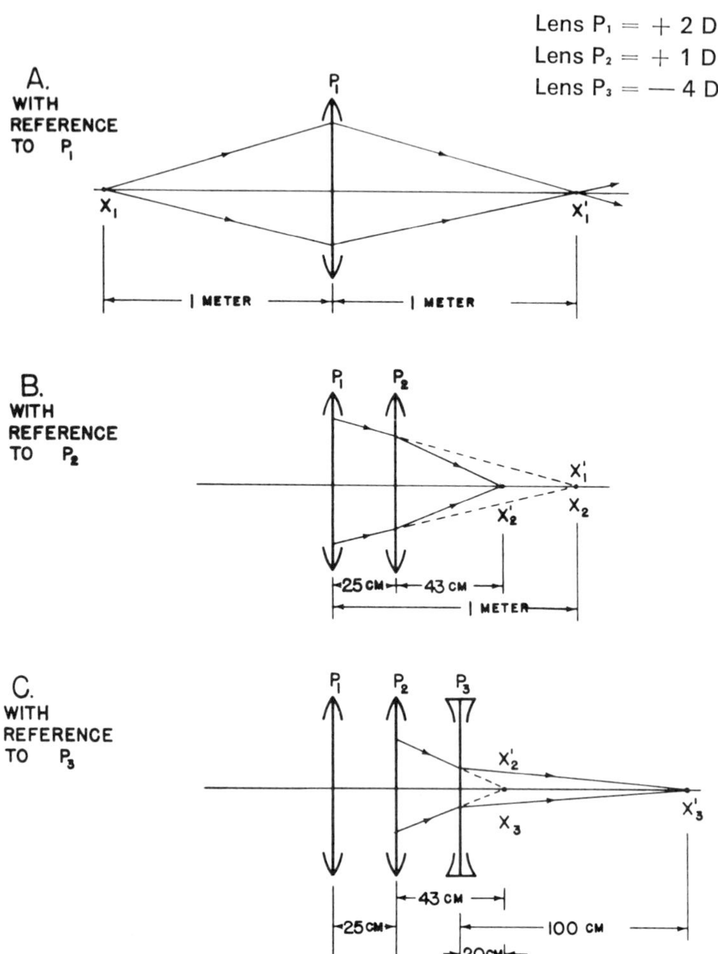


A. *With reference to Lens* $P_1$:

$$U + P = V$$

$$-1 + 2 = +1$$

Therefore, the image is located 1 meter to right of $P_1$ (at $X_1'$).

B. But $P_2$ is situated 25 cm from $P_1$; therefore, the image rays from $P_1$ will strike $P_2$ before the real image is formed further to the right. As far as $P_2$ is concerned, it "sees" incident (object) rays aiming at $X'_1$ which is still 75 cm further on to the right. Hence, the incident vergence of those rays at $P_2$ is $+\dfrac{1}{0.75 \text{ meters}} = +1.33$ D.

In summary, the image ($X_1'$) formed by lens $P_1$ with a vergence of + 1 D now becomes an object ($X_2$) for lens $P_2$ *but* with an object vergence at $P_2$ of + 1.33 D.

*With reference to Lens $P_2$:*

$$U_2 + P_2 = V_2$$

$$+1.33 + 1 = V_2 = +2.33 \text{ D}$$

So the image vergence after $P_2$ is + 2.33 D.

$$\frac{1}{V_2} = v_2$$

$$\frac{1}{V_2} = \frac{1}{+2.33}$$; therefore, $v_2 = +43$ cm, that is, 43 cm to right of $P_2$.

C. Since $P_3$ is located only 23 cm from $P_2$, the rays will strike $P_3$ before the real image is formed. The object distance to $P_3$ is equal to (43 cm — 23 cm) or 20 cm to the right of $P_3$ and thus, the object vergence ($U_3$) presented to $P_3$ is + 5 D.

*With reference to Lens $P_3$:*

$$U_3 + P_3 = V_3$$

$$+5 - 4 = +1 \text{ D}$$

The final image vergence is + 1 D; thus, the location of $X_3'$ — the final image formed by the combination of all three lenses — is a *real* one, 1 meter to the right of $P_3$.

So much for locating by simple algebra the locations of conjugate objects and images relative to single or multiple thin lens systems. One should also be able to determine readily these same relationships by an accurate, *graphical* method. This will be taken up now.

## GRAPHICAL ANALYSIS

We have clearly spelled out that an infinite number of light rays will emerge from each point making up any luminous object. Let us concern ourselves with the rays which leave Point A, the tip of an object which sits on the axis of lens P.

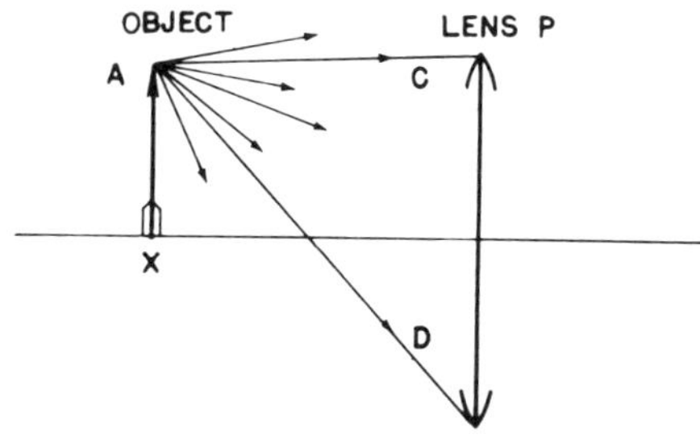


In the diagram, light rays are shown being emitted by point A. Many of these rays will fall on lens P — those emitted between the extreme, "limiting" rays C and D which just barely hit the edge of the lens. Each ray will be bent a different amount, but still towards the same image point. If we knew the exact path taken by any of the rays after refraction, we could follow those and find where any two of them intersected; we would then have localized *the* image point corresponding to object point A. Luckily, we *do* know the precise path of three particular rays; (we have already dealt with these). We know the exact road these rays travel and can utilize them in a simple graphical analysis. These three "known" rays in any optical system are as follows (see next diagram):

1) The *"chief"* ray; that particular ray from A that aims directly for the lens nodal point will continue undeviated.

2) The ray through the primary focal point F; we know that *all* rays *originating* at F must leave the lens parallel to the axis. Thus *any* single ray *from any object* which happens to pass through F will also leave lens P parallel to the axis.

3) The ray parallel to the lens axis; *all* rays which are parallel to the lens axis in the object space must, in the image space, pass through the

secondary focal point F′ of the lens. Thus, if we select *the one ray* which leaves object point A directed parallel to the lens axis, it also must, after refraction by lens P, be bent toward F′.

Given lens P with

N = nodal point
F = primary focal point
F′ = secondary focal point

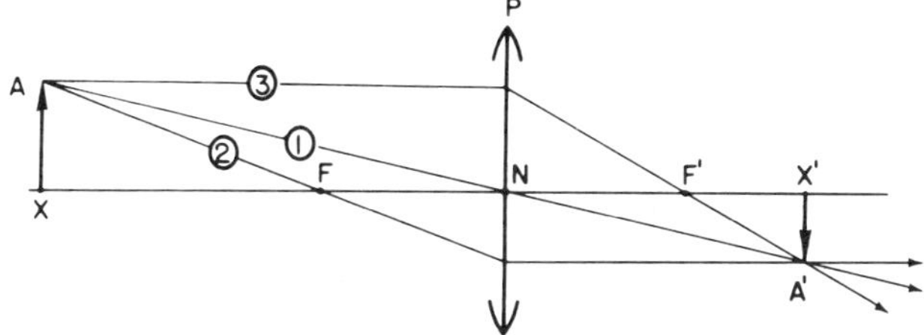


Ray 1 drawn through N is undeviated.
Ray 2 drawn through F must emerge from P parallel to axis.

These two rays cross at A′ and establish it as the image point of object point A. To check, continue with ray 3.

Ray 3 leaves A parallel to axis in the object space and must pass through F′ in image space. We see that all three rays intersect at A′, and firmly fix A′ as that image point conjugate to A. Object point X on the axis is located at the intersection of a perpendicular dropped from point A. X′ is similarly located in the image space once A′ is determined.

With *any* given lens and a given object point, we should easily be able to construct graphically the corresponding image point (and vice versa!) Remember, any two of the above mentioned three rays will suffice. For practice here, you should draw all three.

Another aid: If you find that in drawing rays 2 or 3 that the lens diagram is not large enough in diameter, simply extend the size of the lens with a dotted line (as in the figure below). All refraction of light rays can schematically be considered to take place somewhere in the lens plane, even if there is no actual lens material there! For example, to draw ray 3 parallel to the lens axis, you must extend lens P diagrammatically upward as shown here.

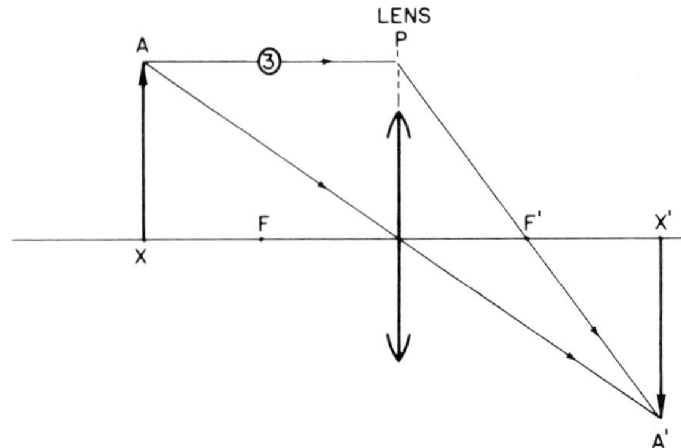


When we are given only an *axial* point X and wish to find the corresponding image point X′, the three ray technique described above will not "work"; all three of the particular rays we have enumerated are still there, but happen to be superimposed upon the line representing the lens axis, and so, we would be unable to locate any *intersection* of rays to find point X′. A slightly different technique can help us, however.

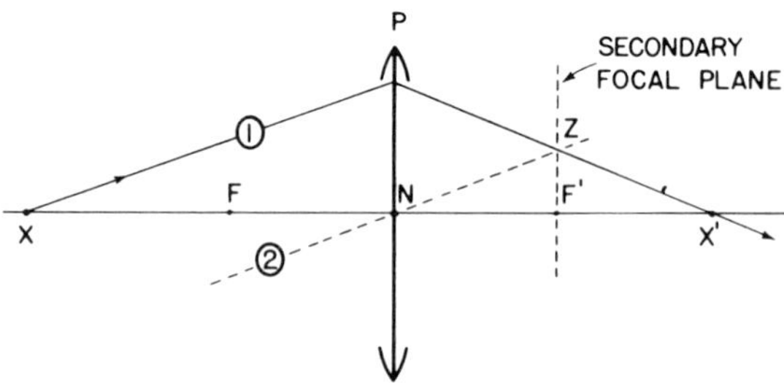


From X draw *any* arbitrary ray 1 which intersects lens P. We know that any and all rays in the object space (even though they do *not* arise at X), if parallel to ray 1, will have to come to a sharp focus somewhere in the secondary focal plane. Let us draw such a ray, 2, parallel to 1, but through N, the lens nodal point. Ray 2 must then be undeviated. (Remember, we can draw this helpful ray even though we know it does *not* denote a *real* ray originating at X. It is simply a logical device to let us fix point Z in the secondary focal plane.)

Since all rays parallel to 1 and 2 in the object space must pass through Z in the image space, ray 1 also must be bent by the lens to pass through Z; thus, ray 1 is given a definite *direction* in the image space. But, we need the intersection of *two* rays to determine image point X′. Can you see where we can find the second ray? You should. That second ray is the one which travels along the lens axis itself; we know both object X and X′ must lie along this ray. Thus, the intersection of ray 1 and the lens axis establishes X′ as the image point of X.

Let us use our same 3 basic rays to indicate graphically the image created by a minus lens with a *virtual* object, a somewhat more complicated, but still very useful exercise.

Given AX — a virtual object on the right lens of P, a *minus* lens. Locate the image graphically:

Solution (see figures below):

To establish object AX on the right of lens P, the rays must be converging as shown.

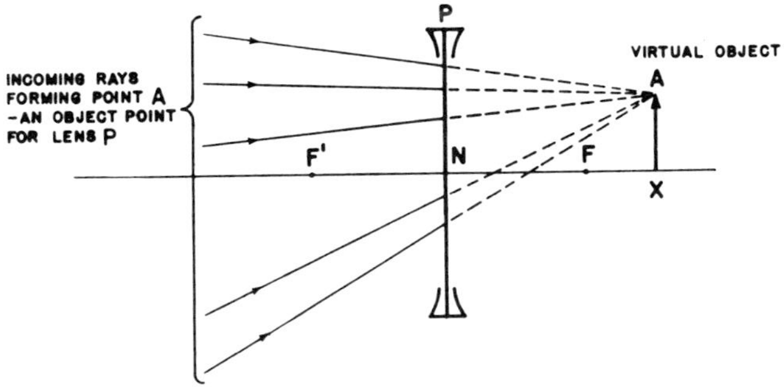


(See next figure.)

1. Of all the rays aiming for point A in the *object* space, select the one ray, 1, which is directed towards A and passes through N. That ray must be undeviated.

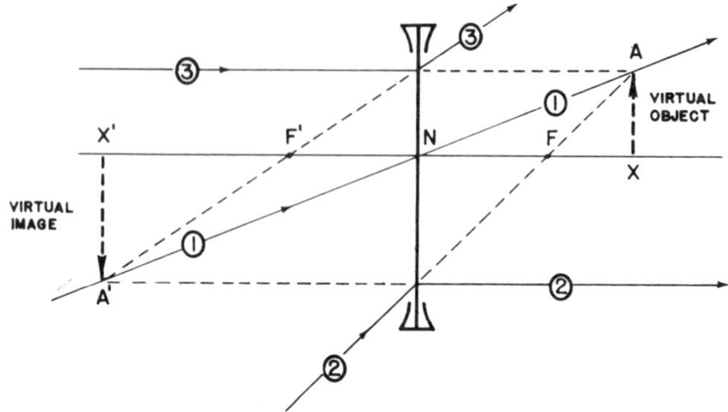


2. Select ray 2 (another of the rays aiming for A) which passes through F, in the "object space" even though it appears on the *right* of the lens. This ray must, after refraction, emerge from lens P as a ray parallel to the lens axis. (Watch this next move since it is a bit tricky!) If the direction of ray 2 *after* refraction were extended backward (now within the image space), it would intersect ray 1 (also in the image space) at A′, which must therefore be the *image* of A. To check this point, utilize ray 3.

3. Ray 3 is selected as that ray aiming for A (in the object space) which is parallel to the lens axis. After lens refraction, this same ray must pass (or seem to come to) F′. Extended even further back from F′, ray 3 also will intersect rays 1 and 2 precisely at A′, and confirms its location.

Rays 1, 2 and 3 after refraction (in the image space) seem to come from A′ (a virtual image) and any eye (or optical system) peering back through lens P to receive these rays will see A′X′ exactly located as diagrammed here.

If one fully grasps the principles underlying the use of the three rays to construct a display of the object-image relationship, he need have no fear in facing *any* construction which might be encountered later.

Try two more construction problems: If you find you are unable to draw these out for yourself or you cannot understand the answers given, please go back and reread this section on graphical analysis.

GIVEN: The lens position and the object position and size.

FIND: The position and size of the image in both problems.

PROBLEM:

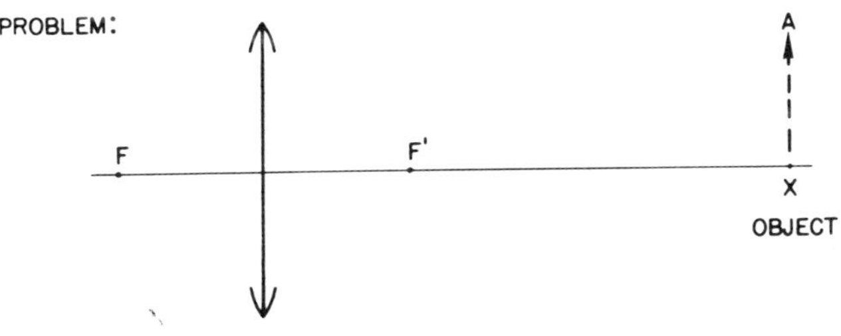

ANSWER:

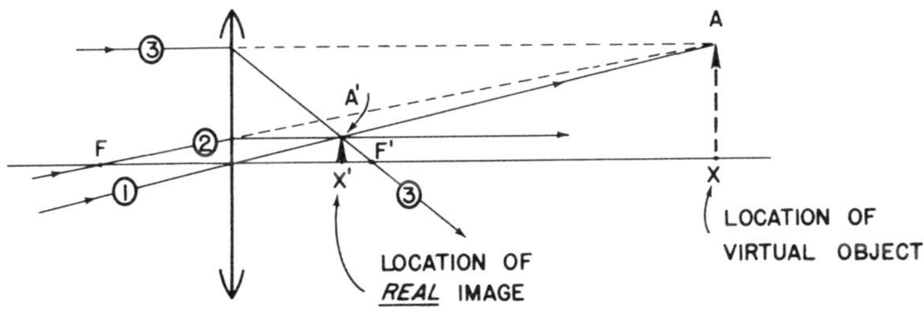


PROBLEM:

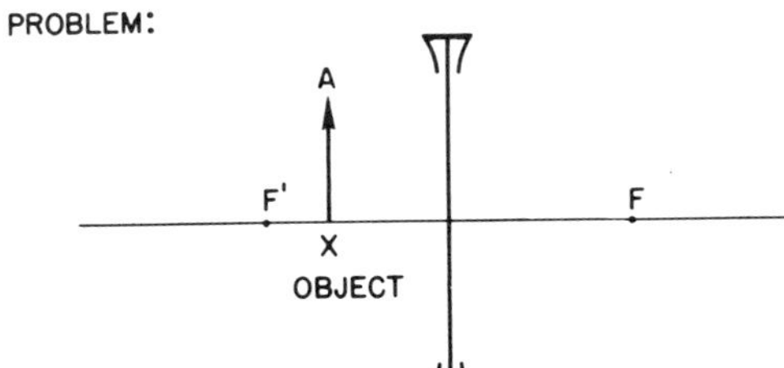

ANSWER:

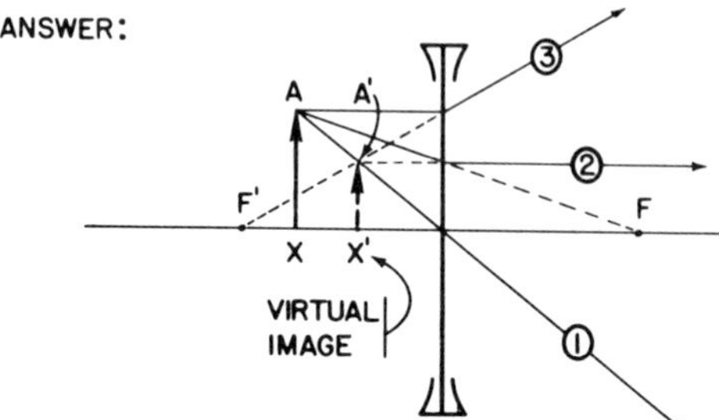


If you wish to construct the rays from an object through *multiple* thin lenses, you must use the same method shown here for *each* lens separately and consecutively. First locate the first image by lens 1. That image then becomes the object for the second lens; this lens forms another image which is "seen" by the next lens and becomes its object, etc. The final image is fixed when all lenses have exerted their own particular influences on the original object rays.

## LINEAR MAGNIFICATION

Though we will get into a more complete discussion later of magnification as it pertains to vision, we should mention *linear* magnification here to complete our image construction section.

So far we have not mentioned anything about the relationship of the linear size of the object to that of the image. That is, when object point A is off the lens axis, how far off the axis is the image? Simple geometry tells us this answer in all cases.

Aside from the lens axis, the only other ray necessary for us to see this relationship clearly is ray 1 — that through the nodal point.

These rays establish 2 similar triangles AXN and A'X'N.

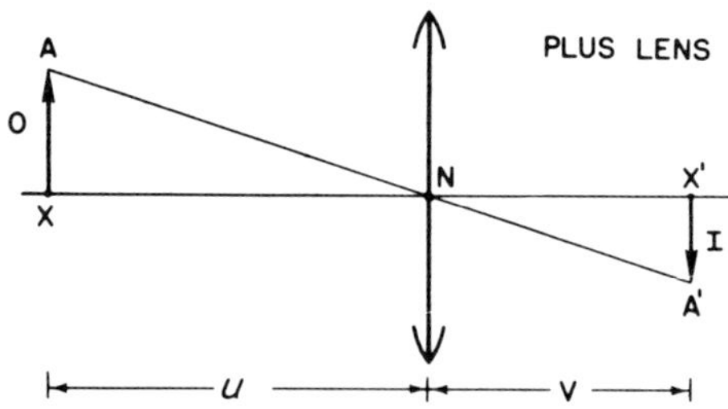

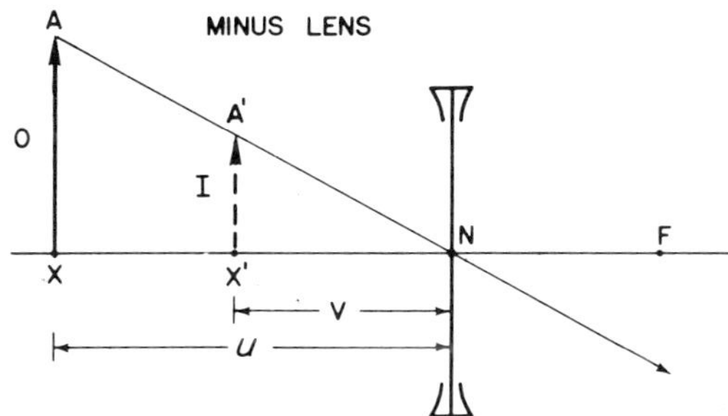


No matter what the power of lens (plus or minus) — no matter where the objects and images are in relation to the focal points — these triangles will always be similar, even, of course, if the object and image are on the same side of the lens (as in the figure above).

You should, therefore, be able to see that the *sizes* of the object and image are always directly proportional to their *distances* from the lens. The LINEAR MAGNIFICATION (M) is hereby defined as the ratio of the size of the image to the size of the object. Refer now to the above diagrams:

$$\text{Image size I} = A'X'$$
$$\text{Object size O} = AX$$
$$\text{image axial distance } v = NX'$$
$$\text{object axial distance } u = XN$$

By definition,

$$M = \frac{I}{O} = \frac{A'X'}{AX}$$

But since the triangles involved are similar,

$$\frac{A'X'}{AX} = \frac{NX'}{XN}$$

Thus;
$$\frac{I}{O} = \frac{v}{u}$$

Since
$$v = \frac{1}{V} \text{ and } u = \frac{1}{U},$$
$$\frac{v}{u} = \frac{U}{V}$$

So, we have three different, but exactly equivalent, ways of expressing the linear magnification M;

$$M = \frac{I}{O} = \frac{v}{u} = \frac{U}{V}$$

M may be greater, equal to, or less than 1, depending on whether the image size is (respectively) larger than, equal to, or smaller than the object size.

That's all there is to it; and it makes no difference whether the objects or images are real or virtual. Continue to use our same sign convention for vergences and distances and you will find that when Magnification Power turns out to be *minus,* this will always indicate that the image is *inverted* compared to the object, while plus "says" the image is upright.

To find the magnification M, you can use either a geometrical construction (to a set scale) or our simple $U + P = V$ relationship. The algebraic expression is obviously the easiest and most direct method to use routinely.

PROBLEM: What is the overall magnification and the actual size of the projected image produced by a 5 cm focal length projection lens using a 35 mm width slide transparency located 6 cm from the lens. The image is sharply projected on a screen.

ANSWER:

To obtain the Magnification, we must know the image distance (or its vergence) at P.

$$U + P = V$$

$$-\frac{1}{.06} + \frac{1}{.05} = V = \frac{1}{v}$$

$$-16.7\text{ D} + 20\text{ D} = V$$

$$+3.3\text{ D} = V$$

$$\frac{1}{V} = v = +30\text{ cm.}$$

So, the screen is 30 cm to the right of the lens; $u = -6$ cm
$v = 30$ cm

a) $M = \frac{v}{u} = \frac{30}{-6} = -5\text{ X}$

Or even simpler, just use the vergences $U$ and $V$ themselves:

$$M = \frac{U}{V} = \frac{-16.7\text{ D}}{+3.3\text{ D}} = -5\text{ X}$$

This means the image is 5 times the size of the object; the minus sign signifies that the image is inverted related to the object.

b) Since the object is 35 mm wide and the magnification is 5 X, the image is 5 *times* 35 = 175 mm wide.

## PRINCIPAL PLANES AND POINTS

I would like now to introduce another set of terms to complete (*not* complicate) our introduction to the nomenclature applied to optical systems. I will expend slightly more space on this subject than it warrants for the level of optics required by students, but so many budding ophthalmologists continue to ask me to explain (not just define) the concept of *principal planes* that I felt it would save time in the long run to do so here.

When we were studying image formation by a series of thin lenses, we showed that we could locate the final image by either numerical or graphical means, but only after processing object rays successively through *each* lens element in the total system. We will now demonstrate diagrammatically a way to simplify the situation, even with a complicated optical set up. (See the next figure). We will then eliminate all the refracting elements and replace them with two theoretical (though mathematically proper) "refracting" planes. The positions of these planes will be determined. These planes will permit us to neglect all the lenses shown in the figure; that is, each ray emanating from object point A will be able to be treated as if it were influenced only by these two planes. These key reference planes are called the PRINCIPAL PLANES — one primary and one secondary. Their intersections with the lens axis are correspondingly called the PRINCIPAL POINTS.

To see how these planes are located, let us begin with a diagram; but don't let it frighten you. I have *purposefully* chosen a rather complex system of seven assorted lenses to demonstrate how these reference planes can simplify the optical considerations.

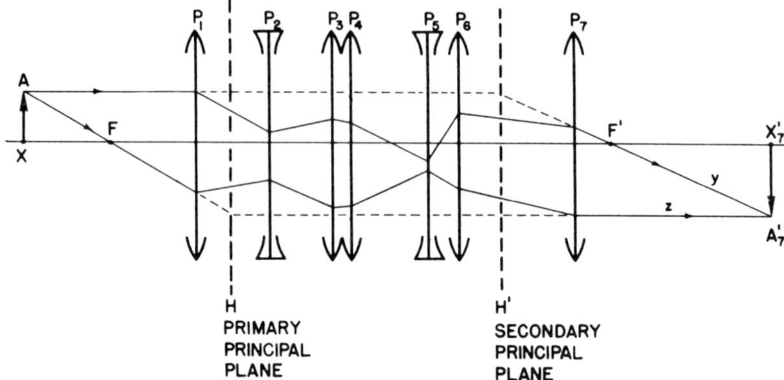


Somehow, by any mechanism we wish (optical bench experiments or actual calculation), we locate F, the primary focal point of the entire system (not just of lens $P_1$), and F', the system's secondary focal point. We succeed in imaging object AX through all seven lenses to its final position $A_7'X_7'$; so, $A_7'$ is fixed. Now, let us draw the one ray from A to lens $P_1$ which passes through F. This ray will then go through the entire optical system. (To know the *exact* path of this or any ray through any such complex system, one would have to go through a process of "ray tracing", determining the influence of each lens on the ray. So, let us not worry about the exact path through the system.) Suffice it to say, after assorted bumps and grinds, the ray leaves the last lens $P_7$ headed directly for $A_7'$. This final ray will be parallel to the axis since it originally went through F, the system's primary focal point. The final ray is shown as ray Z.

For our analysis it will make no difference what actual bending and flexing this ray has been subjected to in its passage through this complex system. If we simply extend ray Z straight back *as if it were uninfluenced by any lens,* it will somewhere intersect the extension of our original object ray which passed through F of the lens system. This intersection point determines a plane H (dotted in the diagram) which we make perpendicular to the lens axis. It should be obvious that our light ray from A (through F) could be considered as being refracted "by some mystical power" located in plane H, whence it would leave parallel to the axis and later arrive at $A_7'$. Any plane so constructed is named the *primary* principal plane.

Back to object point A: Also emitted from A is one ray which leaves parallel to the axis. It also is refracted by the entire lens system and finally emerges from lens $P_7$ aiming towards $A_7'$. This final ray must cross the axis at F'. This latter ray (Y in the diagram) is extended further backward to intersect the extension of our original object ray (the one that left A parallel to the axis); this intersection determines another plane — the *secondary* principal plane — which is a "surface" acting as if the final refraction took place there. This plane is labeled H' in the diagram.

These two planes, then, can be considered and treated as if they replaced all the other optical elements. This is schematically shown in the figure below:

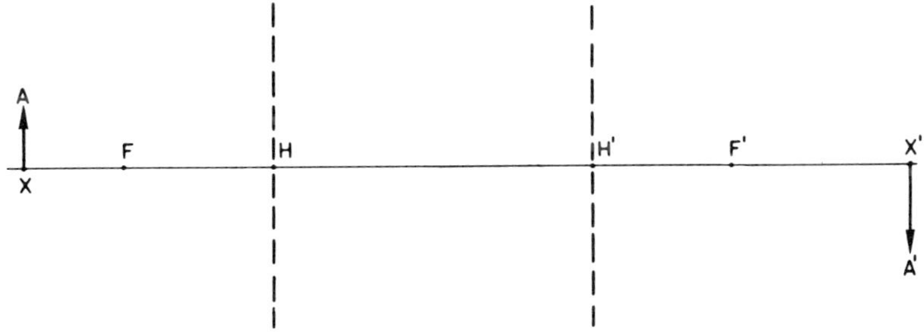


The positions of these planes can be accurately determined by mathematical computations, but these become more arduous as the lens systems become more complex; they *are* relatively easy to establish on an optical bench in the laboratory, if one wishes.

You can see that there is a physical separation apparent between planes H and H'. Here the space seems large; other times the gap is very small. In either case, in geometrical constructions this gap is treated *as if it weren't there,* and any ray from object A which hits plane H, leaves plane H′ at precisely the *same height* — as if the planes were in contact. (You will not usually see any rays drawn within the space HH′).

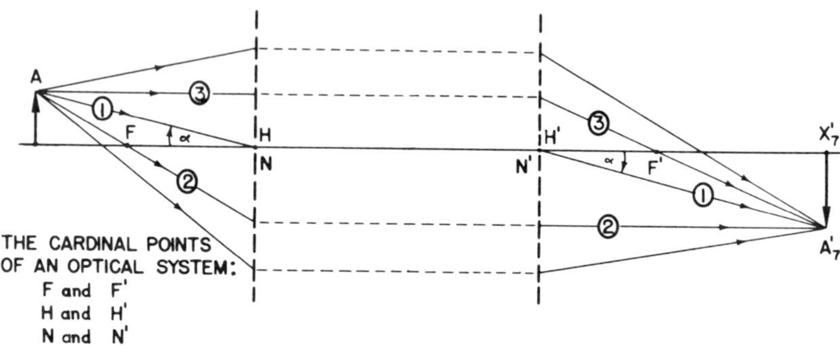


Since any incident ray impinging on plane H will always leave plane H′ at the same height from the axis, we say that the principal planes exhibit the property of *unitary linear magnification.* (Also, these planes are actually *conjugate,* that is, one is the image of the other, optically.)

In the above figure, ray 2 from A through F will arrive at plane H; it will leave from plane H′ parallel to the axis. Ray 3 from A drawn parallel to the axis will leave plane H′ directed to F′ and on to the image $A_7'$. Ray 1 from A, which aims for the axial point of plane H, exits from plane H′ also at the axis *and* at exactly the same angle as the incident ray! This should call to mind the "chief" ray which, when we dealt with the simple thin lens, passed undeviated through the lens *nodal* point. (The close relationship between nodal points and principal points will be discussed shortly. They are *not* identical, though here, where we have labeled H and H′ as the *axial* principal points of their corresponding planes, these same points also represent the nodal points of *this* system; that is, N and N′ happen to be superimposed on H and H′.)

In a complex optical system such as this (actually, in *any* optical system), the principal planes are *the* reference planes; all object and image distances are measured relative to them. So also, the primary and secondary focal lengths are measured to H and H' and are defined as FH and H'F', respectively. But this requires that the position of H and H' be known; if so, distances FH and H'F' will provide a measure of what is called the *true* or *equivalent* focal length — the reciprocal being the *equivalent power.* Though this latter unit is generally conceded to be the *standard* description of a lens' power, it is *not* a clinically useful unit, because the positions of H and H' are not readily found — they are "intangible".

So, when the principal planes are not identified or localized for you, the focal lengths must be measured in reference to some other, more *convenient* surface, like the axial point on the anterior surface of the lens; its distance from F would then be called the "anterior" focal length. Better yet, measure from the axial point on the *posterior* surface of the lens to F', the secondary focal point. *This* distance is called the "posterior" or "back" focal length.

Become familiar with the "back" focal length; *it* is the lens focal length that is implied (when not stated to the contrary) whenever you deal with *ophthalmic* lenses. The clinical instrument known as a lensometer measures this "back" focal length of an "unknown" lens; the instrument's calibrated dial indicates the reciprocal of the "back" focal length, that is, the "back" or *vertex* power. (To add further confusion, "back" vertex power is *also* called the *effective* power of a lens.) In any case, the *vertex* power and not the *true* power is the clinically important one.

Oops! We seem to have drifted off the subject of principle planes for a moment. Back on the track now, I must stress that distances FH and H'F' (the *true* focal lengths) will be equal to each other (as are *both* the focal lengths of any of the thin lenses we have already studied) *if* the media composing both the object and image spaces are of the *same* "refractive index"; (we will get to this term later, in the next section.) On the other hand, if the "refractive index" is *not* the same in both "spaces", the focal length on the side of greater refractive index will be *longer.* More about this later.

## NODAL POINTS

We now come to a place where we can treat the nodal points more specifically. By definition, they are a conjugate pair of axial points (the object and image of each other) which have the following property: any ray striking the primary nodal point leaves the secondary nodal point with an identical inclination to the axis. (For the sophisticate, they are points of *unitary angular magnification.*)

If N and N′ are nodal points of an optical system, and ray 1 strikes N at $\alpha^{\circ}$ to the axis, it will leave N′ at that same angle.

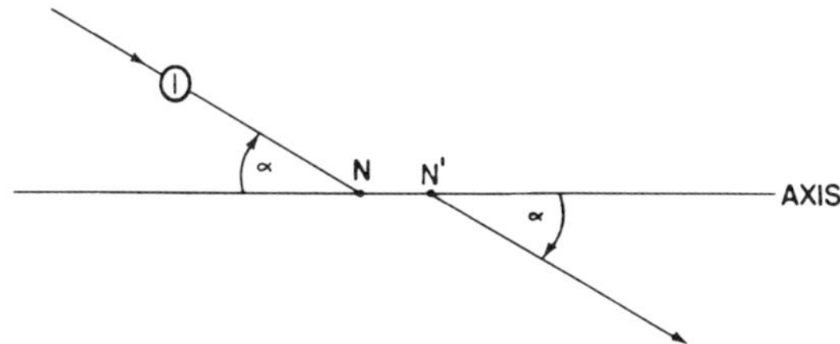


As long as the medium composing the object space and the medium making up the image space are identical — so far, we have considered only *air* in both "spaces" — the primary *nodal* point is located at precisely the same position as the primary *principal* point, and the secondary nodal point at the secondary principal point, that is, they are superimposed; however, if the media are different (as they are for the eye) both nodal points together shift *away from* superposition with the principal points. How much they shift depends upon the media. They will always shift in the direction of the greater refractive index. One principle is always true: The primary focal length of any optical system is always equal to the distance between N′ and the secondary focal point; that is, FH = N′F′. Recall this later when we deal with the eye.

It should now be clear that nodal points (there are no nodal planes) and principal planes and points, as well as focal points, are convenient and important reference positions for *all* optical systems and all join together to describe completely the focusing action of a particular system. The six axial points we have thus far discussed — (two principal, two nodal, and two focal) are called the *cardinal* points of any optical system. (Two *other* pairs of points which you may see listed in other texts are much less important but are also sometimes included as "cardinal" points — these are the *negative nodal points,* and the *symmetrical points.* Forget them for our study.)

In the complex optical system demonstrated here (as well as in the eye), the primary and secondary principal points are physically separated from one another; (so also would be the primary and secondary nodal points N and N'). The actual separation between H and H' and between N and N' is always *identical* for *any* given optical system; (even when the Ns are not superimposed on the Hs). But, this separation does vary with the nature, complexity and linear separation of the elements making up the complete system. For the human eye, the separation between the primary and secondary principal points (and also the corresponding nodal points) is only 0.3 mm.

So, you might ask, when we began fiddling with our single thin lenses, changing the vergence of object rays (using $U + P = V$) and constructing images by drawing those three particular rays, how come we completely ignored H, H', N and N' as reference points? The answer is, we didn't. With thin lenses, all four of these points coincide with the vertex (axial position) of the lens. It is only with more complex systems that individual attention to these points becomes important. Thus, you should now realize that with thin lenses, even though the four individual cardinal points are not obvious, they are there nonetheless, at all times, huddled together and hidden.

## SNELL'S LAW AND REFRACTIVE INDEX

We need to have a few more principles at our fingertips *before* we can indulge ourselves with the "meat" of our course — the optics of the eye, the subject which *really* is what we've been waiting for. This present section, of necessity, is a bit more mathematical, but certainly not offensively so; so, approach it with an open mind.

At the outset, we began by leaping headlong into the concept of vergence and how lenses change that vergence, but we neglected to give you any background in how lenses really work and what governs their activity. As good a place as any to begin is with the basic and fundamental law of optics which makes all lenses and refracting surfaces work to form images — Snell's Law.

This law concerns itself with exactly to what extent each light ray is bent by surfaces which are separated by media differing in *refractive index.* From your early physics courses, you should remember that this *index* is a property of transparent media; as light passes through any medium other than a vacuum, it is slowed down. The index is simply the ratio of how fast light travels in a vacuum compared to its speed in the specific medium. The refractive index

$$n = \frac{\text{velocity of light in a vacuum}}{\text{velocity of light in the specific medium}}$$

Since the denominator of this fraction is always less than the numerator, this ratio $n$ is always greater than 1 for any medium other than the vacuum (or, more practically, air).

Interestingly, the velocity with which light travels in any medium depends not only on the medium itself but on its own wavelength. Each wavelength has its own "private" index of refraction for each particular medium; the index you usually see listed in tables is for the specific wavelength of sodium light, 589 nm ($10^{-9}$ meters). For water, this particular index is $n = 1.333$, for crown ophthalmic glass (1.523), for plastic (1.491), for the lens of the eye (1.42), and for the cornea (1.376).

When a ray of light which is traveling on one medium hits another medium of a greater index of refraction, that ray of light will be slowed down (and vice versa). If it strikes the material perpendicular ("normal") to its surface, though it is still slowed down, it does not change direction but continues on in the same direction, 90° to the plane of the surface. If, however, the ray of light strikes the new material at some angle (inclination) to the "normal", it will be deviated after crossing the boundary. *Snell's Law will tell us how much this ray is bent.*

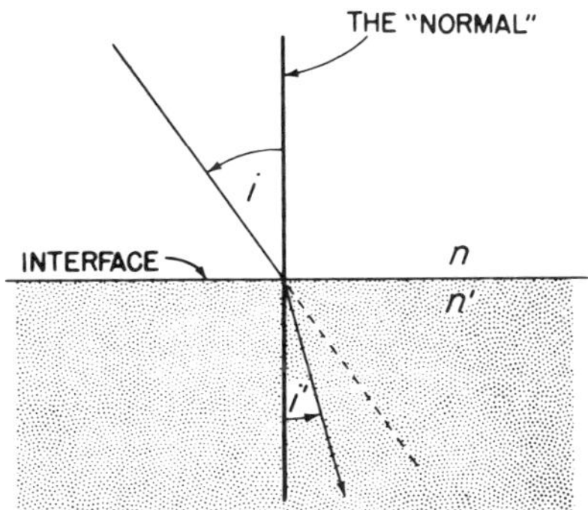


Let $n$ be the index of refraction of the medium on one side of the interface and $n'$ its counterpart — remember that these indices are for a *single* wavelength of light. The angle which an incoming ray makes with the "normal" we will call $i$. After entering the new medium, the ray is bent — *toward* the normal *if* $n'$ is greater than $n$. The angle after refraction is denoted by $i'$. $i$ and $i'$ are called the angles of incidence and refraction, respectively, and are always measured to "the normal", the line drawn perpendicular to the surface at the point where the ray strikes it.

The basic law of Snell — which governs all refraction and forms the basis of how lenses work in their changing the direction of light rays — is as follows:

$$n \sin i = n' \sin i'$$

Simple enough; but, since this represents a trigonometric relationship, let us try to simplify it even further. (I feel it is definitely worthwhile to go through this completely, since you will automatically pick up familiarity with some units you should understand.)

We will begin by looking at how angles are measured.

## ANGULAR MEASUREMENT

In high school geometry and trigonometry, you were given different units with which to measure angles — the degree and the radian.

When studying the field of strabismus, you will encounter three additional ones — the prism diopter (the only important one), the centrad, and the meter-angle. It would be useful to review each of these units briefly:

a) *The Degree* — simply defined as 1/360 of the complete angle around a point.

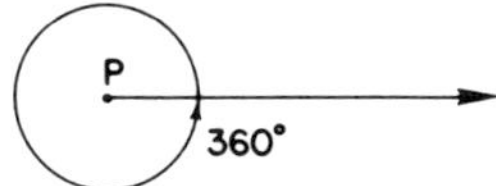


b) *The Radian* — Construct a circle of any radius, and draw two radius arms extending from the center.

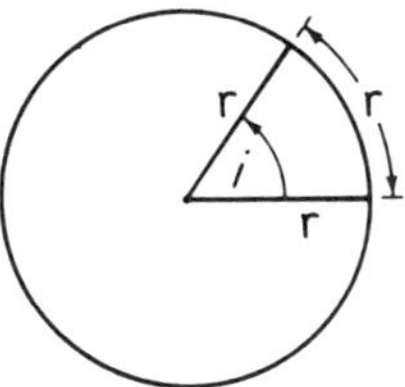


The included angle *i* measures *1 radian* when the actual length of the curved arc subtended is equal to the length of the radius.

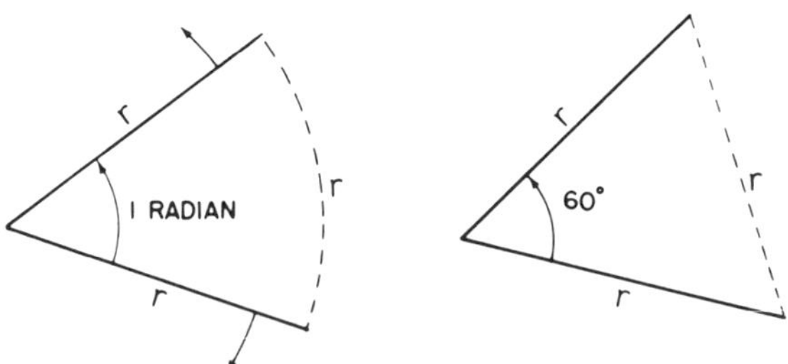

If the circle circumference were made of string and the stiff-radius ends were fixed to it and the included angle was 1 radian, one can see that angle *i* could increase somewhat before the string became taut to a straight line. When this point is reached, *i* would, of course, then be 60°, since the "straightened out arc" would still be equal to the length of the radius. This makes the triangle equilateral (with each angle equal to 60°). So, a measure of 1 radian must be *less* than 60°; let us now see how much less and how we find the exact equivalent.

In any circle there are always a fixed number of radius lengths which (laid end to end) would complete the circumference. That number is $2\pi$; that is, there are $(2\pi)$ radii which make up the complete circumference of 360° (Remember? $C = 2\pi r$). Since *each* $r$ measured along the circumference represents 1 radian of central *angular* measure (the definition of "radian"), and since there are $2\pi$ radii in the complete circumference, there must be $2\pi$ radians of angular measure equivalent to 360°. Since $2\pi$ radians $= 360°$, 1 radian must equal 360° divided by $2\pi$ (approximately 57.3°); and conversely, 1° would equal $\frac{2\pi}{360}$ radians. The reason we are dragging this unit up from the deep dark past will soon be apparent.

c) *The Prism Diopter* — The angle corresponding to an apparent displacement of 1 cm at 1 meter distance.

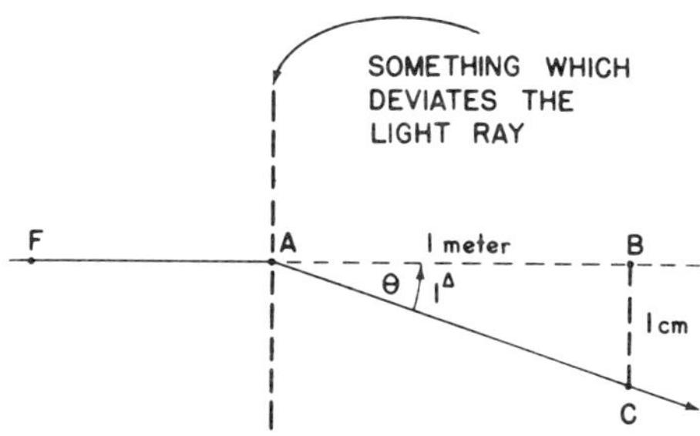


Ray FA (after encountering some optical device) is deflected from its "straight ahead" path by angle $\theta$ towards C. Since triangle ABC is a right triangle, tangent $\theta = \frac{BC}{AB}$.

By definition, when $\theta = 1^{\Delta}$ (prism diopter),

$BC = 1$ cm and $AB = 1$ meter.

Then, $\tan\theta = \frac{.01 \text{ meter}}{1 \text{ meter}} = .01$.

By the same reasoning, if $\theta$ is $5^{\Delta}$, $\tan\theta = .05$.

Now you should see that the converse is also true, that is, for any angle $\theta$, the expression *100 tan θ* equals the angle in prism diopters. So, the conversion from degrees to prism diopters (and vice versa) is tied to the *tangent* of the angle, but, herein lies a problem. Look at Table I, which demonstrates the relationship between $\theta$ in degrees, the tangent $\theta$ (looked up in a standard math table) and the prism diopter:

*TABLE 1*

| *Degrees* | *Tan θ* | *100 Tan θ (Prism Diopters)* |
|---|---|---|
| 1 | .01746 | 1.7 |
| 2 | .03492 | 3.5 |
| 10 | .17633 | 17.6 |
| 20 | .36397 | 36.4 |
| 30 | .57735 | 57.7 |
| 44 | .96569 | 96.5 |
| 45 | 1.00000 | 100.0 |

At small angles, 1° is equivalent to $1.7^{\Delta}$; at larger angles (say between 44° and 45°) 1° is equivalent to $3.5^{\Delta}$! What self-respecting kind of *unit* would change its size along a scale? It should be clear, then, that the prism diopter (100 tan $\theta$) is *not* a true unit; it keeps changing its size compared to the degree (which, of course, *is* a true unit).

Don't get me wrong — the prism diopter is a *useful* "unit", but only for measuring *small* angles. In the typical clinical situation when you are measuring, say, esotropia, you usually deal with angles of up to about 30°. In this range of 0° to 30° we can consider 1 degree *approximately* equal to 2 prism diopters; this assumption is actually quite close to being correct. But, keep in mind that the error introduced by using this approximation with larger angles can be large; the "unit" becomes rapidly meaningless when dealing with angles greater than 45°.

d) *The centrad* ($^\triangledown$) is similar to the prism diopter except that the 1 cm length displacement (at 1 meter) is measured along the *arc* of a circle instead of a straight segment. For small angles the centrad is approximately equal to the prism diopter.

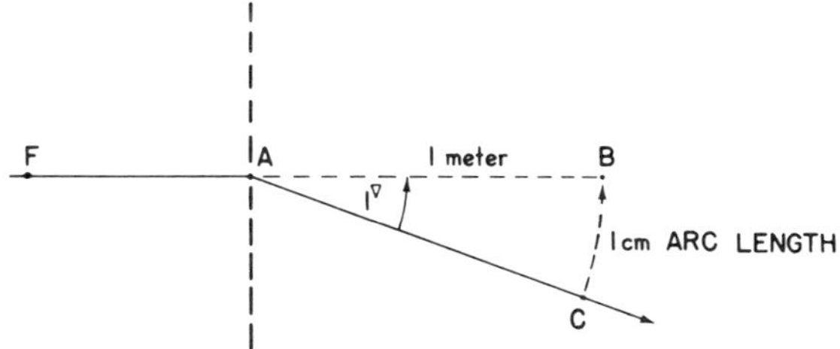


Now, *forget* the centrad; it is outmoded and obsolete.

e) *The meter-angle:* This angular measure is also not a very important one, but it is encountered in the squint literature in connection with the determination of the ratio between accommodation and its synkinetic accommodative-convergence, that is, the $\frac{AC}{A}$ ratio. (The prism diopter is much more appropriate as the unit for this purpose.)

In any case, when a close-up object is being scrutinized by the eyes, the angle (of convergence) between the two visual axes may be given in meter-angles.

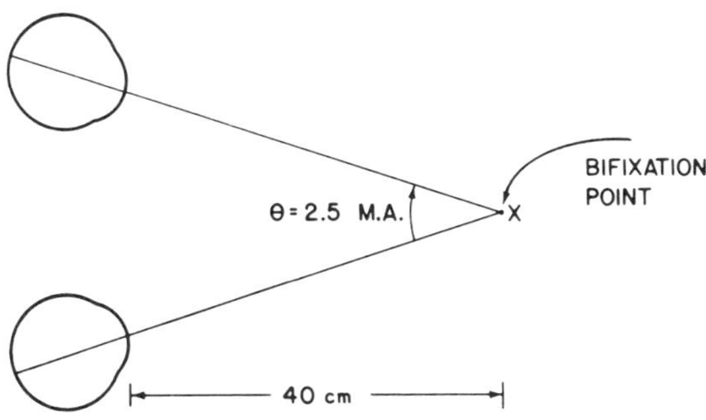

If the eyes converge to point X, angle $\theta$ is automatically expressed in meter-angles when the distance of X from the eyes is given in diopters of vergence. So, if X is 40 cm away, angle $\theta$ is 2.5 meter-angles.

This "unit" also is not a *true* unit since the angle $\theta$ (if it is expressed in degrees or radians) depends on how far apart the eyes are; that is, the true angle *should* depend on the interpupillary distance (p.d.). If the p.d. were greater than that shown here, $\theta$ would actually *have* to be greater; yet it still remains only 2.5 meter-angles! The use of "meter-angles" avoids taking account of the actual p.d., and this turns out to be both the advantage and the disadvantage of this "unit".

We would be quite well off without this term too.

The student may question the need for this digression here into the units of angle measurement. In its defense, not only did I want all readers to understand each unit, but I also hoped that this would allow us to begin at the same baseline of information regarding the *radian.*

We have already noted that central angle *i* (in the figure below) can be expressed in degrees or radians. The subtended arc also can be given in either degrees or radians, its measure being exactly equivalent to that of central angle *i* (that is, we can speak of the *arc* itself as equal to, say, 45° or .78 radians) However, the actual measurement (in inches or meters) of the *length* of arc *i* is obviously dependent on the length of the radius. With the same central angle, the longer the radius, the longer the length of arc *i*. If the angle *i* is 1 radian, arc *i* exactly equals the length of the radius; if angle $i = 0.5$ radians, arc *i* equals one-half the length of the radius. Thus, arc *i* always equals the central angle *i* (in radians) *times* the length of the radius.

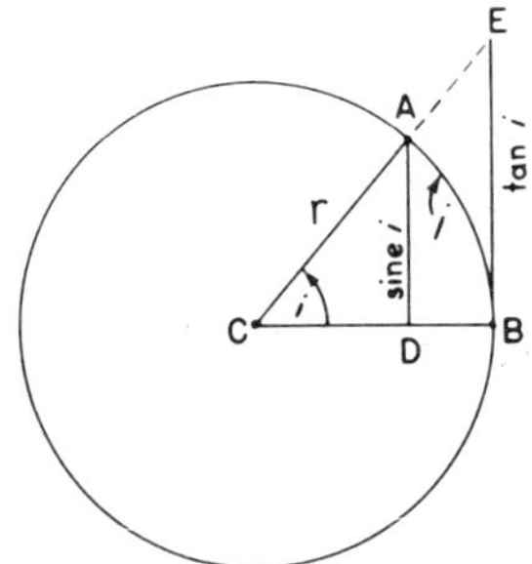

We have drawn the circle above with its center at C and two radii, CA and CB, and constructed two perpendicular lines to radius CB; one is dropped from the end of radius CA, the other at the end of radius CB.

$$\sin i = \frac{AD}{AC} = \frac{AD}{r} \qquad \tan i = \frac{EB}{CB} = \frac{EB}{r}$$

If we assume r = 1 "unit" of any length,

then $\sin i = AD$

and $\tan i = EB$.

In other words, lines AD and EB are the linear representations of the sine and tangent of angle *i*, respectively.

As we have shown, arc AB in the above figure equals *i* (in radians) times the length of the radius. Since the radius is equal to 1, arclength AB is equivalent to angle *i*.

From the diagram, let us now list these three lengths in order of size: AD is shortest, then arclength AB, and longest is BE; thus, by substitution

$$\sin i < i < \tan i.$$

You should now be able to picture what happens as angle *i* decreases; these three become just about equal to each other. So, for relatively small angles, *i* (in radians) is a very acceptable substitute for the trigonometric expressions sin *i* and tan *i*. To show you how little error is introduced by such a substitution, I have constructed Table II.

*TABLE II*

| *i* in degrees | sin *i* (from math tables) | *i* in radians | tan *i* (from math tables) | % error introduced by using *i* (radians) instead of sin *i* | % error introduced by using *i* (radians) instead of tan *i* |
|---|---|---|---|---|---|
| 1° | .01745 | .01745 | .01746 | 0.00% | 0.01% |
| 2° | .03490 | .03490 | .03492 | 0.00% | 0.01% |
| 10° | .17365 | .17450 | .17633 | + 0.49% | — 1.04% |
| 20° | .34202 | .34900 | .36397 | + 2.01% | — 4.12% |
| 30° | .50000 | .52350 | .57735 | + 4.5 % | — 9.35% |
| 45° | .70711 | .78525 | 1.00000 | +11.1 % | —21.4 % |

This table shows the percentage error introduced by substituting $i$ (in radians) for sin $i$ and for tan $i$ for various angular measures. Three points should be clear:

1) $i$ is an excellent substitute for sin $i$ and tan $i$ for small angles, probably up to 20°. Even for angles as large as 20°, $i$ errs by only 2% when substituted for sin $i$ and 4% for tan $i$.

2) $i$ is a somewhat better substitute for sin $i$ than for tan $i$.

3) When one substitutes $i$ for sin $i$, the approximation will be slightly too large; when using $i$ for tan $i$, it will be slightly too small. (This is also evident by looking at the last figure, specifically at the lengths AD, $\widehat{AB}$, and BE; if you compare these lines you will have a *graphical* demonstration of our approximation $i$ for sin $i$ and tan $i$.)

We can investigate this approximation in still another way. When any angle $i$ is expressed in radians, we can find the value of sin $i$ by simply substituting this value of $i$ in a mathematical "trigonometric expansion" which is the mathematical equivalent of sin $i$. (Incidentally, this is how the math tables themselves are constructed.) Don't close your eyes! It's not that scary. That expansion is as follows:

$$\sin i = i - \frac{i^3}{3!} + \frac{i^5}{5!} - \frac{i^7}{7!} + \ldots \quad (3! = 3 \times 2 \times 1)$$

$$(5! = 5 \times 4 \times 3 \times 2 \times 1)$$

Also,

$$\tan i = i + \frac{i^3}{3} + \frac{2\,i^5}{15} + \frac{17\,i^7}{315} + \ldots$$

We can thus find the value of sin $i$ or tan $i$ as accurately and to as many decimal places as we wish. The question is how many of those $\frac{i^n}{n!}$ terms (for sin $i$) must be used to be reasonably accurate. The obvious answer, as we have shown in Table II, is that if we limit ourselves to $i$ of *small* angles, we can leave off all the terms beyond the first one; we make almost no error in stating that sin $i$ is equal to angle $i$ itself (when $i$ is expressed in radians). *This* is the approximation we make in what is called *first order optics.* It is somewhat analogous to using the approximation 3.1 for $\pi$; if you need a more accurate value, you could add another term, as 3.14. Likewise, if you need a more accurate value for sin $i$, you can include the second term $(-\frac{i^3}{3!})$ too.

When we do use the more refined value ($i - \frac{i^3}{3!}$) for sin $i$ in optical calculations, we are using *"third order" optics* — note there is no "2nd order". Comparable terms can also be added for tan $i$ to increase its accuracy. The additional degree of accuracy for both sin $i$ and tan $i$ is necessary to account for some of the lens aberrations. Astronomers may require 5th or even 7th order optics for their accurate calculations of star positions. But, the point of importance for us clinicians is that for everything we will learn here, we really need only the 1st term ($i$) to substitute for sin $i$.

First order optics includes the evaluation of object and image rays — called *paraxial* rays — which lie close to the axis of refracting systems, so that angles of incidence and refraction are relatively small. In spite of the fact that we use large angular diagrams in this book to elaborate certain principles, we must realize we are still only describing *accurately* the action of lenses on the *paraxial* rays.

Later, when we deal with curved refracting *surfaces*, we will substitute $i$ for sin $i$; Snell's Law will become simplified to $n\,i = n'i'$, and correctly so, as long as we do not deal with too large angles and we express $i$ in radians.

## PLANE SURFACE REFRACTION

Snell's Law governs the refraction of light rays. *Without* using any approximations now, we can say, $n \sin i = n' \sin i'$.

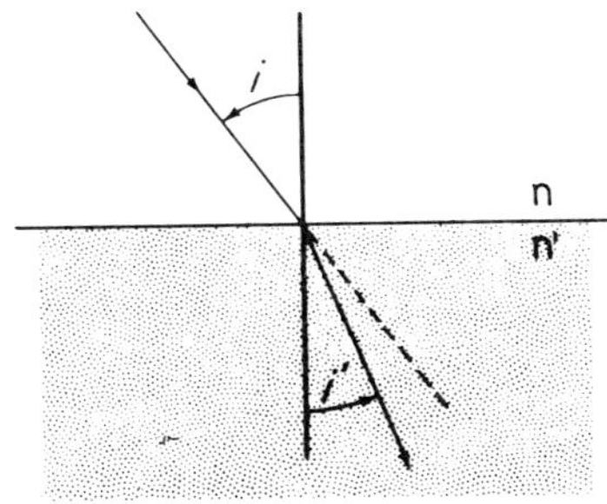

A light ray passing from a less dense medium to a denser one (with a greater index $n'$) is bent *toward* the "normal"; and conversely, light moving the other direction (from $n'$ toward $n$) is bent *away* from the normal. (Remember that all light rays are retraceable.)

## "Critical Angle"

When light emanates from an object point located within the denser medium (figure A below), the stage is set for a peculiar phenomenon to occur.

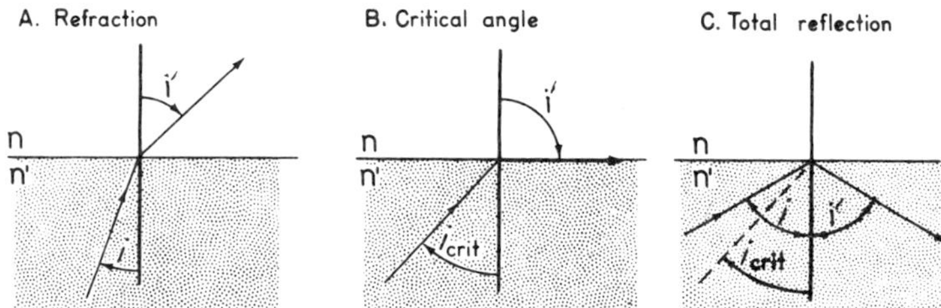


As angle $i$ increases, so does angle $i'$. Angle $i$ will eventually reach a certain magnitude such that $i'$, the angle of *refraction*, will become equal to 90° (figure B above). At this position, $i$ (the angle of incidence) is called the *"critical"* angle. If $i$ now is increased further, even a slight amount, the ray will *not* exit from medium $n'$ at all; it will be totally reflected* internally (as shown in C above).

Since, in this example,

$$n' \sin i = n \sin i'$$

$$\sin i = \frac{n}{n'} (\sin i')$$

By definition, at $i_{crit}$ (the "critical" angle)

$$i' = 90°, \text{ thus } \sin i' = 1.00.$$

Therefore,

$$\sin i_{crit} = \frac{n}{n'}$$

When n' represents water ($n' = 1.33$) and $n = 1.00$ for air,

$$\sin i_{crit} = \frac{1}{1.33} = 0.75$$

$$i_{crit} = 48°.$$

* We will take up "reflection" later.

Hence, the critical angle for water is 48°, but only for a *water-to-air* interface. Similarly, the sine of the critical angle for corneal tissue $= \frac{1}{1.376}$;

$$\sin i = 0.725$$
$$i_{crit} = 46.5°$$

This point has clinical importance. Recall that, if we are to see any object at all, light rays from it must enter our eyes. Consider the situation of the anterior chamber angle; light rays which come from the angle must pass out through the cornea. They do pass and are refracted by the posterior corneal surface; however, because of the particular dimensions of the anterior chamber angle and its distance from the cornea, the light rays which leave there strike the *anterior* corneal surface (an interface with a greater index of refraction *inside* than outside) at an angle of incidence *greater* than the critical angle of corneal tissue ($\angle$ 46°). So, all these rays are reflected back into the eye. Since they are unable to escape, a clinician normally is unable to visualize the anterior chamber angle of a patient. If the cornea happens to be "steeper" than normal (as in a patient with keratoconus), the rays may strike the interface with *less* angular incidence than the critical angle and thus might be able to leave the eye. So occasionally, the chamber angle may be seen by an observer, but not usually. (See A next figure.)

## Diagnostic Goniolenses

In the *typical* patient, the angle can only be visualized with optical help — by "optically" removing the corneal front surface and replacing it with a new surface (or one with a *different* curvature) which allows the light to escape. This can be done with a contact lens whose own index of refraction is substituted for that of the air. This decreases the *difference* in index of refraction across that interface and thus, optically "removes" the original corneal surface. It works by the same principal which causes a glass marble to disappear when it is immersed in a dish of water. The glass ball is easily visible in air because one "sees" the surface due to the marked difference in index between the glass and the air. But, when it is put into water, the indices are so close that the marble's surface is optically eliminated. *Therapeutic* contact lenses* which correct corneal irregularity (as in keratoconus) or even correct simple refractive error also work in the same way, with any necessary prescription "corrective power" being ground onto the front surface of the lens.

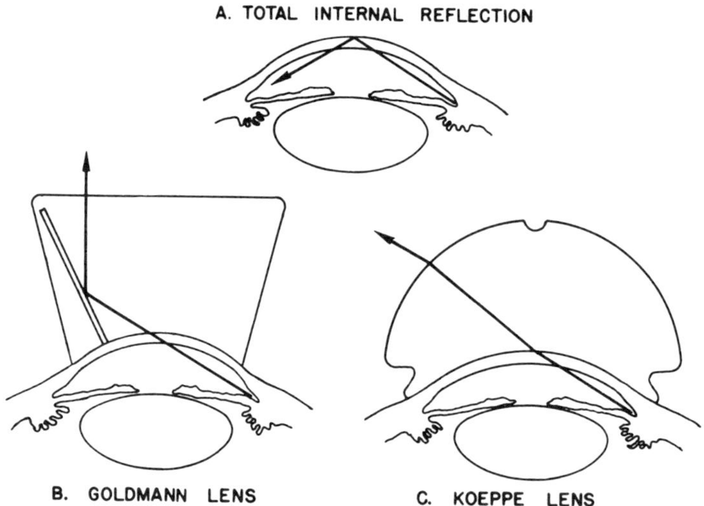


Two main types of diagnostic contact lenses are used for gonioscopy, the viewing of the chamber angle: a Koeppe contact lens allows *direct* viewing of the angle (see Figure C; and a Goldmann type lens incorporates a mirror so that an examiner sees a *reflected* or *indirect* view of the anterior chamber angle on the opposite side (that is, an upper mirror allows view of the lower angle). (See figure B above.)

The presence of a critical angle is not always a disadvantage; it *can* be used constructively. For their useful operation, many optical instruments require that a light beam change its direction. An ophthalmoscope is a good example. It requires "something" to bend the light

* The subject of therapeutic corneal contact lenses is so broad that it would be out of place in this book and will not be discussed further here. Excellent texts (such as *Corneal Contact Lenses* by Girard, Soper and Sampson, Mosby, 1970) are available for consultation and reference in this important clinical area.

beam which runs parallel to the handle and change its direction by 90°. This allows you to shine the beam into a patient's eye while you are observing the fundus. Another example is the typical binocular telescope ("binoculars"); this instrument would be much longer in dimension were it not for the fact that the light path is "folded" inside. The changes of light direction in both these instruments is accomplished by prisms which totally *reflect* light (as would a mirror, but more efficiently — with *less* light loss).

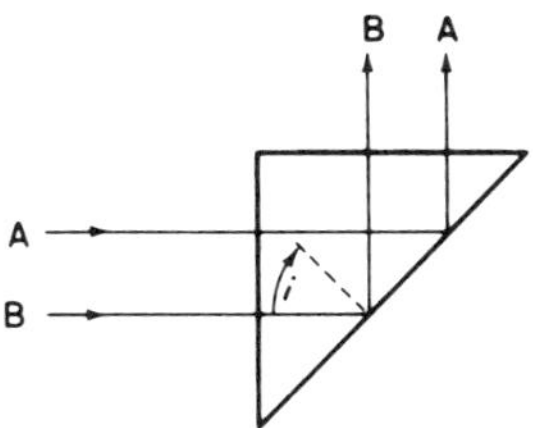


This total reflection is shown in the accompanying diagram above. Since the angle of incidence *i* exceeds the "critical angle" of this prism material, the light rays are totally reflected internally. For special purposes, many other prism forms are available which not only totally reflect light but also twist and reorient the image as well. (e.g. Porro and Abbe prisms.)

There are many other useful applications of total internal reflection, for example, "fiber optic" bundles.

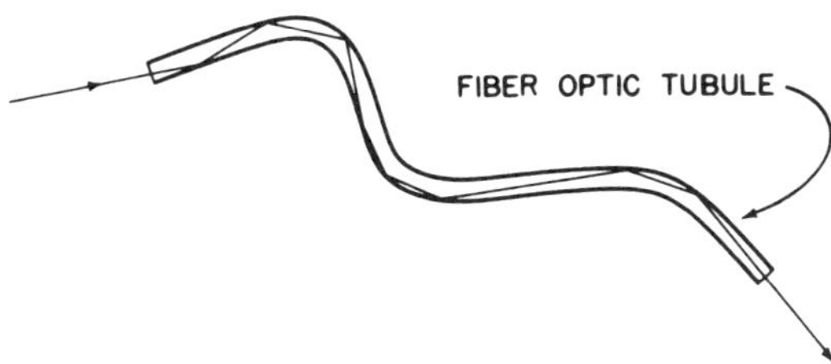

Thin flexible tubes of transparent material capture light at one end and keep it "trapped" inside by total reflection until it reaches the other end, where it is allowed to exit. The beauty of these "light pipes" is they can be bent into unusual shapes and the light will follow the bends, seemingly in defiance of the principle of *linear* propagation of light. Not only can the light emitted at the end of such a "pipe" be used as an illumination source for dark recesses, say, of the inside surface of the stomach; if properly constructed, *optical* fiber bundles (created by thousands of very fine, perfectly aligned, closely packed tubules) can allow sharp visualization and photography of these same, hitherto unseeable areas. New endoscopic applications for these fiber optic bundles are continually being discovered.

You should also know that the visual cells (rod and cone) outer segments "trap" light via this same mechanism of total internal reflection and thus permit more efficient utilization of light entering the eye.

## Prismatic Deviation

So far, we have explored the refraction of a *single* light ray by a plane surface separated by two different media. Now let us follow a number of these rays which leave an object point.

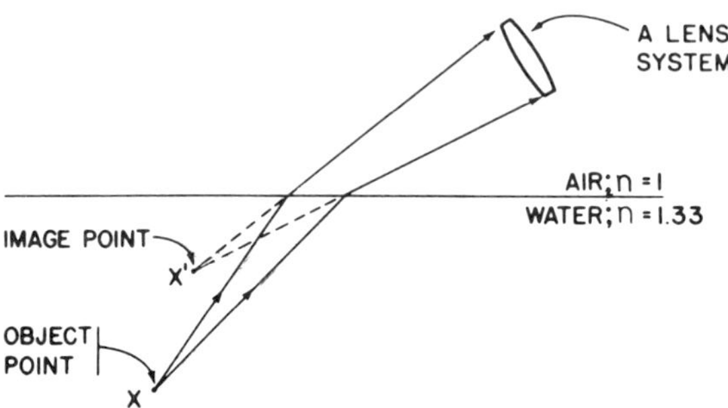


Object point X is immersed in water. The light rays from X are shown; each is bent appropriately (Snell's Law) as it leaves the water. These rays will seem to emanate from point X' (in a straight line path)

to any optical system (camera lens or a human eye) which intercepts them. Point X′ will seem closer and not as deep in the water as the original point X, and accounts for the "displaced" position of objects which are under water. The greater the viewing angle $\theta$ makes with the water, the greater the apparent displacement. But, even if the view is directly perpendicular to the surface, the image point X′ will still seem closer than the object really is. (You should now be able to diagram this yourself.)

This is our first exposure to "displacement" caused by a plane refracting surface. It leads us to consider refraction and displacement by a two surfaced plane-parallel, such as a slab of glass with parallel faces.

Refraction by two plane-parallel surfaces:

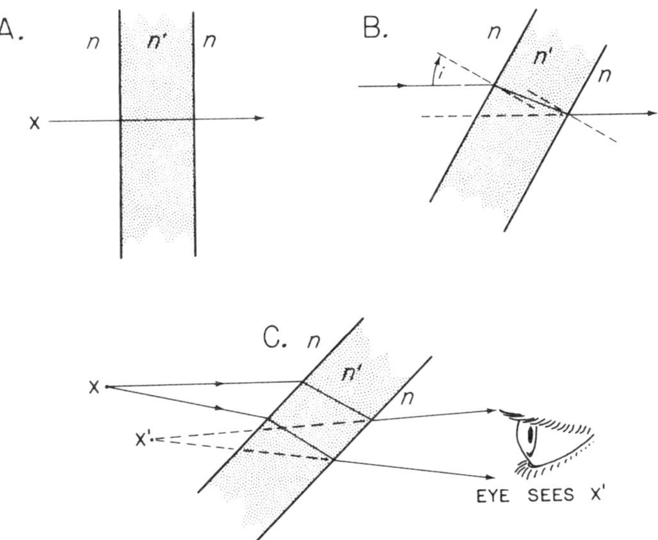


A) If the sheet of glass has 2 parallel faces and a ray strikes along the "normal", there is no deviation of this ray.

B) If the glass plate is tilted so that the "normal" makes an angle $i$ to the incoming ray, the ray will be bent towards the normal inside the glass and then away from the new normal as it exits again.

C) To an eye looking through the glass at rays from point X, the object will appear closer and displaced, as if it were situated at X'. *This* is prismatic deviation (or displacement). To an observing eye, this type of deviation will appear to be much more pronounced for *near* objects than for distant ones.

Refraction by two plane *non*-parallel surfaces: If the sides of the glass are not parallel, we have a "prism"; there will always be a deviation imparted to any incoming ray.

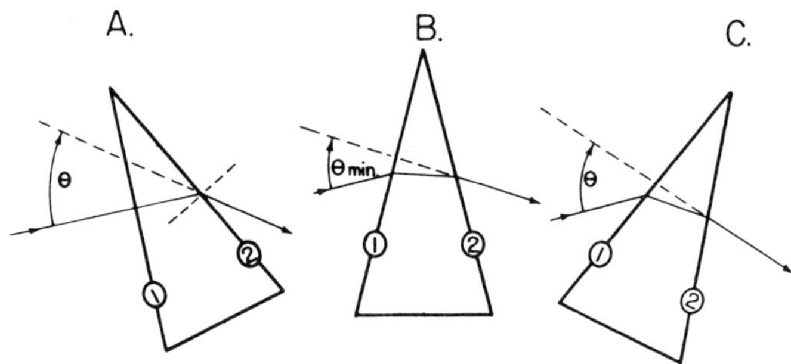


A) If a ray strikes the first face along its normal, it will continue without angulation until it strikes the second face, when it will be bent away from *its* surface normal; as shown above, the total angular displacement is $\theta$.

B) $\theta$ will be *minimal* for any given prism when the ray travels symmetrically through the prism, so that the angle of incidence on face 1 is equal to the angle of refraction from face 2.

C) If the prism is pivoted further so that the ray proceeds through the prism asymmetrically again, the deviation of the ray will *increase* from the minimum shown in B.

In general, then, the capability or effectivity of a prism in bending rays is indicated by its "power", which is usually measured in prism diopters. The "power" is that which causes the *"minimum"* angular deviation, though we should now realize that the prism's power is *not* absolutely fixed and varies a little as you tilt or pivot it. (The variation with tilt is greater with prisms of higher power. From this you can draw the following clinical conclusion: if you desire an accurate measurement of a phoria or a muscle weakness, especially if you are using higher powered prisms, you must hold the testing prisms squarely — not tilted — in front of a patient's eye.)

Aside from the effect of tilting the prism, for most practical purposes, you should consider the induced deviation for any clinical prism as *constant;* that is, (see figure below) if you sight an object X through the prism, there will be an immediate but fixed apparent displacement of that object to X'. If you continue to sight while moving the prism up and down along its base-apex dimension, that object will "stay put"; it will *not* appear to move.

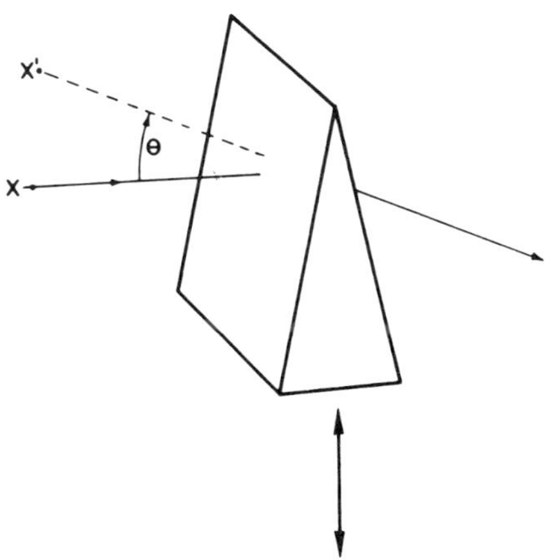


The fact that an object does not appear to move in this situation tells us that there is no *spherical* power incorporated along with the prism power; apparent "movement" *would* occur if there were refractive (vergence) power present. We will later learn how to "hand neutralize" unknown corrective lenses by using this "movement" technique.

Another point about prisms: when you look through one, you should also discern that distortion is introduced — straight lines may look curved and parallel lines may appear to diverge. Actually three *different* types of distortions are introduced by prisms and each has its own characteristic optical basis.* Sufficient for us is simply knowing that prisms induce distortion in addition to the object displacement.

In another short digression, we initially mentioned that the refractive index of any material varied not only with the material itself but with the *wavelength* of the incident light. Thus, if light composed of many wavelengths falls onto a prism such as that shown below, each of the component wavelengths is refracted a *different* amount; the shortest wavelength (blue light) is bent most, and red, the longest wavelength, the least. This is the chromatic dispersion you saw when you played with prisms in nursery school. It is an important occurrence and is the physical basis for the "chromatic aberration" of all optical systems.

Practical use is made of this "aberration" in an instrument called the spectroscope (for physico-chemical analyses) and clinically, in the *"Bichrome"* or "Duo-chrome" test, to be explained later.

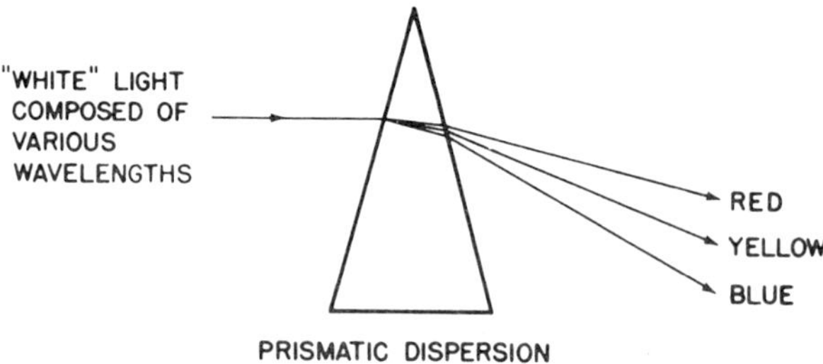


PRISMATIC DISPERSION

The extensive foregoing discussion then, explains how prisms work and how Snell's Law governs ray deviation and object displacement. In summary, the basic prism "power" is derived from two factors: how steeply the flat faces of the prism abut and the index of refraction of the prism material itself. (The greater the index, the greater the overall power.)

---

* (See Optics, by Ken Ogle, Thomas, 1968, p. 74-5)

## Ophthalmic Prisms

Realize that in using prisms in ophthalmology, the eye must pivot (about its center of rotation) to follow the displaced images. Though this direction of displacement is always toward the prism's apex, *prism* positions are typically given (just to confuse you) by their *base* locations! A 10 prism diopter *base*-down prism would displace an image upward, so that it appears off-axis by 10 cm for each meter of distance, (or 25 cm for a distance 2.5 meters away). The eye must rotate upwards to fixate the displaced object. (See diagram.)

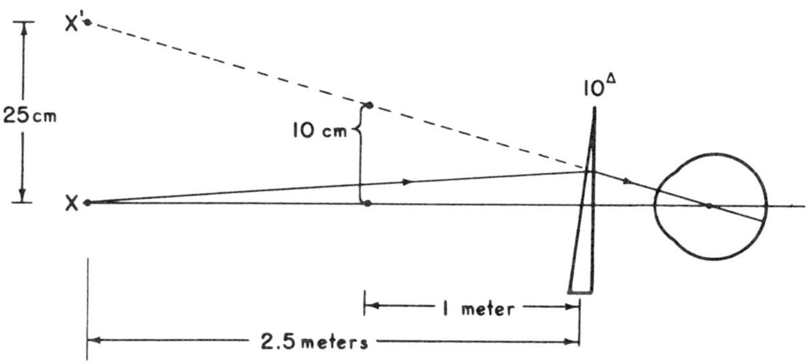


Ophthalmic prisms come in various forms:

1) single, loose prisms of varying powers.

2) bar prisms — a graded series of prisms mounted on a stick. Usually they are constructed so that when the bar is held vertically, they introduce a horizontal prismatic effect. Bars introducing vertical prisms are also available.

3) A "Risley" prism — which is of continuously variable power. This is constructed with two 15$^\Delta$ prisms mounted so that by turning a single knob, you rotate both prisms equally, but in opposite directions. This maintains a resultant prism activity in one primary direction. When the prisms are fully aligned (with their bases pointing in the

same direction), their two powers are additive and will yield $30^{\Delta}$ (see A below). If they are rotated so that the apex of one falls on the base of the other, the resultant prism power is $0^{\Delta}$. (See B.) Partial rotations yield the "in-between" powers (see C). This tool is extremely handy and is supplied commercially on almost every type of refracting unit made for clinical purposes.

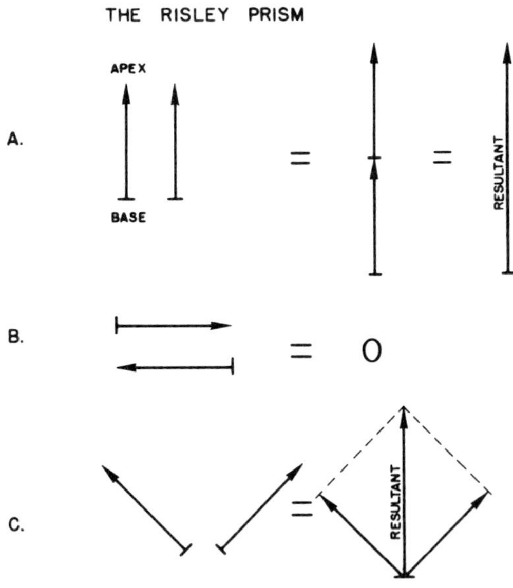


4) Fresnel prisms — to be covered later.

We have now covered sufficient detail to understand the properties of refraction by plane surfaces and how Snell's Law supervises this activity.

## CURVED SURFACE REFRACTION

Let us now investigate how curved surfaces might act on light rays.

*PLEA:* Do not allow this one derivation to scare you away. Bear with me and do make an effort to go through it. It is *not* tough despite the use of a few mathematical symbols. Actually, you may find it is fun to learn exactly how your useful formula, $U + P = V$, is derived. However, those readers who find even simple mathematics abhorrent can proceed to skip over to page **71**.

To see what happens at a curved surface, two points must be resurrected from your geometrical background:

a) the *exterior* angle of any triangle equals the sum of the opposite two interior angles:

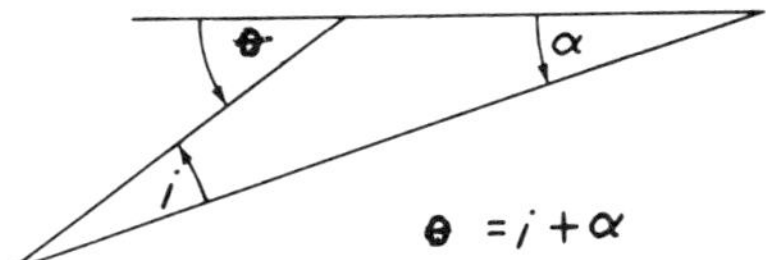


b) The "normal" (perpendicular) to a spherical surface is any line which passes through the center of curvature of that surface. If C is the center of curved arc AB, the line shown is considered "normal" to that surface:

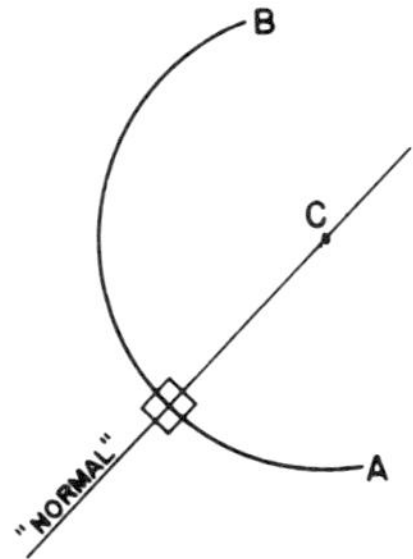


Now, let us take a brief glimpse at refraction at the curved surface. Incoming ray 1 strikes point P on a spherical surface whose center is at C. (See next figure.)

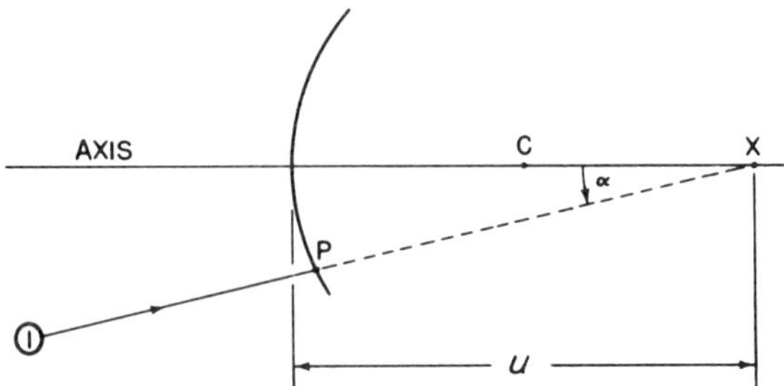


If this ray did not hit the surface, it would continue to travel and intersect the axis at point X. The distance to X (measured along the axis) is object distance *u*, and the angle this ray 1 makes with the axis is $\alpha$.

To point P from C, we draw a line; this is "normal" to the surface. The "normal" makes angle $\theta$ at C, and incoming ray 1 makes angle *i* with this "normal". Angle *i* is the "angle of incidence"; it is equal to its *vertical angle* (created by two intersecting lines), shown below as one angle of triangle PCX.

Since $\theta$ is an exterior angle of PCX,

$$\theta = \alpha + i$$

So,

$$i = \theta - \alpha$$

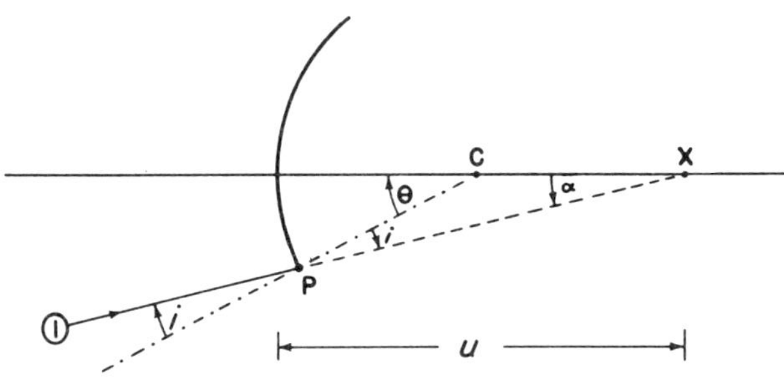


Ray 1 will be refracted by the surface and bent toward the "normal"; the "angle of refraction" is now $i'$ (next figure.)

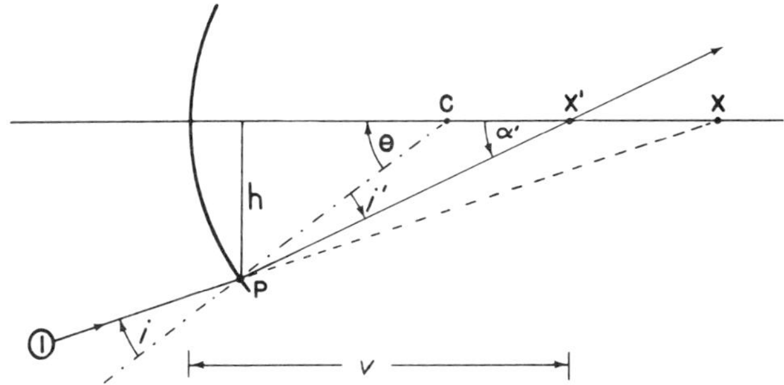


This refracted ray will continue and intersect the axis at X′ making angle $\alpha'$ with it, and at an image distance *v* from the spherical surface.

$\theta$ is also an exterior angle of triangle PCX′, so

$$\theta = i' + \alpha'$$

and $$i' = \theta - \alpha'$$

We have shown, then, that

$$i = \theta - \alpha$$
$$i' = \theta - \alpha'$$

In our simplification of Snell's Law for small angles,

$$n\, i = n'\, i'$$

So, substituting for *i* and *i′*,

$$n\, i = n\,(\theta - \alpha) = n'\, i' = n'\,(\theta - \alpha')$$

We will call the following, *FORMULA 1:*

$$n\,(\theta - \alpha) = n'\,(\theta - \alpha')$$

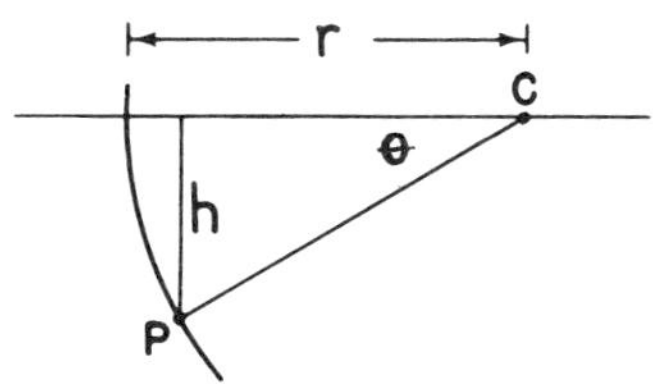


The distance from C to the curved surface is, of course, the radius *r* of that surface. Assume point P is at a distance *h* from the axis. When P is relatively close, that is, when *h* is short, $\theta$ will be a small angle, and $h$ will just about be equal to the length of the arc subtended by $\theta$. So, in radians,

$$\theta = \frac{h}{r}.$$

From the previous set of diagrams, and using the same reasoning, you can see that

$$\alpha' = \frac{h}{v}$$

and $$\alpha = \frac{h}{u}$$

To eliminate consideration of the actual angles themselves, ($\theta$, $\alpha'$, and $\alpha$), we can substitute the equivalent expressions for the angles into FORMULA 1:

$$n\,(\theta - \alpha) = n'\,(\theta - \alpha')$$

$$n\,(\frac{h}{r} - \frac{h}{u}) = n'\,(\frac{h}{r} - \frac{h}{v})$$

Factoring out $h$,

$$h\,n\,(\frac{1}{r} - \frac{1}{u}) = h\,n'\,(\frac{1}{r} - \frac{1}{v})$$

Dividing by $h$,

$$n\,(\frac{1}{r} - \frac{1}{u}) = n'\,(\frac{1}{r} - \frac{1}{v})$$

Multiplying this out,

$$\frac{n}{r} - \frac{n}{u} = \frac{n'}{r} - \frac{n'}{v}$$

Regrouping,

$$\frac{n'}{v} - \frac{n}{u} = \frac{n'}{r} - \frac{n}{r}$$

Since $$\frac{n'}{r} - \frac{n}{r} = \frac{n' - n}{r}$$

Then $$\frac{n'}{v} - \frac{n}{u} = \frac{n' - n}{r}$$

Or $$\frac{n}{u} + \frac{n' - n}{r} = \frac{n'}{v}$$

EUREKA! This is the general relationship for a single spherical refracting surface. It ties together object vergence $(\frac{n}{u})$, image vergence $(\frac{n'}{v})$, and the refracting power of the surface $(\frac{n'-n}{r})$. Starting to look like something familiar? It should!

$$\frac{n}{u} = U$$

$$\frac{n'}{v} = V$$

$$\frac{n'-n}{r} = P^{*}$$

$$U + P = V$$

When we talked about our simple, thin lenses and their vergence powers, the refractive indices of the media on both sides of the lens were identical — $n$ and $n'$ were equal to 1.0. If, however, the media are not air, we *must* take account of the indices, since they influence the object and image vergences; $U$ does *not* equal $\frac{1}{u}$; it equals $\frac{n}{u}$, and $V$ does not equal $\frac{1}{v}$, it equals $\frac{n'}{v}$.

Originally we declared that the power of a lens P was a certain number of diopters; we should now see that the *total* lens power of such a lens depends on its *two* surface powers, each in turn is dependent on its own radius of curvature and a *difference* in index of refraction. Suffice it to say that with *thin* lenses, the powers of each of the two surfaces simply add together algebraically. A + 5 D front surface and a + 3 D back surface sum to a + 8 D of total power. Similarly, a + 7 D front surface and a − 1 D rear surface would yield + 6 D of total power. (This latter lens is like most ophthalmic corrective lenses in that it is meniscus-shaped, that is, it has a concave surface facing the eye.) So, in general, $P_{total} = P_1 + P_2$. This is for thin lenses. With thicker lenses, however, there is a third, *thickness* factor $[\frac{d}{n}(P_1)(P_2)]$ which must be *subtracted* from the sum of the

* NOTE: To maintain our light convention, $r$ is considered *plus* if the light (moving from left to right) strikes a *convex* surface and minus if it hits a concave surface.

surface powers to arrive at the true "equivalent" lens power. You'll *not* be dealing with thick lenses in this course.

Recall I said you had to know only a *few basic formulas?* Well, here's another of them:

$\frac{n' - n}{r}$ is the refractive power of *any* surface. Fix it in your memory. It tells you that when $r$ is small, the refractive power of the curved surface is high; if $r$ is large, then the power is low; if it is infinitely large, the refractive power is zero — a flat plane is just such a surface, with no vergence power.

PROBLEM:

a) What is the refractive power in air of a polished convex surface of a glass rod ($n = 1.5$) of radius 10 cm?

b) What happens to the power of that same glass rod after immersing it in water ($n = 1.33$)?

ANSWER:

a) $P = \frac{n' - n}{r} = \frac{1.50 - 1.00}{+ .10} = \frac{.50}{.10}$

$P = + 5$ Diopters

b) $P = \frac{1.5 - 1.33}{.10} = \frac{.17}{.10}$

$P = + 1.7$ diopters; thus, the refracting power of the rod has decreased by placing it into water.

When we spoke of thin *lenses,* $P = \frac{1}{f} = \frac{1}{f'}$;

Single surfaces also have focal lengths, and we can use the same reciprocal relationship to power to determine those lengths as we did above, but we must also take into account the refractive indices of the object and image spaces.

$$P_{surface} = \frac{n}{f} = \frac{n'}{f'}$$

This means that if an air-to-water interface has $+ 5$ Diopters of surface power, the $f$ in air is $\frac{1}{+5} = 0.20$ meters or $+ 20$ cm; however, the $f'$ in the water (the position of sharp focus of an object located at infinity) $= \dfrac{1.33}{+5}$ = 26.5 cm to the right of the surface.

This shows that, if the medium on one side of a surface is air, the secondary focal length in the denser medium is *n′ times* the primary focal length: $f' = n' f$; this will prove to be a useful relationship.

Summary: In any calculations dealing with lens and surface refraction, do not forget that any object and image vergences depend not only on the actual *distances* from the reference plane, but also on the refractive indices of the object and image spaces. We will further explore these concepts when we study the model eye.

PROBLEM:

An object point X is located *in* a water tank and is 40 cm away from the end of that tank, which has a *concave* surface of 5 cm radius. Air surrounds the tank.

a) Where is the image point formed?

b) How long are the primary and secondary focal lengths of the refracting surface?

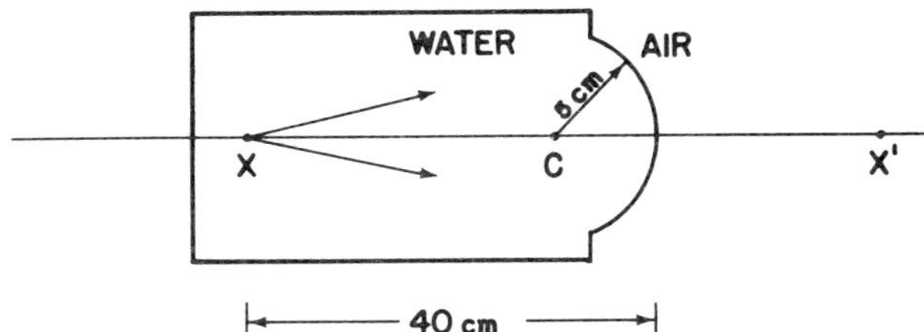


ANSWER:

a) $U + P = V$

*U:* The rays are diverging from X as they strike the curved surface, therefore, the vergence is minus at the surface.
The index or refraction of the object space is

that of water or 1.33. Thus, $U$, the object vergence $= -\frac{n_{water}}{u} = -\frac{1.33}{.40 \text{ meters}} =$ $-3.33$ D.

*P:* The difference in index of refraction referred to in $\frac{n'-n}{r}$ is always the index of the *image space* minus the index of the *object space:* Here, $(n_{air} - n_{water}) = 1.0 - 1.33 = -0.33$. $r = 5$ cm $= (.05$ m$)$ but is *concave* to the light traveling from left to right, and therefore *minus.* $r = -.05$ meters.

So, $\frac{n'-n}{r} = \frac{-0.33}{-0.05} = +6.6$ D.

(Thus we see that a *concave* surface has *plus* refractive power when the index of the image space is less than the index of the object space.)

*V:* The vergence of the image rays $= \frac{n'}{v}$; *$n'$ here* is that of the *image space* which is *air:* thus, $n' = 1.0$

$V = \frac{1}{v}$.

Now, solve equation for *V:*

$$U + P = V$$

$$-\frac{n_{water}}{u} + \frac{(n_{air} - n_{water})}{r} = \frac{n_{air}}{v}$$

$$-\frac{1.33}{.40} + \frac{1-1.33}{-.05} = \frac{1}{v}$$

$$-3.33 + 6.60 = \frac{1}{v}$$

$$+3.27 = \frac{1}{v}$$

$$v = +30.5 \text{ cm}$$

There will be a sharp image of X at X′, which is located 30.5 cm to the right of the refractive surface (which is the water tank side).

b) The primary focal length $= \frac{n_{water}}{P} = \frac{1.33}{+6.6}$

$$f = +20 \text{ cm in water.}$$

The secondary focal length $= \frac{n_{air}}{P} = \frac{1.00}{+6.6}$

$$f' = 15.1 \text{ cm in air.}$$

(Both are measured from the axial vertex of the spherical surface.)

(See Appendix A for a discussion of the influence of the refractive indices of the object and image spaces on the linear magnification.)

## ASTIGMATIC REFRACTION

Compared to other icebergs in the sea of optics, astigmatism always seems to offer the student a bit more of a challenge — an obstacle to otherwise smooth sailing. The teaching technique which will serve us as radar to neutralize this impediment has in the past proved to be a most useful one; it allows a firm grasp of the subject *now* while encouraging easy recall *later.* (In my opinion, it is probably Paul Boeder's greatest contribution to optics pedagogy). It presents one particular way of looking at the relationship of the refractive power of a cylindrical lens to the *orientation* of the astigmatic line images formed. But that's getting ahead of our story; we will come back to this shortly.

In our elucidation of $U + P = V$, we learned that $P$ was the power of a spherical surface, that is, one with *one* radius of curvature. If we neglect the lens aberrations of 3rd order optics, that spherical surface can form a point image for each object point. In this type of point-for-point correspondence, the image is said to be stigmatic (point-like). However, not all surfaces are spherical and those that are not, do not usually form stigmatic images. One such surface is called *toroidal* — a particular type of *non*spherical surface which forms non-stigmatic images, but does so in a certain way. It is only for *those* images that we reserve the term astigmatic (*not* point-like).

A toroidal surface is exemplified by the surface of a doughnut. It has *two* fixed radii in contrast to only one for the spherical surface and can be completely described by those two radii. Let us diagram a complete doughnut (torus):

A. WHOLE

B. VERTICAL SECTION C. HORIZONTAL SECTION

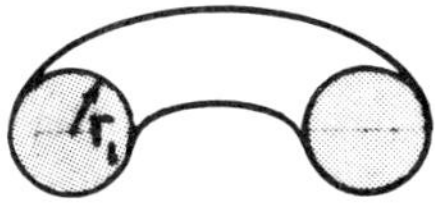


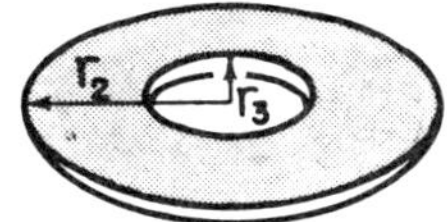


If we cut it in half vertically, we will see two small circles along the cut face (figure B above). Each circle can be described by a radius, $r_1$.

Compare this, instead, with the torus sliced in half horizontally (figure C above). The cut face now also shows two circles, an outside one of radius $r_2$ and a smaller, inside one of lesser radius $r_3$; we will ignore the latter one. All this about doughnuts is nice if you're hungry, but what we're interested in is a description of its outside surface. It is the shape of that front surface that is called toroidal. It is this specific surface which, when treated as a refracting surface (by definition), forms *astigmatic* images.

We can describe the outside surface of the torus as one which is created by the rotation of two radii of unequal length, each in a plane at right angles to the other.

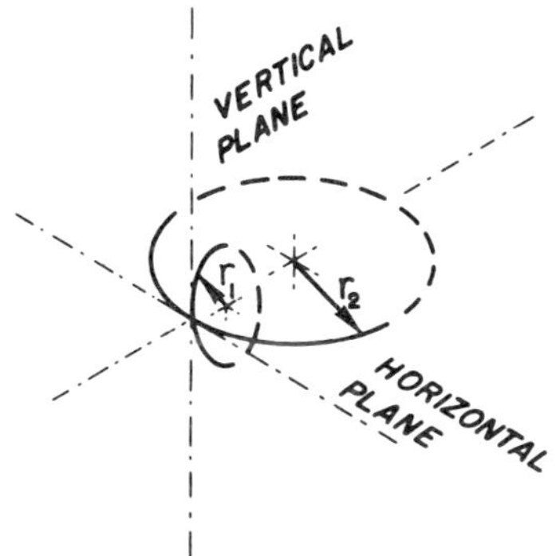


As shown in the figure, $r_1$ lies in the vertical plane while $r_2$ is in the horizontal plane. $r_1$ and $r_2$ do not have the same center. Their respective centers, however, do lie along a line called the axis of rotation (which is *not* the cylinder axis).

Let us look at each of the two planes drawn in the figure. The curve representing $r_1$ is a smooth circular outline present in the vertical plane. $r_2$ and its corresponding circular section are located in the horizontal plane. Each of these should be considered separately, completely independent of one another.

Look at the curve generated only by $r_1$ in the vertical plane. If it happens to separate two media of different refractive indices, we can calculate its surface power. So, in the vertical plane,

$$P_1 = \frac{n' - n}{r_1}$$

and for any object vergence $U$ we can find the corresponding image vergence $V$.

$$(U + P_1 = V_1).$$

We can do the same for $r_2$ in the horizontal plane, and find $P_2$, and calculate $V_2$. So, any object vergence $U$ which impinges on the toroidal surface generated by $r_1$ and $r_2$ will be influenced differently by $P_1$ and $P_2$, each of which act in separate planes which are at right angles to each other. We are forced to the conclusion that for any given object vergence, a toroidal surface must create two separate images. It is this composite image which is called *astigmatic.*

We can locate the position of certain parts of the image formed by our toric surface by using $U + P = V$ as just shown. But what does this peculiar astigmatic image look like? To find this out, let us investigate the optical effect of a toric surface *other* than that of the *doubly*-curved doughnut; let us use a straight vertical cylinder (as that shown in the accompanying diagram A below), which also happens to be a toric surface, but one with *one* of the radii, $r_2$, as *infinitely* long. (If $r_2$ were slightly *shorter* than "infinitely long", you should see that we would have a cylinder which was part of a very large doughnut with a tremendous hole [Figure B].)

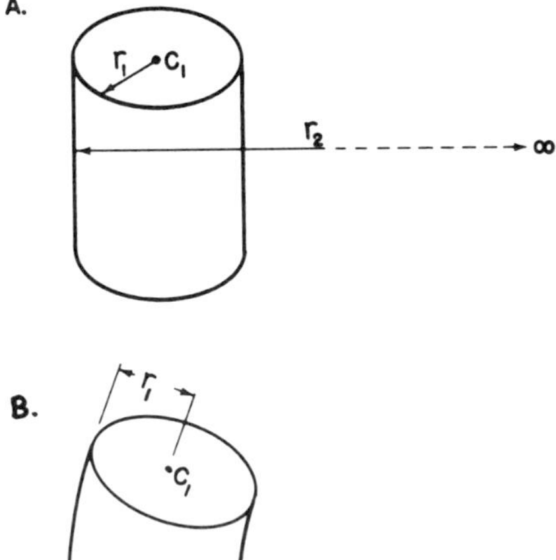


In any case, slice off a thin vertical piece of the cylinder (shown in A above) and it will look like that shown in diagram A below. Then slice the vertical slab into horizontal segments; these will look like a stack of thin, plus lenses (figure B).

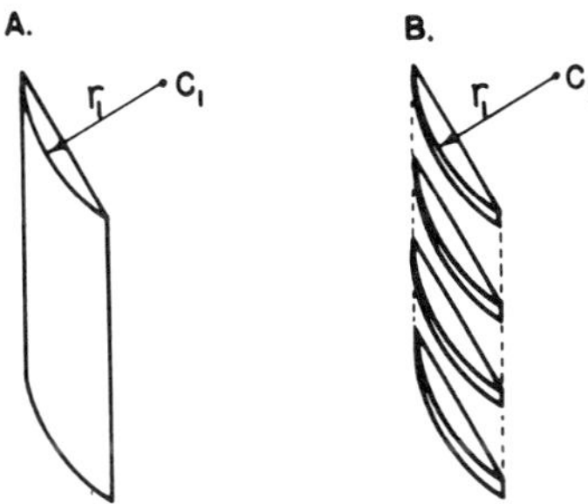


Each one of these thin, plus lenses will have the same refracting power, $P_1$*, say, + 5 D. If we put an object point 50 cm in front of this stack, *each* thin single "lens" would form a point image (see figure below). Where is this point image (X') located?

$$U + P_1 = V$$
$$-2 + 5 = +3$$

So, each point image X' will be 33 cm away. Since every one of our slices will also form its own point image of X, the final composite image of X will have to be a series of X' positions oriented in a straight vertical line. We say that our complete right angled cylindrical lens forms a real vertical line image at X'.

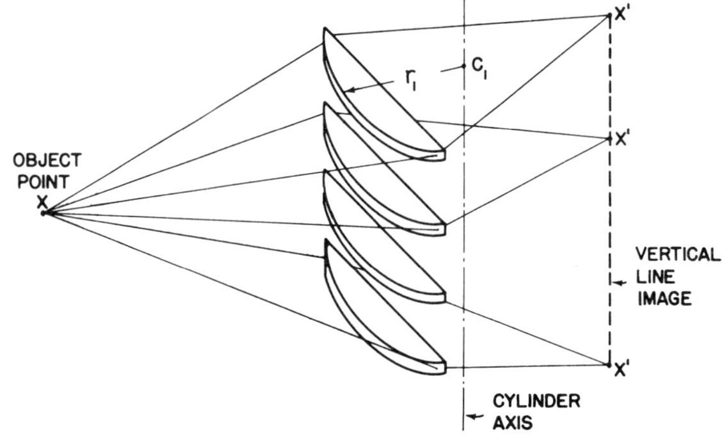


* $P_1$ refers to the power in the horizontal plane (where $r_1$ is located) as shown in the diagrams above.

Note the vertical line drawn through the center where $r_1$ originates. *That* dotted line is called the *cylinder axis.* The distance from that line to the cylinder surface *is* $r_1$ and, of course, is one of the variables which gives refracting *power* to that surface. This power of $P_1$ diopters lies only in the *horizontal* meridian as shown. The axis, however, is *vertical.* In ophthalmologic jargon, we say that $P_1$ is the power *associated with* the vertical axis (at 90°), or, in shorthand, $P_1 \times 90°$. So, tying these concepts together, we have learned that the cylinder power $P_1 \times 90$ creates an image which is a *vertical* line. (Keep the association *"vertical line image* and *axis 90°"* sharply in mind.) Such a vertical image line is produced for each and every object point.

Let's look at another right cylinder but of a different, longer *r.* Again, shave off one face and turn it so that its axis (the locus of all the $r_2$ *centers*) runs horizontally. Further, slice this piece vertically so that our slab is now composed of a series of plus lens segments stacked side by side (as shown).

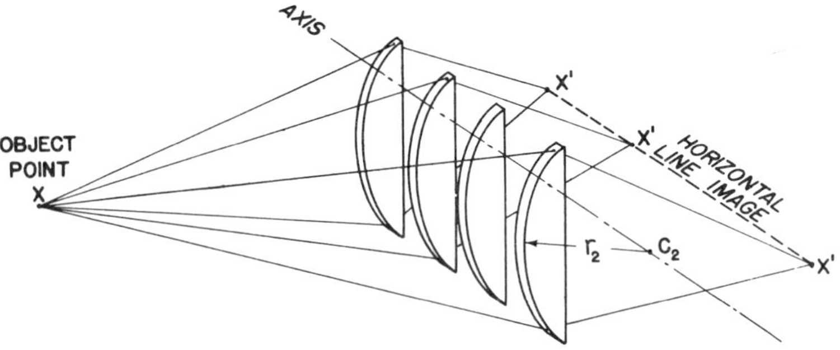


When point X is imaged, *each* of these lens segments will act to form a *point* image X'; when all are considered together, a horizontal line image is composed.

Since we chose $r_2$ to be longer than $r_1$, $P_2$ will be of *less* power than $P_1$ in the last example; let's arbitrarily say $P_2$ is + 3 D, that is, each vertical slice (representing $P_2$ in the vertical — 90° meridian) has plus lens power of + 3 D.

With object point X located 50 cm from this cylindrical lens, $U = -2$ D.

$$U + P_2 = V$$
$$-2 + 3 = +1$$

Thus, the location of the horizontal line image is 1 meter away from the lens. And again we see that power $P_2$, though it *acts* in the vertical meridian, is associated with a *horizontal* axis (that is, $P_2 \times 180°$) and so, focuses a *horizontal* line image.

In *both* the right cylindrical lenses considered so far (the $+ 3 \times 180$ and the $+ 5 \times 90$ we looked at initially), we dealt with the power in only one of the meridians — that exerting + 3 D and + 5 D, respectively. Any and every cylindrical lens will, however, always have *two* "major" meridians — one with maximal spherical power and another (at right angles to it) of minimal power. Each of these major meridians will form a line image, so, there will always be two line images formed by any cylindrical lens. In our examples above, we showed you the location of only one of them. Where is the other hidden? To find out, let's examine our second lens ($+ 3 \times 180$), shown in the last diagram, and specifically look at the meridian which is responsible for imaging the *vertical* line — that is, the horizontal meridian.

We had sliced this lens into vertical segments to study the vertical meridian, so now let's cut the slices horizontally to examine the horizontal meridian, (see figure below). Each one of the horizontal slices of this segment would yield a piece of material with parallel faces having no curvature and, therefore, no refractive power; thus, we can conclude that the horizontal meridian (180° meridian), which also happens to be the direction of our cylindrical lens *axis,* has no power.

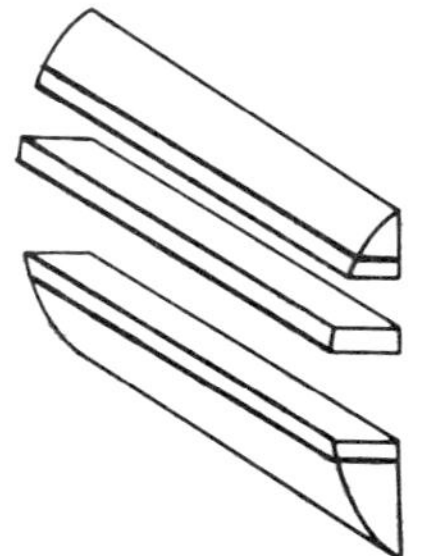

Now, locate the vertical image line with the object point still kept at 50 cm.

$$U + P = V$$
$$-2 + 0 = -2$$

Thus, the vertical line image, which we know must be present with this cylindrical lens, is optically located *at* the same position as the object point (but, in the image space).

## The Maddox Rod

This somewhat surprising finding forms the basis for one of the most useful of optical tools — one which you will probably use every day — the Maddox Rod. So, pay attention and please follow through the next few optical diagrams. These will enable you to understand how this instrument works. We will use the same right cylinder $P_2 \times 180$ as above and diagram the object and image rays originating from object point X.

In the series of diagrams, object point X lies on the optical (*not* cylindrical) axis of this lens. Select 4 rays emitted by point X towards the 4 corners of our right cylinder whose axis is horizontal — first, rays 1 and 2 (diagram A). These rays lie in a vertical plane on the far side of the lens and are converged to $X'_{1,2}$ by the lens. In the image space, these rays will remain in the same vertical plane as they were originally.

In diagram B, rays 3 and 4 lie in another vertical plane on the *near* side of the lens and are also converged, but to a point $X'_{3,4}$. We have already seen that the *composite image* of X is a horizontal line. $X'_{1,2}$ and $X'_{3,4}$ are the point extremes of this horizontal line image.

All converging rays (in the image space) can be extended backwards. Let us do so from $X'_{3,4}$ (diagram C). These extended rays (as shown) will form a triangular shaped, vertical plane with the apex at point $X'_{3,4}$ and bounded by *image* rays 3 and 4. These rays will lie in the same plane as the *object* rays 3 and 4, since objects and images always lie in the same plane. Object point X will, of course, also lie in this plane. (Diagram D shows a *side* view of this vertical plane, with pertinent positions labelled).

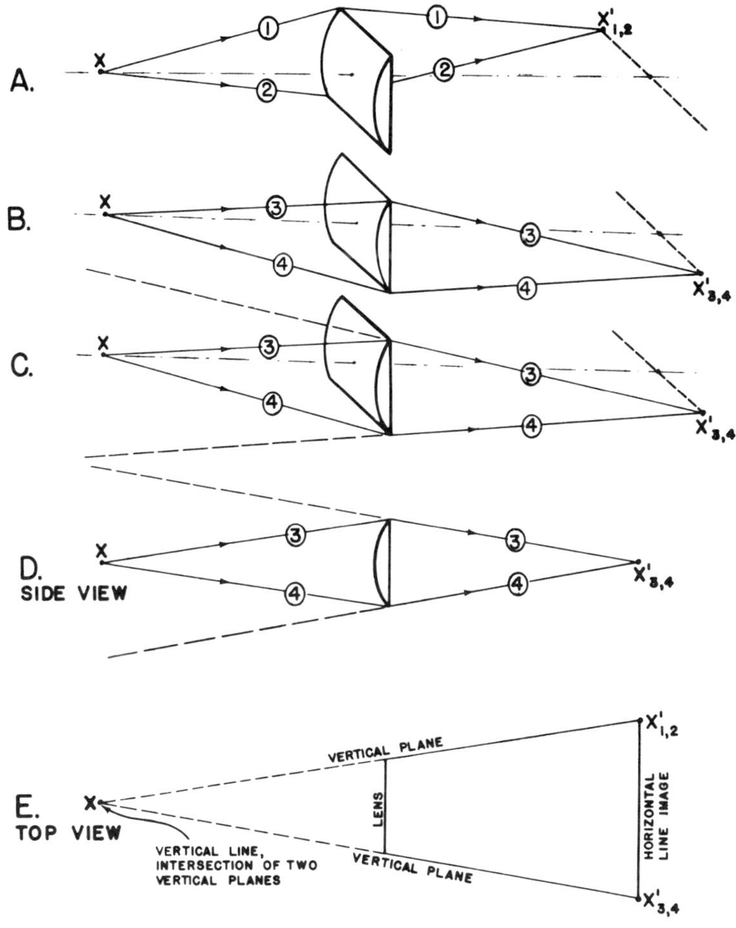


Follow this same procedure for object and image rays 1, 2 and point $X'_{1,2}$. This will create a different vertical plane, but also will include object point X. These two vertical, triangular planes will intersect in a vertical line located at the same position of object point X.

You can see this somewhat more easily if you look down on the diagram from the top (diagram E).

If we now draw the full optical diagram (see next figure), you should be able to understand how each of the rays is projected. Note the virtual vertical line image, which must be located at the same position as object point X (but in the *image* space). This vertical line image is virtual and cannot be focused on a piece of paper; however, it can be photographed or seen, but only when looked at *through* the cylindrical lens towards object X.

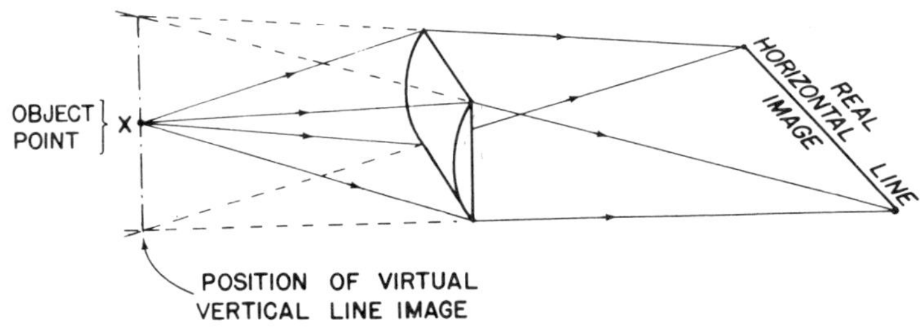


This rather long preamble is necessary for you to understand the Maddox Rod: this instrument is simply a high-powered plus cylinder. If held in front of an eye which is looking at a point source of light, and if the cylinder axis is horizontal, the lens will create a real *horizontal* line image close behind the lens. However, since the eye is *very* close to the lens, the horizontal line cannot be seen; (it is too close to be focused upon). The eye then will see the virtual, *vertical* image line apparently located at the object point. (Usually, the object light is distant, but it may also be nearby, as it was in the previous diagrams — in either case, the vertical line image will always be located wherever the object light is).

In practice, instead of using a single, high powered cylindrical lens, a *series* of high powered cylinders are stacked side by side with their axes aligned as shown below. This will make the vertical line image appear longer, and therefore, more visible to a patient.

MADDOX ROD

The Maddox Rod is used to measure extraocular muscle balances and phorias. When it is presented to one eye, the rod makes a *line* image visible to that retina while the point light source itself is seen by the other; this "breaks fusion" between the two eyes and permits them to settle into their tonic (relaxed) positions. Thereby, any tendency to ocular misalignment can be detected and, if present, measured. This is done by determining (in prism diopters) the angular separation which appears to exist between the two dissimilar images.

This detailed elaboration of the imaging of an object point by a cylindrical lens should have made the following quite clear: even a lens which has no power in one of its primary meridians will form *two* line images and each line image will be parallel to one of those primary meridians.

## The Astigmatic Cross Diagram

So far we've dealt with the real image line formed by a simple cylindrical lens. We paid more attention to the axis than the power since it was the *axis* that had the same orientation as the image line. However, the lens *power* belonging to that axis is what actually places that line at some definite location within the image space. So, let's now discuss not the image line but the cylindrical lens itself.

When I talk about a cylindrical lens, say $P_2 \times 180$, I mean one which has a *maximum* refractive power of $P_2$; though it functions by putting the *horizontal* image line into some position, that *power* is located in (and acts in) the vertical meridian, that is, at 90°. Note that I said "meridian". I can talk about a certain meridian because

I wish to identify one specific *direction.* This direction does not automatically associate with any *particular* lens power; it *could* be zero if it happens to represent the location of the axis, or it might represent any other sectional power of that lens. In *this* lens ($P_2 \times 180$), the vertical meridian (*at* 90°) represents the plane which contains power $P_2$.

This same information can be written as "$P_2$ acts in the vertical meridian", or "$P_2$ is located in the vertical meridian" or more simply "$P_2$ @ 90°" — all these are equivalent expressions. Furthermore, I can draw a "cross" diagram (which portrays the two major lens meridians) and label $P_2$ along the vertical meridian (see below).

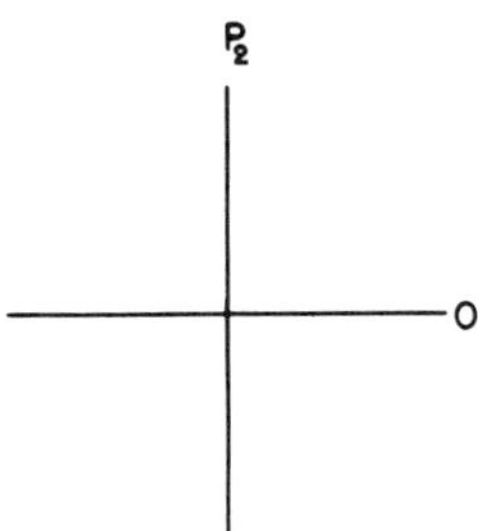


This lens' *axis* (× 180°) is located in the horizontal plane of the lens and is shown in the diagram as being present along the horizontal meridian with "zero refractive power".

Thus, any lens can be shown on a "cross", with two major meridians labelled. A *spherical* lens would show both of these meridians labelled with identical powers. Any "cross" which shows "zero power" along one of its major meridians must clearly be a simple *cylindrical* lens and that meridian must be the *axis.*

Let us now use the "cross" diagram to represent the *powers* in each of the primary meridians of the cylindrical lens we have already discussed, ($P_2 \times 180$). (See figure A below).

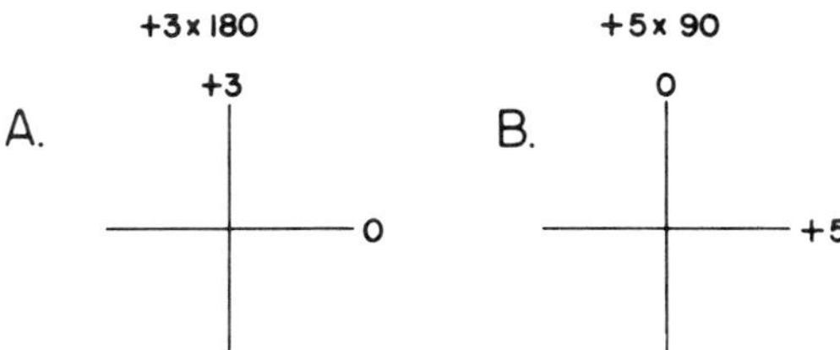


Since the vertical meridian has + 3 D of power, it is labelled so, and the horizontal, 0 D. We should, just by looking at the "cross", be easily able to write the formula of this simple cylinder as $+ 3 \times 180°$.

If you will recall, our *first* cylinder lens example was oriented with the axis situated vertically ($P_1 \times 90$). This lens, of formula $+ 5 \times 90°$, would be drawn on the "cross" in the manner shown in B above. Each of these lenses is a simple right cylinder with no refractive power given by the axis meridian. We have already considered how each lens imaged a point object 50 cm away. To reiterate, the $+ 5 \times 90°$ cylindrical lens imaged the point as a vertical line 33 cm from the lens. The $+ 3 \times 180$ lens created a horizontal line image 1 meter away. In addition, we should now know that when each of these lenses is considered separately, it also forms another, but *virtual* line image which is located at the same position as the original object point and oriented perpendicular to the cylinder axis.

Without too much difficulty, we should now understand what happens if we put both cylindrical surfaces *simultaneously* onto *one* piece of glass. Instead of a right cylinder with a straight vertical face (no curvature), we would now have one which has *some* curvature vertically, as well as horizontally. We can still slice the lens into horizontal segments and obtain pieces with power $P_1$ just as we did with the right cylinder. When, however, we slice this glass vertically, the segments will now *have* power $P_2$. Such a combination lens will act like our two separate right cylindrical lenses — the $+ 5 \times 90$ and the $+ 3 \times 180$ — in forming their line images as already discussed, but this lens will do so simultaneously. (However, there will be no virtual image lines at the object position as with the Maddox Rod.)

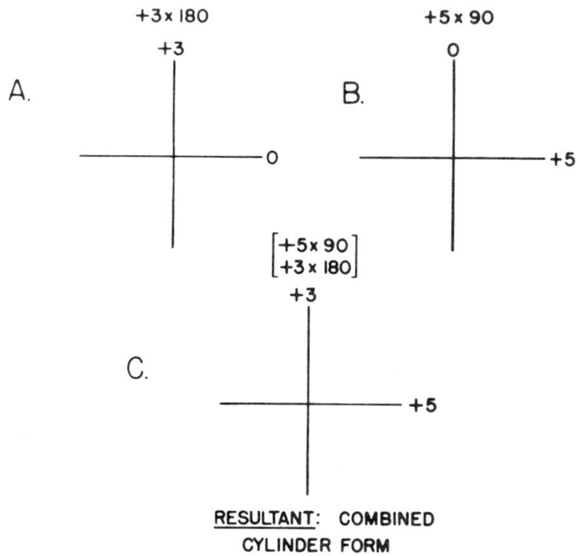


The diagram of these two separate cylindrical lenses (shown in A and B above) can be combined on a resultant "cross" (C above) to yield a single lens in which the power acting in each of the major meridians is simply added algebraically. This resultant represents the cylindrical combination which is written $+ 5 \times 90$ *combined with* $+ 3 \times 180$, or $\left[ \begin{matrix} +5 \times 90 \\ +3 \times 180 \end{matrix} \right]$.

Conversely, any "combined" cylindrical lens can always be redivided into two simple cylindrical lenses, as those shown above. (You will get more experience with this later). But remember, each cylindrical lens will be responsible for creating a focal *line* for every object point, and each of the lines will have the same orientation as the *axis* associated with the power which imaged it. This is *the* most important practical point about cylindrical lens imagery, and you must know it backward, forwards and should be able to recite it even if you are awakened from a sound sleep; once again, it is the meridian of the *axis* that determines the orientation of the particular line-image created: So,

$P_1$ × *90* forms a *vertical* line image for each object point

$P_2$ × *180* forms a *horizontal* line image for each object point.

## The Circle of Least Confusion

Again back to our example with the object located 50 cm away, the + 5 × 90 forms the vertical line image 33 cm from the lens and the + 3 × 180 forms the horizontal line image 1 meter away. What about the character of the image located in between the two focal lines?

The following diagram should help to answer this question:

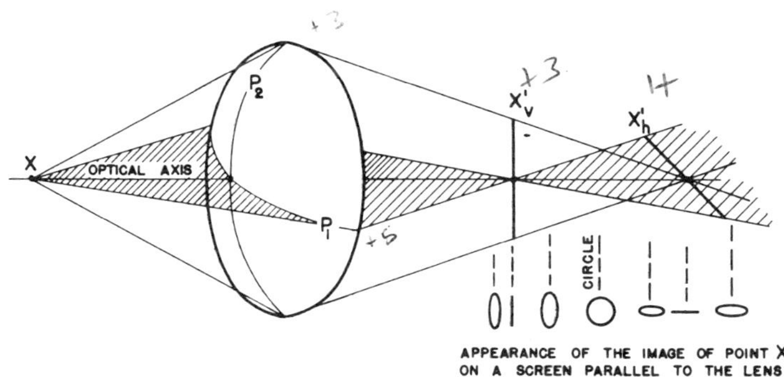


APPEARANCE OF THE IMAGE OF POINT X ON A SCREEN PARALLEL TO THE LENS

We are showing the same "combination of cylinders" lens we have been talking about, but here it is shaped more like a conventional "lens".

$$P_1 \times 90 = +5 \times 90$$
$$P_2 \times 180 = +3 \times 180$$

From object point X, only four rays are drawn to keep the diagram simple: two are in the horizontal plane (which is lightly shaded). These rays proceed to the horizontal edges of the lens. The other two rays shown are in the vertical plane. $P_1$ (with its *axis* at 90°) *acts* in the horizontal plane but focuses the vertical line at $X_v'$ as shown. (Remember, × *90* means we will be dealing with a *vertical* line image). Similarly × *180* means we will have horizontal line imagery with power $P_2$ placing the horizontal line into position.

In the figure above, we are given a bright object point and an image by the lens. If we place a screen parallel to the lens plane in the image space at position $X_v'$, we will see a sharp, bright vertical line there (surrounded by some dim, diffuse light); if the screen is moved to position $X_h'$, we will have a horizontal line. In between these two line images, other shapes are present; we will find bright spots of light shaped as shown in the diagram — vertically oval when in the proximity of the vertical line and horizontally oval when near the horizontal line image. At one discrete screen position between the two line images, we will find a perfectly circular spot; remember that this spot of light represents the image at that particular screen position of only *one* object point. Since the spot is here circular, it more closely represents a desired "point" image than does either *line* image or some *oval* image. This image is called "the circle of least confusion" and represents the axial position which provides an image which is less "blurry" overall than any other.

The exact position of the "circle of least confusion" is *always* one-half way *dioptrically* between the two line foci (that is, with a vergence which is the *average* of $X_v'$ and $X_h'$). Clinically this is a most important vergence.

In *linear* distance the "circle" is always somewhat closer to that image line which is nearer the lens; for example, with the object at 50 cm, $X_v'$ had an image vergence of + 3 D and $X_h'$ an image vergence of + 1 D. The "circle of least confusion" has a vergence of $\frac{1}{2}(+3+1) = \frac{+4}{2} = +2$ D and, thus, is located 50 cm from the lens. But, $X_v'$ was 33 cm away and $X_h'$ was 100 cm away; so you can see that the "circle" image is closer to (only 17 cm from) the vertical line, and 50 cm from the more distant, horizontal line. It will *always* be closer to that image line which is most proximal to the lens. But, *vergence-wise,* the "circle of least confusion" is always halfway between, and the lens power which is responsible for imaging this circle is called the "spherical equivalent" power; that power is one half way dioptrically between the maximum and minimum meridional lens powers. *That* is the item you must remember. We will encounter the spherical equivalent later in reference to "cross-cylinders".

The entire image which stretches between the image lines (including the "circle") is known as *Sturm's interval* and is often referred to as such. (J.F.C. Sturm investigated astigmatism in 1838).

## Images of Extended Objects

So far, we have spoken only of the representation of object *points* in the astigmatic image. What about the images of *full* objects? Let us schematize first again; for a *point* object:

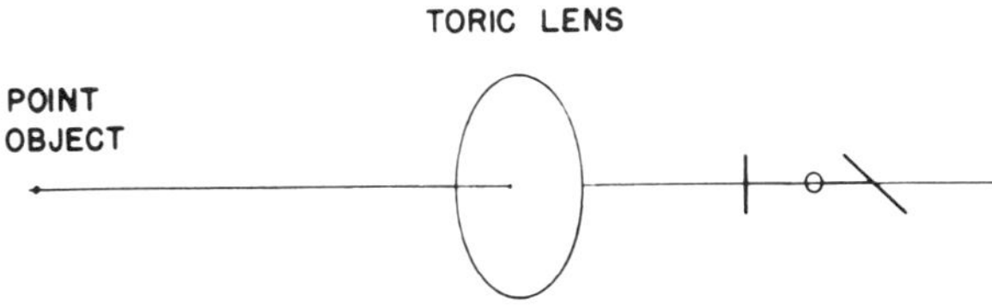


Now, instead of considering only one object point X, as above, let's put a letter E at that same position.

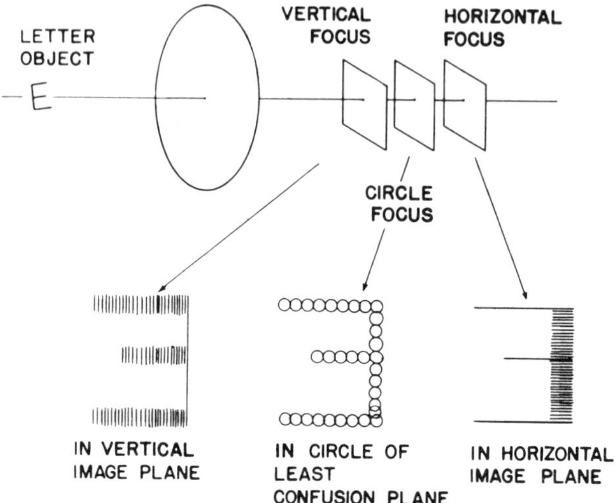


Every point making up the E will form a vertical image line in the vertical image plane, where the image of the E looks like that shown on the left. (Notice that it's upside down, as are all real images of an upright E). The vertical, right-hand edge is sharp because all the individual vertical line images fall superimposed on each other.

In the horizontal image plane (shown on the right), every object point in the E will give rise to a horizontal image line. Again, the horizontal bars on the E are sharp since each horizontal image line falls on top of the adjacent one, reinforcing it.

In the plane of the circle of least confusion (center above), the images of *each* object point are *circles;* this image more closely represents the true shape of the actual object E than do either of the other two images.

If the astigmatism were such that the two cylindrical axes were not oriented vertically and horizontally but were at some other positions, each image line (for every point in the object) would still be parallel to the corresponding axes — let's say 45° and 135°. The images might look somewhat like those below; notice that neither the vertical nor horizontal parts of the E are clear, but, the image in the plane of the circle of least confusion would, of course, look exactly like that shown above.

At this point we will skip learning what actually determines the *length* of the image lines and *size* of the "circles of least confusion". This will be taken up later in a discussion of the use of the stenopeic slit in clinical refraction.

## Differential Motion of Image Lines

Without any difficulty whatsoever, we should now be able to find the locations and orientations of the image lines. This is particularly easy to do when we express the cylindrical lens as a combination of two cylinders, ($P_1 \times 90 \supset P_2 \times 180$). So, let's place a new cylindrical lens in front of the above combination. This new lens will add directly to the combination lens in a particular way. Assume that its axis is *parallel* to either one of the others; (if it is not, the cylindrical addition becomes more complicated — we will broach this latter subject later on.) For the moment, then, we will hold our new cylindrical lens $P_3$ so that its axis is at 180°. Now $P_3 \times 180$ will add only to that power already present which has its axis also at 180°; that is, $P_3 \times 180$ adds only to $P_2 \times 180$ and yields

$(P_2 + P_3) \times 180$; the $P_1 \times 90$ is *not affected at all.* Since $P_2 \times 180$ was the power influencing the position of the horizontal line image, the new $P_3 \times 180$ will move *only* that same horizontal line.

$P_3$ can be either plus or minus. We are already acquainted with a *plus* cylinder × 180 (A below); A $-P_3 \times 180$ lens is called a *minus* cylinder × 180 and would look like diagram B.

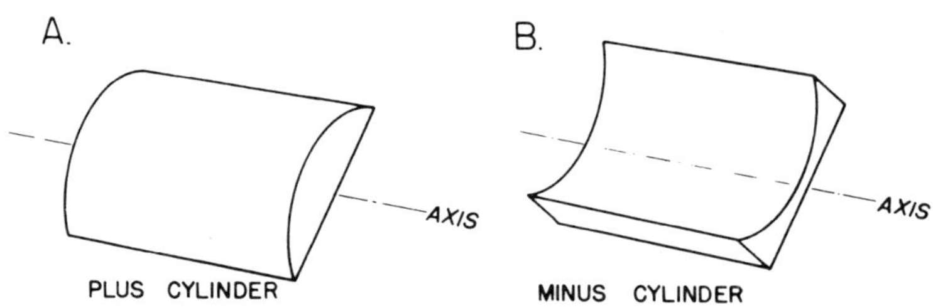


Either of these two cylindrical lenses would move only the pre-existing horizontal line, but each would move it in the opposite direction: the *plus* cylinder would "pull" that horizontal line image closer to the lens itself since it adds plus vergence to the image line; the *minus* cylinder will decrease the image vergence of that line and "push" it further away. (Notice, we *cannot* automatically say that either lens will move the horizontal line closer to or further from a *vertical* image since that depends on where that vertical line is situated!)

Let's look at a problem: If I show you a cylindrical lens combination where the image-lines are at the positions shown in figure A below, you should quickly be able to tell that the power associated with axis 180° ($P_1 \times 180$) must be more *plus* than power $P_2$ *with* its axis at 90°. You know this because the horizontal image line lies closer to the lens than the vertical one.

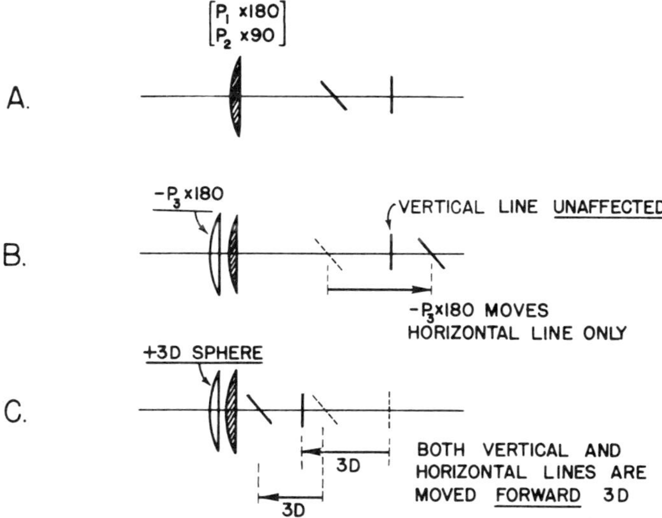


If I place a $-P_3 \times 180$ in front of (and in contact with) this lens (figure B above), I will move the horizontal line back away from its present position and towards the vertical line. (Figure B shows the horizontal line being moved beyond the position of the vertical). You can see that if $-P_3$ is of just the "proper" strength in minus power, it will move the horizontal line exactly to the location of the vertical. This would completely "collapse" the conoid of Sturm, and thus eliminate all the astigmatism present. At that moment, the sum of $P_1$ and $P_3$ (both with their axes at 180°) will equal $P_2$ with its axis at 90°; that is, the *total* power in both the 90° and 180° meridians will be equal. It should be obvious that this total combination will be equivalent to a *spherical lens.* (A + 7.50 D spherical lens can also be written as $+7.50 \times 90 \supset + 7.50 \times 180$).

So we have a beautiful mechanism for moving whichever image line we wish. Use simple cylindrical lenses oriented with the axis parallel (aligned in the same direction) as the image line you want to move. (Remember, plus cylinders move the line closer while minus cylinders push it away). The *other* line will stay put!

If instead of adding a simple cylinder ($P_3 \times 180$) we add another "combined" cylindrical lens ($P_4 \times 90 \supset P_5 \times 180$), all we have to do is add the powers associated with identical axes to obtain the correct *resultant* lens power:

| | |
|---|---|
| When lens | $P_1 \times 180 \supset P_2 \times 90$ |
| is added to | $P_5 \times 180 \supset P_4 \times 90$ |
| you obtain | $(P_1 + P_5) \times 180 \supset (P_2 + P_4) \times 90.$ |

The combined powers associated with each axis will together move their corresponding lines.

Instead of adding *cylindrical* power to a "combination" cylinder lens, let's add only spherical power. Since the spherical lens will add equal powers to both meridians simultaneously, both image lines will be moved *equally,* but only in a *dioptric* sense. (See figure C above). Thus, if + 3 D sphere is added to $+3 \times 180 \supset +2 \times 90$, the resultant would be

$$(+3+3) \times 180 \supset (+2+3) \times 90$$
$$\text{or} \quad +6 \times 180 \supset +5 \times 90.$$

Although both lines would be moved toward the lens an equal *dioptric* distance (each by + 3 D), the *linear* distance moved by the more distant line must be greater than that moved by the closer line. This simply shows the effect that proximity (to a reference plane) has on vergence, something you learned in the first few pages of this book. Look at the following figure:

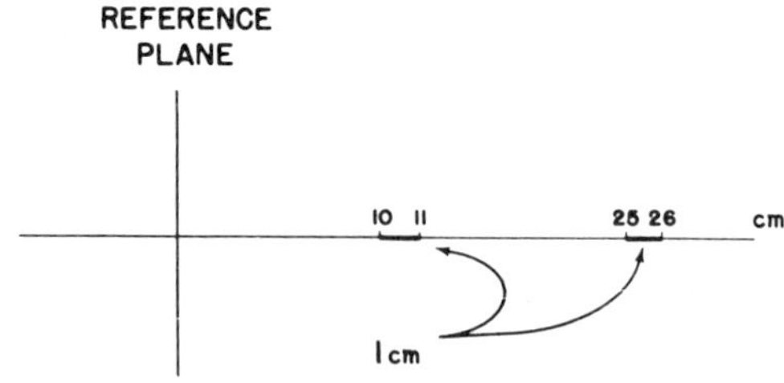

A 1 cm linear distance between 10 and 11 cm corresponds to a dioptric change of 10 — 9.1 or 0.9 D, while that same 1 cm distance between 25 and 26 cm corresponds to a dioptric change of 4 — 3.85 or 0.15 D. The addition of a full *diopter* of plus *at the reference plane* will "pull" both sets of points forward but will exert a greater movement effect on the more distant pair of points.

## Transposition of Cylinders

We now have sufficient background to understand the different ways we might describe a cylindrical lens combination.

On the "cross" diagram, we showed that a + 3 × 180 ⊂ — 1 × 90 would look like the following with the resultant lens shown below:

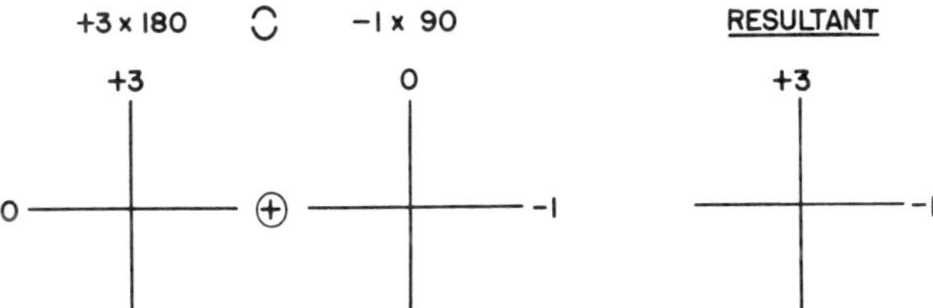


This resultant is *read* as "+ 3 @ 90° (or + 3 D in the vertical meridian) combined with — 1 @ 180°."

The resultant is the *key* way to describe any lens — just note the dioptric power of the two major meridians.

We have expressed the above lens as a combination of two simple cylindrical lenses (+ 3 × 180 and — 1 × 90). This same resultant can also be divided into other combinations — just as long as the algebraic sum of meridian powers yields the same powers shown in the major meridians of the resultant lens.

If we wished, we *could* subdivide the same resultant shown above into any number of other lenses; for example, the following three:

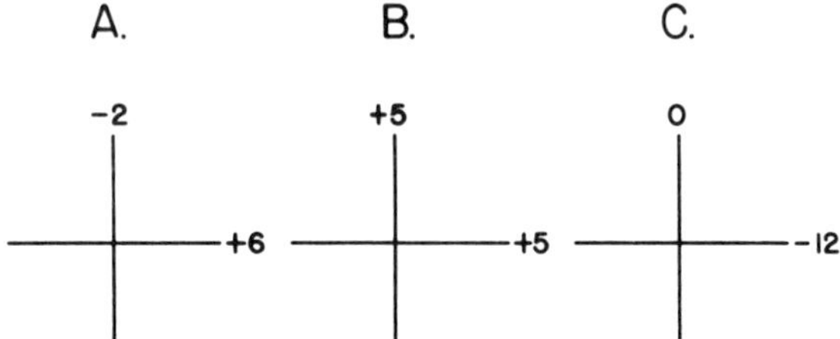


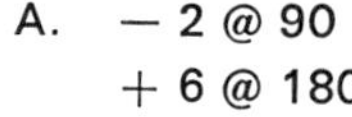

| A. | − 2 @ 90 | B. | + 5 @ 90 | C. | 0 @ 90 |
|---|---|---|---|---|---|
| | + 6 @ 180 | | + 5 @ 180 | | − 12 @ 180 |

Since the algebraic sum in each of the major meridians is $\begin{bmatrix} +\ 3\ @\ 90 \\ -\ 1\ @\ 180 \end{bmatrix}$, the resultant of this 3-lens combination is still the same as the one given in our last example. However, let's get back to a more practical "factoring" of the resultant. It is conventional to divide the resultant into two lenses, one of which is a sphere, the other a simple cylinder. We can determine the spherical component contained in our resultant by subtracting, from *both* meridians, the power shown at either of them. For example, the two meridional powers here are + 3 D and − 1 D. We can remove + 3 D of sphere from the resultant by subtracting + 3 D from both the vertical and horizontal meridians; we would then have the situation below:

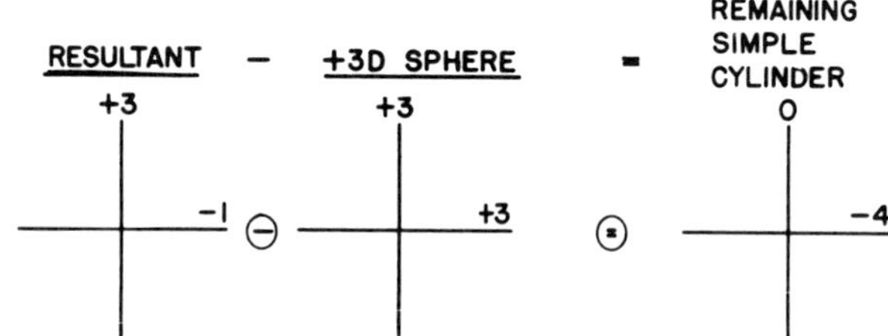

It is clear, then, that our same resultant can be considered to be composed of a $+3$ D sphere combined with a $-4 \times 90$ simple cylinder. (To check, just add the two major meridians algebraically; you should obtain the original resultant's major meridional powers).

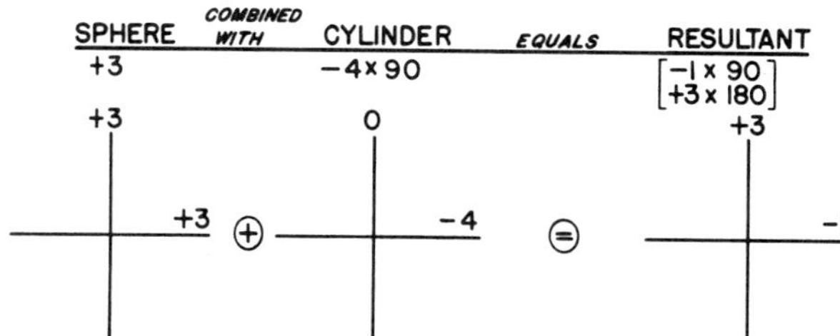


Written longhand, these same two lenses can be stated in a *spherocylindrical* form:

$+3$ D sphere combined with $-4$ D cylinder axis 90°

or better, $+3 \subset -4 \times 90°$;

or even shorter, $+3 -4 \times 90°$.

This is a spherocylindrical shorthand for precisely the same resultant lens which also represented our original "combination" of two cylinders. This form is called the *minus* cylinder form since the cylinder sign is *minus.* It is obtained from the resultant by choosing to extract a sphere of such power so as to leave one meridian remaining with minus power.

Return again to our resultant; we could, instead of extracting a $+3$ D sphere, have separated out a $-1$ D sphere — representing the other meridional power.

This would give us the two following lenses:

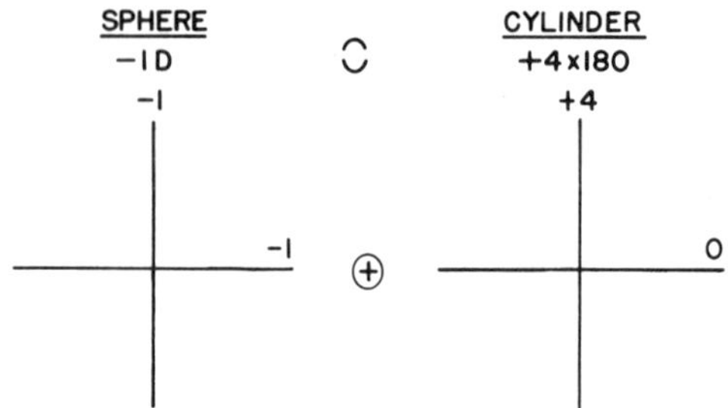


(Again, a check by adding each of the corresponding meridian powers would yield the same powers evident on the resultant).

These lenses also can be written in a spherocylindrical form:

— 1 D sphere combined with + 4 D cylinder axis 180°,

or — 1 + 4 × 180°.

This is called the *plus* cylinder form.

Both plus and minus cylinder forms are convenient ways to write down the corrective lens prescription for a spectacle lens. Both expressions represent single lenses that would focus light identically on an optical bench or for a patient's eye. However, only the *powers* of the major meridians are explicit in these prescriptions. The actual *shape* or surface curvatures of such a combination lens is *not* denoted by these formulas; many *different* shapes or forms of proper prescription lenses *could* provide exactly the same refractive powers for each meridian. So, don't think you are prescribing a specifically shaped lens with your notation. You are only indicating specific powers, which will be identical for any of the ways in which you can write the prescription.

The plus and minus cylinder forms can easily be interconverted. If a minus cylinder lens is + 7 — 3 × 45, it can be changed to *plus* cylinder form; to obtain the new sphere, algebraically add the *sphere* and *cylindrical* powers — (+ 7) + (— 3) = + 4; the new sphere will be + 4 D. The cylinder sign is simply changed to the opposite sign (— 3 D becomes + 3 D) and the axis is rotated 90° — thus 45° becomes 135°. The plus cylinder form of this lens is + 4 + 3 × 135.

The great advantage of both the plus and minus spherocylindrical forms is the *immediate* accessibility of the amount of cylinder and its axis; however, if you want to get some insight to the positions and directions of the *image lines* (which we *will* when considering the clinical aspects of cylinders), the spherocylindrical form is not as useful as the third form — "the combination of the two cylinders". It is this latter form which gives us more complete flexibility.

To find the third prescription form requires a minimum amount of arithmetic; and this always seems to be a pitfall leading to possible errors.

My own favorite way to change the lens to this form is by way of the "cross". Any errors are much less likely with this simple graphical means than with an algebraic one.

To convert from the spherocylinder $+4 + 3 \times 135$, first jot down both sphere and cylindrical "crosses"; then form the resultant by adding the corresponding meridians and finally, write each meridian power as a separate simple cylinder.

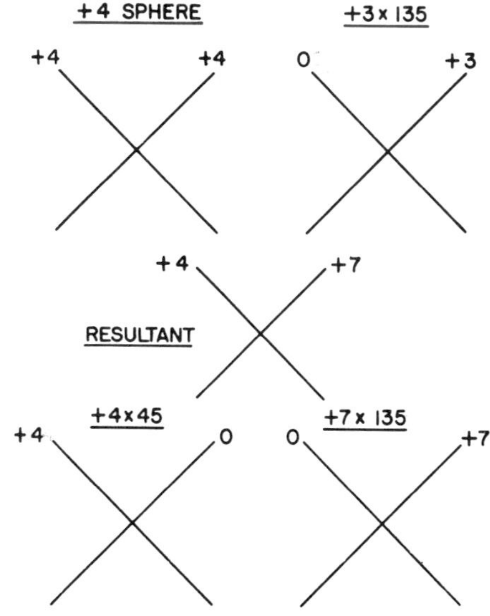

This is the required conversion:

$$\begin{bmatrix} +4 \times 45 \\ +7 \times 135 \end{bmatrix}$$

Again, you can check by adding the powers in the major meridians; you should arrive at the resultant lens.

Remember we are still talking about *one* lens — we are *describing* it as a combination of two cylindric lenses only to make it easier to visualize its action on light. This lens would take an object point at $\infty$ ($U = 0$) and form two focal lines; one is located $+\frac{1}{4\text{ D}}$ or 25 cm from the lens and oriented at *45°* (just like the axis); the second line is located $\frac{1}{7\text{ D}}$ or 14.3 cm from the lens and oriented at 135°.

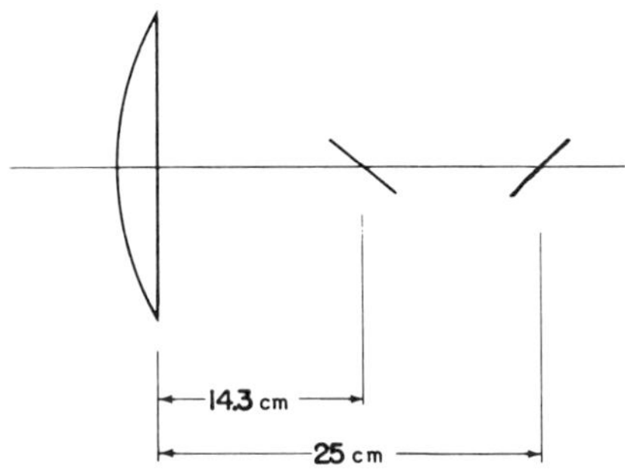


Now, I think the beauty of this way of expressing a spherocylindrical lens should be apparent, since it enables you to visualize the focal lines clearly. However, you should, at the drop of a hat, be capable of transposing cylindrical lenses into *any* of the three forms shown.

Here, we should point out what the usual clinical axis-convention is. You will use it every day when refracting a patient, and you will just have to memorize it. The 0° — 180° is the horizontal meridian.

0° begins at the patient's *left* ear and rotates counterclockwise when you are facing the patient; this holds true for *both eyes.*

Two sample meridians are shown below. Each designates one particular axis.

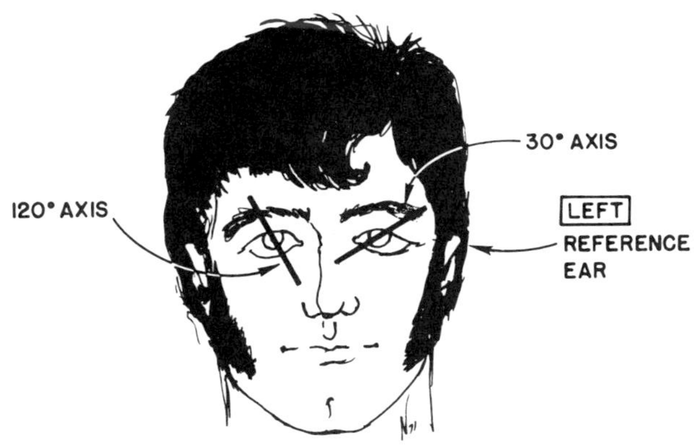


## Meridional Powers of Cylindrical Lenses

We have just seen that cylindrical lenses vary in their refracting power, going from a maximum dioptric power in one major meridian to a minimum power at a second major meridian which is always perpendicular to the first. But what about the powers in between?

The power gradation going from maximum to minimum meridional power is *not* a straight line change; the power gain moving from the *axis* meridian (which *is* minimum) to the maximum one increases by the *sine*$^2$ of the angle away from the axis.

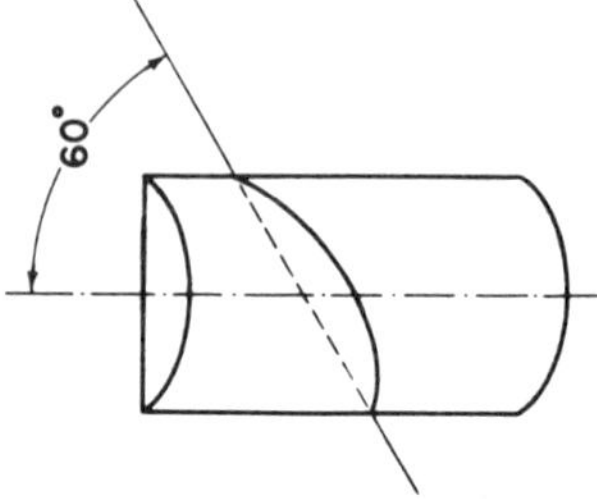


In this 3 D cylinder, the power in the meridian *60°* to the *axis* is

$$3\ (\sin^2 60°) =$$
$$3\ (.866)^2 =$$
$$3\ (.75) = +\ 2.25\ D$$

If there is an accompanying *sphere* in the lens, *its* power must be added to that found in the cylinder's meridian to obtain the *total* power exerted by that specific meridian. Try one for yourself:

PROBLEM:

What is the dioptric power in the 45° meridian of the spherocylinder combination $+\ 2 - 5 \times 90$?

ANSWER:

*Sphere:* + 2.00 D in all meridians

*Cylinder:* The angle between the cylinder axis and the required meridian is 45°.

$$-\ 5\ (\sin^2 45°) =$$
$$-\ 5\ (.707)^2 =$$
$$-\ 5\ (.5) = -\ 2.50\ D \text{ at } 45°$$

Therefore, the *total* dioptric power (at 45°)

$$= +\ 2.00 - 2.50 = -0.50\ D$$

For *this* lens there happens to be an easier way, which depends on a fact: For any spherocylindrical lens, the power in the meridian 45° to the axis (that is, halfway between the maximum and minimum meridional powers) is always the spherical equivalent of that lens — half of the cylinder added to the sphere ($\frac{1}{2} \times (-5) + 2 = -0.50$ D). We will study much more about "spherical equivalent" later.

Knowing this sine² relationship will help you determine the dioptric power in *any* meridian of *any* cylindrical lens. This may be useful to you when dealing with bifocals later, since you must know the total refractive power at a particular point in a lens to be able to calculate the *prism* power induced there.

The table below shows some selected meridional powers (in diopters) of a 1 diopter cylinder, axis 180°. (These power figures denote the $sine^2$ of the angle between the specific meridians mentioned and the cylinder's axis.) The meridional powers of cylinders of higher power than 1 D are directly proportional to those shown.

| *Meridian* | *Power* | *Meridian* | *Power* |
|---|---|---|---|
| 0° and 180° | .000 | 50° and 130° | .587 |
| 10° and 170° | .030 | 60° and 120° | .750 |
| 20° and 160° | .117 | 70° and 110° | .884 |
| 30° and 150° | .250 | 80° and 100° | .970 |
| 40° and 140° | .413 | 90° | 1.000 |
| 45° and 135° | .500 | | |

With this exposure to cylindrical lenses and astigmatic imagery under your belts, you should be well equipped to handle clinical astigmatism and refraction. We will delve deeper, when appropriate, into further aspects of this fascinating field.

## REFLECTION

Somehow we're going to have to squeeze in another subject area that seems slightly out of place; yet it is important, so here goes.

When we first started out, we showed how the vergence of light rays emanating from some object was influenced by lenses, each type of lens adding or subtracting its own particular vergence power to that of the incoming light and thus creating an image. There are *other* surfaces which (like lenses) can also change vergence. These accomplish this feat *not* through the *refraction* of rays (via Snell's Law) but through the *reflection* of rays.

Reflected rays can form an image (which also may be real or virtual) in much the same manner as refracted rays, and reflection should be considered as a "special case" of refraction — that is, subject to many of the same rules we have laid down for refraction. The difference is that *reflected* rays are bounced off a *smooth* (flat or curved) surface in an absolutely characteristic manner, with the angle of incidence *i* always equal to the angle of *reflection r* — this is called *regular* or *specular* reflection.

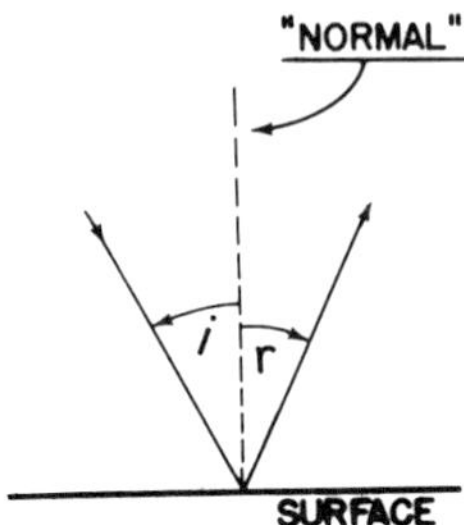


If, on the other hand, the surface is rough or rippled, the reflected rays may bounce off at some unpredictable angle. When rays incident at one angle are bounced off in an unexpected way, we say *diffuse* reflection has occurred. Our discussion here will be limited to *regular* reflection by mirror-like surfaces.

Regular reflection from a flat, *plane* mirror is easy to understand. The plane mirror adds no vergence power to influence the object vergence of the incident rays; the plane mirror serves only to *reverse the direction* of those rays. Every ray from any object will always be reflected from the surface by our rule — equal angles of incidence and reflection. Normally, *any* two of the rays which strike a mirror will pin-point the image; actually, with plane mirrors, only *one* ray is necessary, as we will soon show.

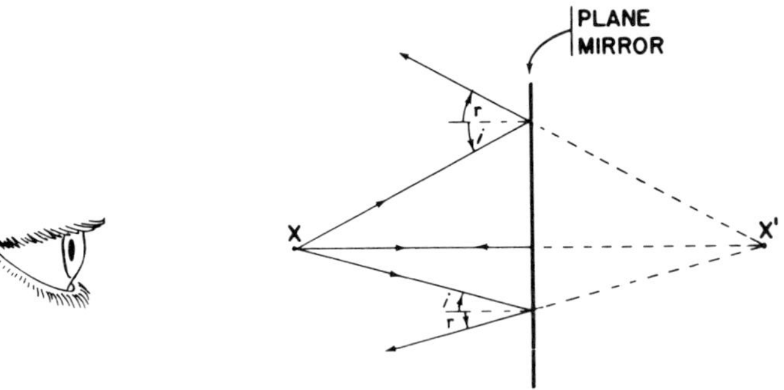

In the above figure, we have selected any two rays from X. Each is reflected off the plane mirror so that the angle of incidence (to the "normal") equals the angle of reflection. These rays, diverging from X, will be reflected by the mirror and will be diverged further. An eye or camera aimed toward the mirror will, of course, see object point X, but will also see image point X′. The diverging reflected rays will appear to have arisen at X′.

With the plane mirror, X′ will always seem to be situated at a distance behind the mirror which is identical to the distance that X is in front of it. (The distance would be measured along a line interconnecting X and X′, a line which is always perpendicular to the plane of the mirror.)

Interestingly, the mirror itself does not have to be physically located *directly* between X and X′. Point X′ can still be found exactly as described above, that is, along the perpendicular to the mirror plane, even if the real mirror must be extended as shown below:

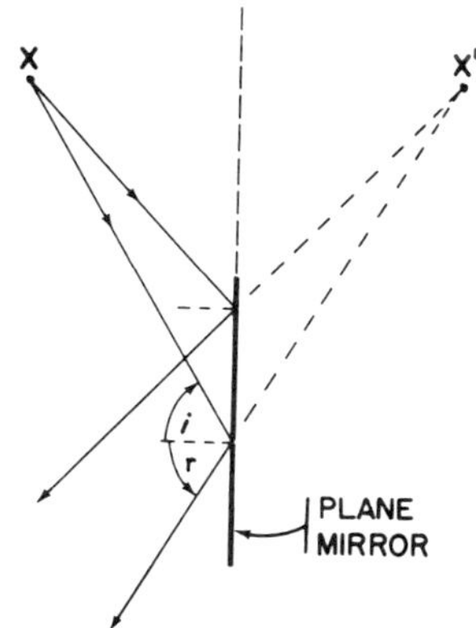


Our old standby $U + P = V$ will just as well apply to mirrors as to lenses and refractive surfaces. Let us first look at a plane surface and show the two comparable situations — one for "refraction" and one for "reflection". An object has been placed 40 cm in front of each surface:

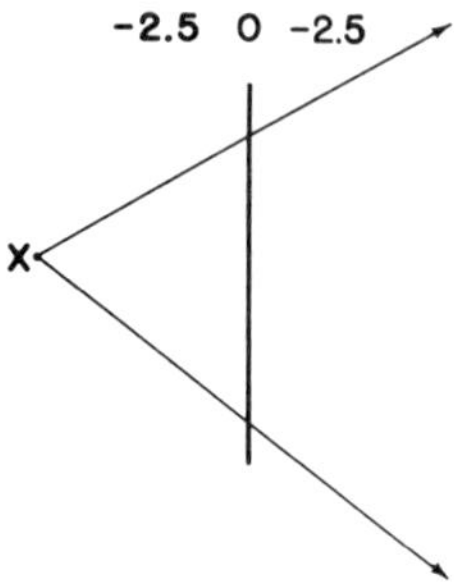


a) A plane surface with no *refracting* power is 40 cm from object point X, so

$$U = -2.5 \text{ D}$$
$$P = 0$$
$$U + P = V$$

Therefore, $V = -2.5 \text{ D}$

As you should know by now, this signifies that the rays in the image space are diverging from a point 40 cm to the *left* of the reference plane, that is, where the object is.

b) Now consider the influence of a plane *mirror* in place of our "refraction" plane:

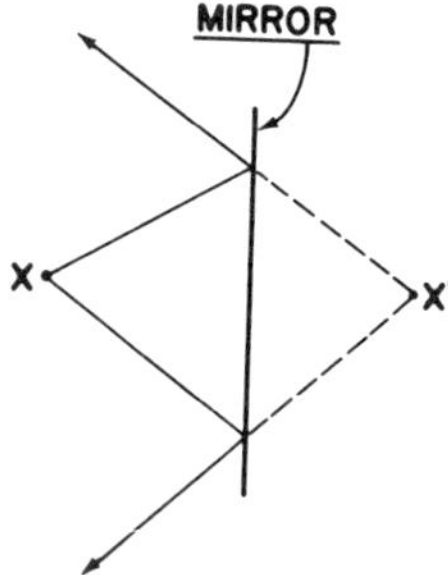


$$U = -2.5 \text{ D}$$
$$P = 0$$
$$V = -2.5 \text{ D}$$

Realize that the mirror has *reversed* the direction of the incident light rays, but still, the $-2.5$ D of image vergence here also has to signify divergence in the image space (meaning *after* being influenced by the optical device), and from an "image" point located 40 cm on the "opposite" side of the mirror. So, *minus* image vergence created by any *mirror* must mean the same as it does with lenses, that is, divergent light and a *virtual* image (apparently located *behind* the mirror).

So, for the $U$ and $V$ factors of our equation $U + P = V$, the sign convention we established for object and image vergences dealing with refraction still continues to hold true for reflection. But what about the sign conventions for $P$? To determine that, let us continue with our reflection-refraction analogy, but now with curved surfaces, which also have inherent reflecting as well as refracting power.

We learned that the *refractive* power of a surface is $\frac{n' - n}{r}$; we studied how this influenced object vergence $U$. Mirrors can also change vergence and thus have a certain *reflective* power, but this power influences object vergence differently — it is a different commodity. Since only the *surface* is involved, reflecting power does *not* depend on the index of refraction of the mirror material, however, the radius of surface curvature certainly *is* important in creating reflecting power; in fact, it is the *only* variable! (Our originally stated sign convention for $r$ will continue to be the same for *reflecting* surfaces as it was for refracting; that is, if a surface is concave to the incident light, $r$ is *minus* and, if convex, $r$ is *plus*.) We will express the fact that the light direction is always reversed by any mirror by routinely making the surface power *minus.*

Let us now derive an expression for reflecting power while maintaining as much similarity to the important concepts of refraction as we can.

We can find an axial image point which represents an axial object point located at infinity. We can call this point F′ as we did with lenses. There is also an axial object point situated so that its image is at infinity — that is F, the primary focal point. Since the mirror reverses the light, image space is folded over onto object space; this folds F′ back into exact super-imposition onto F. Thus, though we can speak of both F and F′ separately, they are located at the same spot on the mirror axis (though each is in a different "space", of course!)

Derivation of a mirror's reflecting power:

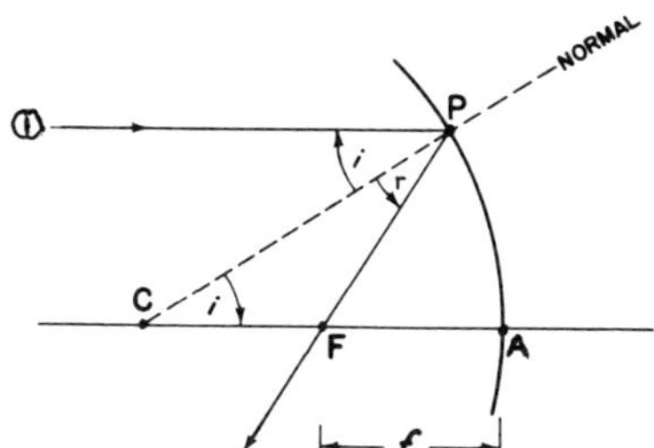


Here comes another easy derivation; again, you can, if you wish, skip over it; however, I urge you to stay with me and try to understand it.

When incoming ray 1 is parallel to the axis, it must be reflected from P and cross the mirror axis at F, the focal point of the mirror, since we are making refraction and reflection analogous. Distance FA then would be considered the focal length *f*.

CA is the radius of curvature of the mirror surface. The "normal" to the mirror surface at P is created by a line drawn through the center of curvature C of the surface.

Ray 1 makes angle *i* (the angle of incidence) with the "normal". The law of reflection states angle *r* must equal angle *i*. Angle FCP must also equal angle *i*. (Geometry: ray 1 and the axis are parallel lines intersected by line CP; then, the alternate interior angles must be equal.) Thus, angle FCP = angle CPF, triangle FCP is isosceles, and the sides CF and FP are equal.

If we assume that ray is a paraxial ray and lies very close to the axis, FP will be approximately equal to FA. (Draw ray 1 close to the axis and see.) Since CF = FP, CF must also be approximately equal to FA. But CA is the "radius of curvature" *r* of the mirror surface, and

FA is the focal length $f$ of the mirror; thus, the focal length $f$ is equal to one-half $r$, or $f = \frac{r}{2}$. Then, $\frac{1}{f} = \frac{2}{r}$.

We already agreed that the power of the mirror should be *minus* to represent the fact that any mirror will reverse the direction of all incident rays, so,

$$P = -\frac{1}{f} = -\frac{2}{r}$$

Thus, we will define here that the reflective power $P$ of any mirror is $-\frac{2}{r}$. Note that this is a general expression which includes, of course, the concave mirror, already designated as having a *minus* radius of curvature. So, in the case where one wishes to determine the reflective power of a concave mirror, one will naturally utilize a *minus* value in the denominator, and this will automatically generate the proper sign for the reflective power $P$, which here will be *plus.* Concave mirrors will always turn out to be *plus* mirrors and are comparable to plus lenses, since both *add* vergence to that presented by incoming object rays; convex mirrors will be *minus* and comparable to minus lenses. (See also Appendix B.)

To summarize then, for $P$ (in our expression $U + P = V$), we can fully substitute $-\frac{2}{r}$ (the minus signifying the mirror's property of reversing the light direction) and $U + (-\frac{2}{r}) = V$. The power term of a *reflecting* surface is $(-\frac{2}{r})$ — $r$ in meters, of course — just as the power term of a *refracting* surface is $(\frac{n' - n}{r})$. Otherwise, the other factors and sign conventions are exactly comparable.

Let's go through an example of locating an image by a mirror:

PROBLEM:

a) Where is the image of an object located 50 cm from a *convex* mirror, whose radius of curvature is 20 cm?

b) If the object is 5 cm high, what is the comparable size of the image?

c) Sketch a diagram of the optics involved.

ANSWER:

a) $u = -50$ cm; $U = -2$ D

$P = -\frac{2}{r}$; since this $r$ is that of a *convex* mirror, $r$ is plus.

$$P = -\frac{2}{+0.20} = -10 \text{ D}$$

$$U + P = V$$

$$-2 - 10 = -12 \text{ D}$$

$$v = -\frac{1}{12} = -8.3 \text{ cm}$$

The image is located 8.3 cm *behind* the mirror and is virtual. Remember, this means that to an observer looking at the mirror, the image will seem like it is situated 8.3 cm behind it.

b) The image size compared to the object size is called the magnification (M), which is always proportional to the respective distances of the image and object from the mirror (just as with thin lenses).

$$M = \frac{\text{image size}}{\text{object size}} = \frac{\text{image distance}}{\text{object distance}} = \frac{-8.3 \text{ cm}}{-50 \text{ cm}}$$

$$\text{image size} = \frac{v}{u} \text{ times (object size)} = \frac{-8.3}{-50} \times 5 \text{ cm} = +0.83 \text{ cm}.$$

Thus, the image size is + 0.83 cm; as with our previously stated convention for magnification, when M is plus (as here), the image is upright.

c) Diagram:

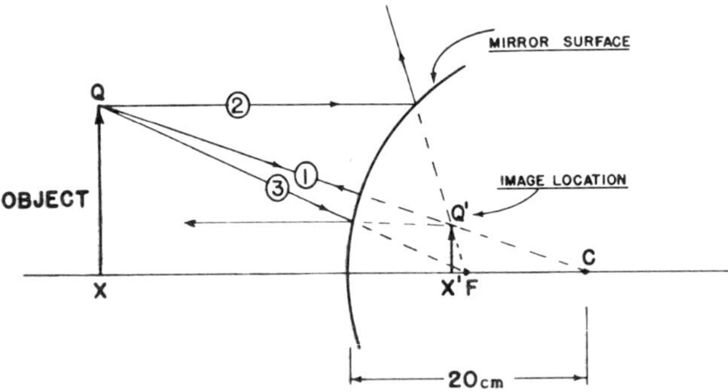

Of all the rays issuing from Q, we select our three known ones.

1. Ray 1 is directed toward *C*, the center of curvature of the surface. Since this is "along the normal", that light ray will be reflected back exactly along this line. This reflected ray is the equivalent of the "undeviated" one (in refraction) through the *nodal* point. (With thin *lenses,* the nodal point was seen to be at the axial vertex; with a single curved surface, however, the nodal point happens to be at the center of curvature of that surface, just as here!)

2. Incident ray 2 is directed parallel to the axis and must be reflected through F'.

These two rays, of course, cannot pass through the mirror; they are reflected by the mirror but *appear* to originate at *Q'*, behind the mirror surface, where the extensions of rays 1 and 2 intersect.

These two rays are all we need *but* there is another "known" ray.

3. Ray 3 is directed towards F (which, with a mirror, is at the same position as F'). It will be reflected back as parallel to the axis. When extended back "through" the surface, it also will appear to originate at Q'.

PROBLEM:

What is the reflecting power of the front surface of the cornea assuming $r = 7.7$ mm?

ANSWER:

$$P = -\frac{2}{r} = -\frac{2}{+.0077} = -260 \text{ D}$$

This is a real "wow" of a mirror. This reflectile property of the cornea is the optical basis for the clinical instrument which measures the corneal curvature — the keratometer.

Note that the reflective power of the cornea is so great that an object does not have to be very far away from the corneal surface before it is considered to be at "optical infinity" for the cornea. Then, the image, of course, will be at F'. For example, F' of the cornea is at $\frac{7.7}{2}$ or 3.85 mm behind the front surface. An object located only 4 *cm* away has an object vergence *U* of — 25 D. *V* would then be — 285 D and the image would be 3.51 mm behind the surface, and thus is *very* close to F. I know it's difficult to conceive that an object distance of only 4 cm is at "optical infinity" for the cornea acting like a mirror, but it is true.

CLINICAL POINT:

For a surface to do its job well as either a refractor or reflector, it must be smooth. Any tiny ripple or loss of perfect regularity will show up in a degraded image.

Take a cornea for an example: a small dense opacity located within the corneal stroma will probably not disturb any image formed by the cornea *if* it does not disturb the surface. (You'd probably be surprised how much corneal opacity can be present and still permit good acuity when the surface stays smooth.) On the other hand, a perfectly clear cornea with 1) only a tiny abrasion of the epithelium, or 2) minimum epithelial "bedewing" from increased intraocular pressure, or 3) surface irregularity from keratoconus, will play havoc with its image forming properties.

You can detect corneal surface irregularity clinically by following the movement of the corneal reflection created by a pocket flashlight. Since the surface's mirror-like qualities are normally unblemished, the reflection (reflex) should be sharp, bright and clear. In moving the flashlight, you will note that the image will move evenly and undistortedly across the surface, as you hold your viewing eye motionless and watch the reflection (which seems located "inside" the anterior chamber). If, however, any surface irregularity is present, the reflection will display some distortion, sometimes subtly and sometimes obviously. It is a good practice to examine the corneal reflections closely whenever a patient has reduced vision that cannot be corrected with spectacles. (Later you'll be told about an even more sensitive detector.)

Sometimes, you will not be sure whether or not an acuity problem stems from the corneal surface irregularity which you have detected. If it does, spectacle correction should *not* improve the acuity significantly; but, if you optically replace the irregular front surface curvature with a contact lens, the vision should then improve, and significantly so. This is an important clinical diagnostic point.

You have now completed your initial exposure to some of the basic principles of optics. However, you haven't learned them all! Later, we will pick up other pieces of geometrical optics (the study of how light is influenced by mirrors, prisms, and lenses). We *will* continue to expand, when appropriate, on the principles of thin lenses, cardinal points, multiple lens systems, and geometrical constructions. So far, we have not even mentioned the *physical* nature of light and the really important optical phenomenon of diffraction. *A modicum of familiarity with this area is mandatory for any clinician,* so a simplified summary of elementary Physical Optics is presented beginning on page 327.

Finally, with no further delay, let us get underway by immersing ourselves in a study of Visual Optics — that branch which deals with the eye.

# VISUAL OPTICS

# GENERAL PRINCIPLES

The eye can only be considered as a *moderately* complex optical structure — certainly not as difficult to deal with optically as are many types of camera lens systems. The focusing power of the eye is dependent upon its many curved surfaces, each separated by media of different indices of refraction. By far, the most important surfaces are those of the cornea (front and back), and lens (front and back, *plus* many in between!). You should also know that *some* refractive power is even ascribed to the anterior face of the vitreous and the concave, curved surface of the foveal pit. However, the anterior corneal surface stands above all as the monarch in importance for the overall refractive power of the eye. (If you want to get "picky", it is not really the *corneal* surface but the "tear film" which first encounters any entering light rays; thus, the "tear film" exerts the most influence of any of the surfaces of the eye. Its curvature is, of course, obligatorily that of the anterior corneal surface.)

## Model Eyes

To help us in our study of the refractive surfaces of the eye, we must make use of a standardized model. In creating such a model through a study of post-mortem and living eyes, many prominent scientists established a schematic eye whose size, dimensions, and optical properties seem close to those of the average living eye. The most important of these models was presented by Gullstrand, the only ophthalmologist to be honored by a Nobel Prize. Using a number of reasonable assumptions and measurements, he established the dimensions and located the optical cardinal points. These can be found in any standard textbook.*

The schematic eye, though useful for a better understanding of how the eye works optically, is simply too cumbersome a tool with which to begin. A much more useful model is created by further simplifying this Gullstrand schematic eye to a structure which has

---

* Ogle, K.: OPTICS
C. C. Thomas, Springfield, 1968, pp. 156-7.

only one refracting surface and a unified intraocular medium of one refractive index. This model eye has the same overall dimensions of Gullstrand's, but the anterior surface curvature must be different, since we are replacing all the intraocular structures as well as the original cornea by a single refracting surface. It should thus be obvious that this new eye (called the *reduced schematic eye* or simply, *reduced* eye), must have a front surface power which is considerably greater than that of the corneal dimensions given by the Gullstrand eye or indeed, any real cornea.

The diagram below gives the dimensions of such a *reduced* eye. It is this eye which we will use in our discussions of refractive error and later, magnification.

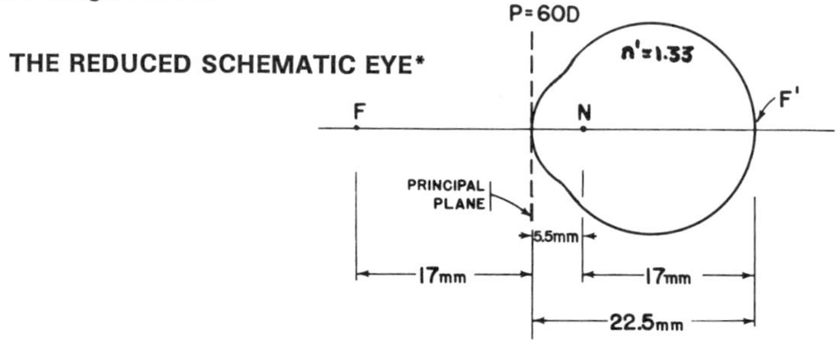


* The RSE constants given above for the 60 D eye have been slightly fudged. Since $P = 60$ D, the *true* focal length $= \frac{1}{60} = 16.67$ mm; and this has been rounded off to an easier-to-handle figure of 17 mm. This little "white lie" will necessarily beget another, which asserts erroneously that the axial length is 22.5 mm.

How's this derived? Well, based on our assumptions of $P = 60$ D and $n' = 1.33$, the radius of curvature $r$ of the "corneal" surface, (in this eye, the distance between the surface and N), is accurately calculated as you have been taught (pp 71-2):

$$P = \frac{n' - n}{r}\ ; \quad 60 = \frac{1.33 - 1.00}{r}\ ; \quad r = \frac{0.33}{60} = 5.5 \text{ mm}$$

Since the distance from N to F' is always also equal to $f$ (here, 17 mm and a *very* key number), the total axial length of the RSE will be $5.5 + 17$ or 22.5 mm. This is the length given in the diagram above, but remember, it presupposes the approximation "17 mm" for $f$. The *actual* axial length is $5.5 + 16.67 = 22.2$ mm, which *should* also be the $f'$ of this RSE. Let's check this:

$$f' = \frac{n'}{P} = \frac{1.33}{60} = 22.2 \text{ mm. It checks.}$$

Thus, in the diagram above, I have sacrificed some slight accuracy to maintain as "sacred" the figures of 60 D and 17 mm, the latter of which is especially important for the following discussion of Visual Size. Later (on pages 131-139), I will switch back and use the "accurate" axial length of 22.2 mm instead of the "fudged" 22.5.

## Visual Size

The anterior focal point of our reduced eye is 17 mm from the "cornea" ($\frac{1}{60\text{ D}}$). The distance from the central, axial point of the corneal surface to the *secondary* focal point of the system (F') (which should be on the retina) is $n' f$ or 1.33 times 17 or 22.5 mm. The distance from N, the nodal point of the eye, to F' is *also* equal to the primary focal length $f$ (see previous diagram). This is so for *any* optical system. Here, it is 17 mm and that particular distance should be cemented in your brain.

All rays directed at N will pass through the cornea without deviation by it, so N forms the apex of all angles subtended at the eye by any object (called the angular object size) as well as the corresponding angular size of the retinal image. Thus, knowing that the distance from N to F' is 17 mm will allow you to determine the actual sizes and angles subtended at the eye by any object — a test target on a tangent screen, a particular Snellen letter, or even a 1 mm lesion on the retina.

A few examples will point this out:

PROBLEM:

What is the angular size (in degrees) of the optic nerve head (1.5 mm in diameter)?

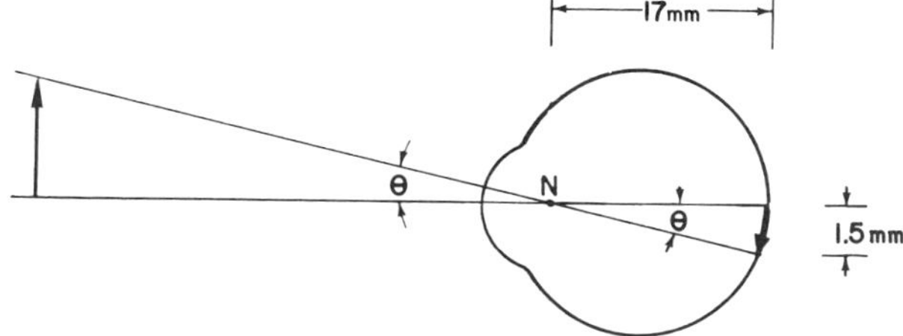


ANSWER:

$$\theta = \frac{1.5}{17} = .088$$

But this answer is in *radians.* To convert it to degrees, remember

$$1 \text{ radian} = \frac{180^\circ}{\pi} = 57.3^\circ$$

Then, $.088 \text{ rad} = .088 \times 57.3 = 5.0^\circ$

So, the optic nerve head subtends an angle of 5.0° and would project a blind area of this same angular subtents onto a tangent screen. If that screen were 1 meter away from the cornea, the *size* of the blind spot would be determined simply "by proportion" of comparable sides of similar triangles — (the *true* distance from N to the tangent screen would be 1000 + 5.5, but the 5.5 is so small compared with the 1000 that it can be safely neglected):

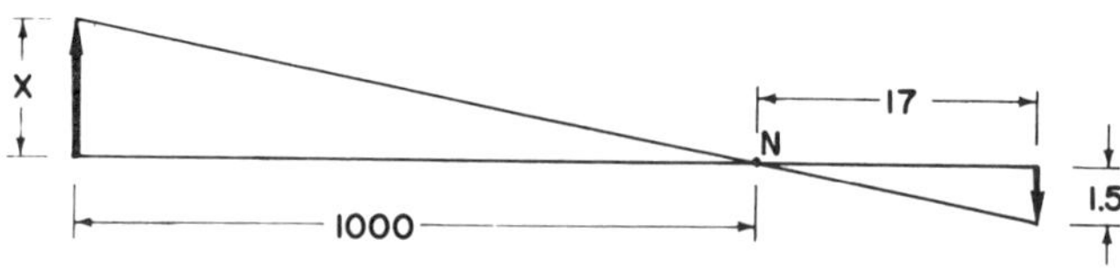


$$\frac{1.5}{17} = \frac{x}{1000}$$

$$x = 88 \text{ mm}$$

Thus, the size of the blind spot on a 1 meter tangent screen is about 88 mm in diameter.

PROBLEM:

a) What overall size should we make a 20/200 letter on a Snellen acuity chart if it is to be used at 20 feet?

b) What is the actual size of the corresponding retinal image?

ANSWER:

First, you must know that the element to be discriminated in any test letter, say the "break" in a letter "C", is standardized. The angular size of the "break" for a 20/20 letter is 1 minute of arc. The *total* size of any test letter is always 5 times the angular size of the "break". So, a 20/200 letter would be 10 times as large as a 20/20 letter and, therefore, a 20/200 letter has a "break" of 10′ angular size, and its *overall* size must be 50′ subtended at the eye's nodal point.

a) In the sketch below, our similar triangles are labeled with what we know:

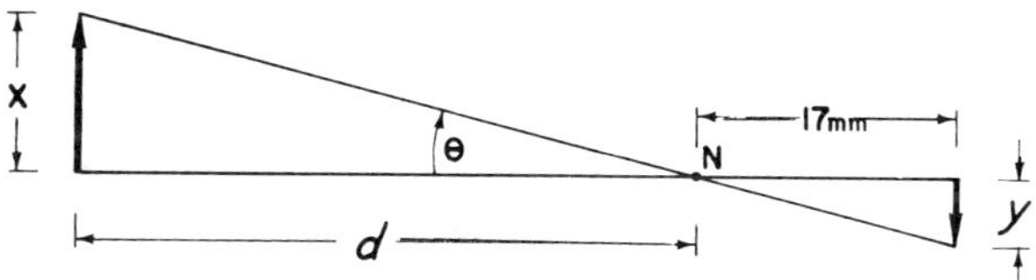


$x$ = the height of the letter

$d$ = 20 ft = 6000 mm

$\theta$ = 50 minutes

To solve this problem we must have $\theta$ in radians rather than in minutes of arc.

$$50 \text{ minutes} = \frac{50}{60} \text{ degrees}$$

$1 \text{ degree} = \frac{\pi}{180}$ radians (for most of us that can't remember the decimal conversion factor!)

Therefore, $50 \text{ minutes} = \frac{50}{60} \cdot \frac{\pi}{180} \text{ radians}$

$$\theta = .0145 \text{ radians}$$

Now, $\theta = \frac{x}{d}$

$$x = d \cdot \theta$$

$x = 6000\ (.0145) = 87$ mm, the true height of the 20/200 letter.

b) y = the retinal image size (see sketch above).

$$\frac{x}{d} = \frac{y}{17}$$

$$y = \frac{17 \cdot x}{d} = \frac{17\ (87)}{6000} = \frac{1479}{6000} = .246 \text{ mm}$$

The retinal image size of the 20/200 letter is 0.246 mm.

So, if you keep the 17 mm figure in mind, you should have no difficulty with these types of determinations.

## ACCOMMODATION

The normal eye has the ability to form sharp images and place them on the retina, and moreover, this ability is not limited to one, fixed object distance; the eye can *change* its power, within certain limits, to accommodate to a shift in the object distance. Say an eye were a *fixed* focus instrument and set for a perfect focus for an object at infinity (figure A below); since, as we have shown, objects and images always move in the same direction, if that object approached the eye, the corresponding image would move backward off the retina (figure B below). (We can speak of this movement happening *optically* even though we know that *physically* the opaque structures at the back of the eye would preclude it). To compensate for the increased divergence at the eye as the object approaches, the *total* eye power must increase (or the eyeball would have to elongate) to maintain a focused image on the retina. It should come as no surprise that the increase in eye power is by far the more important mechanism and serves to "pull" the sharp image onto the retina (figure C below). This increase can be as much as 18 additional diopters in a young eye, which you know already possesses about 60 diopters. This variable power (from 0 to 18 D) supplied "automatically" by the ocular lens is called *accommodation.*

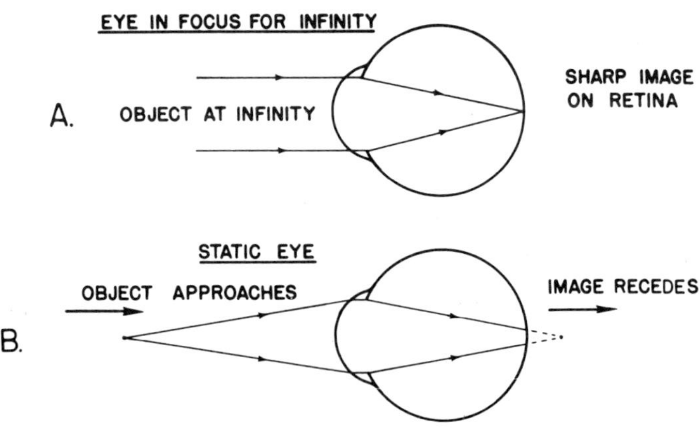

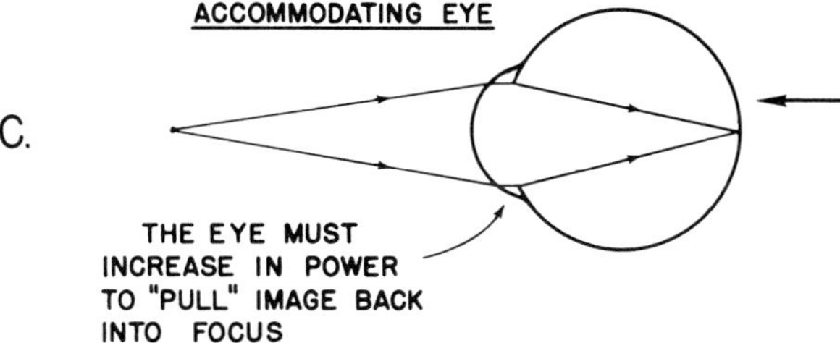


In the sketches above, we are "stretching" somewhat to compare the "reduced" (lensless) eye to its dynamic human counterpart — it is done here for simplicity. We will also assume that all of the increase in power of our model eye occurs at its front surface instead of somewhere within. The error in doing so is not great nor is it critical here.

The availability of 18 D accommodation in a child would allow objects to be brought close enough to present a divergence of up to − 18 D to the eye and still allow the image to be kept in focus on the retina. Since − 18 D corresponds to a $u$ of $-\frac{1}{18\text{ D}}$ or − 5.5 cm, the object could approach as close as 5.5 cm and still maintain retinal conjugacy. We call the 5.5 cm position the *NEAR POINT OF ACCOMMODATION* — that point on the visual axis which is conjugate to the retina when accommodation is *maximally* active.

This remarkable capacity of the eye to accommodate is gradually lost with increasing age. It is one of those physiologic functions that begins to be taken away from us immediately after birth, and it is just about completely gone by age 70 with an almost straight-line loss in the intervening years. The accommodation loss (like taxes and aging) is relentless, predictable and inevitable. It results in *PRESBYOPIA* ("old-age" vision) and is based on rigidification of the ocular lens; it becomes clinically evident by a recession of the "Near Point of Accommodation".

*TABLE III*

ACCOMMODATION LOSS WITH AGE (Donder's Table)

| Age in years | 1 | 5 | 10 | 15 | 20 | 25 | 30 | 35 | 40 | 45 | 50 | 55 | 60 | 65 | 70 | 75 |
|---|---|---|---|---|---|---|---|---|---|---|---|---|---|---|---|---|
| Total accommodation (overall amplitude) in diopters | 18 | 16 | 14 | 12 | 10 | 8.5 | 7.0 | 5.5 | 4.5 | 3.5 | 2.5 | 1.75 | 1.00 | .75 | .25 | .00 |

The amount of accommodative loss only becomes symptomatic to each of us in our early forties, since it is only then that the *remaining* accommodation approximates the amount required for average *reading* distances. At that time, the required task demands the expenditure of a high percentage of the accommodative power that an individual possesses, leaving no reserve "in the storehouse".

An example should make this clear: reading at a distance of 33 cm normally requires an accommodation of 3 D. If an individual is 20 years old, he can easily supply that 3 D from his "storehouse" of say, 10 Diopters. However, if his accommodative reserve has been depleted by age or disease, he may only be able to supply the 3 D and no more; thus, he will be operating at his full capability with no reserve capacity. Since he cannot maintain this maximal effort for very long, he will tire quickly and the image of what he is reading will blur on his retina.

He can help the situation somewhat by moving the reading material further away, placing it beyond his *near point* of accommodation and lessening the accommodative demand. At some point, however, this becomes self-defeating since an increase in the object distance also will *decrease* the image size on the retina. (Remember our magnification relationship; if $u$ increases relative to $v$, the image size will decrease). Thus the increase in object distance can make it *harder* for him to read a given print size. So, any patient entering the presbyopic world will try to move the reading material further away to reduce the accommodative demand, but the distance to which he can move it will be limited by the following: 1) the minimal size of print which is legible by the individual and 2) (more practically) the length of his arms!

When a patient comes to you with such complaints, what can you do to help? First suggest that he increase the illumination on his reading task. This will certainly help, since one's ability to resolve fine image detail increases with an increase in light in the retinal image.* However, this will almost assuredly have been discovered empirically by the patient himself and he will already be using strong illumination. What else do you have to offer? The answer is clear. When a short object distance makes an intolerable demand on the accommodation supply and the presbyopic eye itself cannot increase its own power, the optical lens manufacturer is ready and eager to step in. For a small charge, he will supply (to your prescription) an accessory lens to be worn in a spectacle frame (also at a small charge). This lens will replace some or all of the accommodative power required for close work.

For average reading distances, say around 40 cm, 2.5 D accommodation is required of every eye. The early presbyope will seldom require more than about 1.0 D help in his "reading spectacles"; his own eye will easily supply the 1.5 D balance. In the best attitude of self-reliance, the typical patient will want to use as much of his own accommodative power as possible. (Later, we will see optically why it is to his benefit to do so.) But, by the time he is about 62, his own total accommodative power will have grown so feeble, he will likely need all the lens help he can get to read at 40 cm; then, he will probably require the "full help" of a 2.50 D lens. More about presbyopia and its correction later on, but you should understand now why a typical 70 year old presbyope does not require any *more* than 2.50 D of add at 40 cm. If, however, he *has* to read at a closer distance (say 10 cm) because some macular lesion demands that he have a larger retinal image, then he *will* obviously require a greater add than 2.50 D (here it would be + 10 D) to see with at this ultra-close distance.

* Rubin, M. D. and Walls, G. L.: *Fundamentals of Visual Science*, C. C. Thomas, Springfield, 1969, p. 173.

## REFRACTIVE ERROR

If an eye is to serve its beholder well, it must provide sharp imagery; to do this, there must be a "proper", coordinated match between the *power* of the various refracting surfaces and the *length* of the eye. When a perfect match exists for distant object vergences, we say *EMMETROPIA* is present; if not and a mismatch occurs, a *refractive error* (or *AMETROPIA*) supervenes.

What is surprising is *not* that the eye power and eye length are often inappropriately coupled; the real mystery is why a mismatch is *less* frequent than chance. If we plot a frequency distribution curve of refractive error as it exists in the population, we will find that the distribution is not "normal" as it *is* with height, weight, head circumference, etc. There is a *much* greater frequency of emmetropia (or *near* emmetropia — since the actual mean is about 0.25 D on the hypermetropic side) than one expects to find in a truly randomized "normal" distribution. (See graph.)

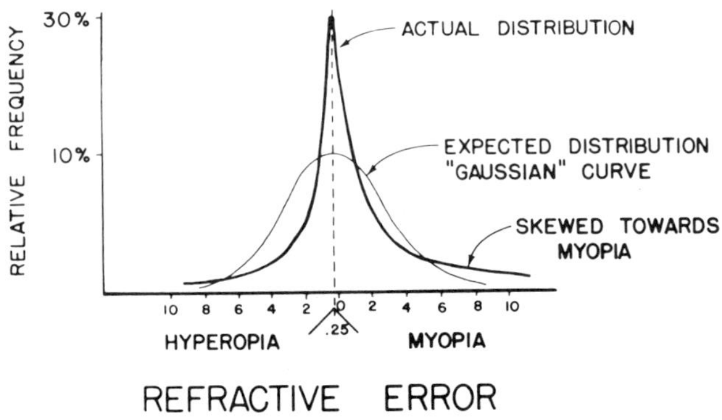


You might be surprised to learn that the individual components of eye refraction — corneal curvature, lens power, axial length (neglecting degenerative elongation) *are* all "normally" distributed. Some "thing" must happen to coordinate these individual parameters, so that when the eyeball grows too long, the corneal radius tends also to be greater than usual and therefore, of less power — this balances the increase of the eye length; conversely, the shorter eyeball tends to be coupled with a steeper cornea than "normal".

So, it seems clear that some wonderful mechanism keeps eyes close to emmetropia. Arnold Sorsby calls this tendency, the "emmetropization process." Alas, sometimes it fails, and ametropia enters the picture (and provides job opportunities for optometrists, ophthalmologists, and electronic refraction machines)!

An important point to remember is that an emmetropic eyeball (with no refractive error) can be a big eyeball or a little eyeball; the absolute size (or its correlate, absolute power in diopters) is immaterial — just as long as the power is exactly proper for the length as it exists, no refractive error will be present. Thus, a small turtle's tiny eyeball (perhaps 6 mm long) and that of a horse (perhaps 60 mm long) both can be emmetropic and allow sharp vision when their refractive components are sufficiently strong to compensate perfectly for their lengths; in these instances the dioptric powers would be 270 D and 23 D, respectively. So, you cannot tell the degree or direction of ametropia by knowing the dioptric power alone. What you must know is how well that power correlates with the corresponding axial length.

The "average" *human* emmetropic eye is roughly 60 D, though of course, it may still be emmetropic yet of greater or lesser power than this. In our optical diagrams of the eye, we will assume the emmetropic reduced schematic eye is 60 D in power, is composed of one ocular medium of $n' = 1.33$ (water), and has only one refracting surface of radius 5.5 mm — a convenient, though distorted model of a real eye.

When emmetropia exists, light from a distant object will be focused onto the retina so that a clear image is present there. That is, each object point is represented by an image point which will lie on the retinal surface. A distant point located straight ahead along the visual axis would be imaged at the secondary focal point F', which will be on the retina in the emmetropic eye (as shown in A below):

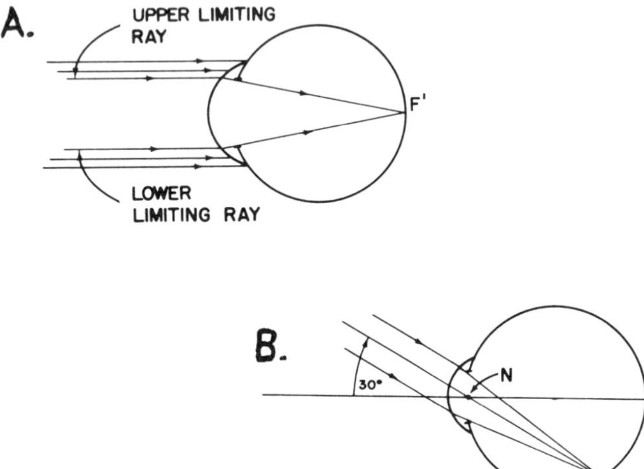


(In both of the above diagrams, the upper and lower most rays signify those which just barely pass the upper and lower edges of the pupil. Any rays which are more peripheralward cannot enter the eye and cannot help form the retinal image. These two extreme rays are called the "limiting rays". The axial line is shown for orientation.)

In Figure B above, if the rays from the object at infinity arise from an off-axis point (say 30° above the axis), the particular ray which goes through the eye nodal point will be undeviated and will, therefore, locate the position of the *image* point. That image will be in the retinal plane only if the eye is emmetropic. (Actually, the image point will *have* to lie somewhere along this ray no matter what the power or refractive error of the eye. That is what makes this ray so valuable.) The nodal point in our reduced eye must be located exactly at the center of curvature, 5.5 mm from the front surface, and any ray which passes through a center of curvature must strike that surface perpendicularly — the angle of incidence at the surface is then *zero* degrees — and is therefore undeviated by it.

The distance between the surface and the retina can be easily calculated if you don't happen to remember it; $P$ of the surface is 60 D and $n' = 1.33$; $U + P = V$

Since the object is at infinity

$$U = 0.$$

Image distance $v = f'$ (the secondary focal length).

$$\therefore \quad P = \frac{n'}{f'}; \; 60 = \frac{1.33}{f'}; \; f' = \frac{1.33}{60} = 22.2 \text{ mm*}$$

Therefore F′ (which is *on* the retina in an emmetropic eye) is 22.2 mm from the surface.

If an eye is not emmetropic, it may either be too strong in power or too weak, or too long in length or too short. The "too" is a relative term; it refers only to what *would* be required to make that eye "correct". Whenever one wants to express the degree of ametropia present, he has only to state how much *power* difference there is "away from" emmetropia for *that* eye — *not* a difference from "normality" or from some "average" power or from some other arbitrary standard.

## Myopia

If the power of an eye is too strong *for its size,* we say that the eye is myopic or "nearsighted." A myopic eye, then, has its F′ located somewhere in front of the retina — the more in front it is, the greater the myopic refractive error.

Say an eye is 5 D myopic; this means the dioptric components are relatively too strong by this amount *or* the eyeball is too long by an amount which causes 5 D of myopia. Either possibility exists and simply knowing that 5 D of myopic error exists does not tell us which parameter is at fault (or whether some proportion of the error is contributed by both) or what the *total* power of the eye is. No; all we know is that there is the 5 D of error, and that it errs by being *relatively* too strong — the actual power and the actual length may be any value. These points must be well ingrained.

* Reread footnote, page 120.

## The Far Point Plane

It doesn't matter whether the myopic fault is with too great a refractive power or with too long an eyeball; in either case, when an object is at infinity, F′ of the eye is where the sharp image is located and that is in front of the retina, somewhere in the vitreous. We can move that sharp image backwards onto the retina by bringing the object closer to the eye, since as the object approaches the eye, the image plane will move in the same direction. (This obeys our law about object and image motion.)

As the object comes closer, it will reach a certain object plane position so that its image will fall squarely on the retina. That particular object position is called the *FAR POINT PLANE* of that myopic eye; its axial point is known as the *FAR POINT.* All rays leaving an object point in that plane (and entering the eye) will form sharply focused image points on the retina. By definition then, the far point plane is that object plane which is conjugate to the retina when the eye is *not* accommodating. (This definition holds true for *any* type of ametropia.)

When the object rays leave the myopic far point plane, they will have a certain divergence. This divergence of object rays is necessary to compensate for the "overpower" of the myopic eye and enables it to create sharp retinal imagery. The specific amount of the divergence required is *equal* to the amount of "overpower", that is, it is a quantitative measure of the existing myopia. Thus, all we have to know is where the far point is and we *know* the amount of myopic error. So, find the axial position where a myopic patient can just see details of an object clearly, measure the distance from the eye and convert it into diopters and *Bingo!* If a far point plane is located 23 cm in front of an eye, that eye must be $\frac{1}{.23\text{ m}}$ or 4.35 D myopic.

With a real patient, you have to be careful in trying to determine myopic error by locating the far point plane in this way since you may stimulate accommodation with your test target. As any target is brought steadily closer, the eye will be able to see finer detail more easily; so, you might be led astray in thinking you had not yet reached the far point, when , in reality, you may have passed it and are now only forcing the eye to accommodate (if it can) in an effort to keep up with the approaching target. Since you are attempting to locate that far point and our definition of *far point* specifies that *no* accommodation should be active, you must make sure the patient does *not* accommodate during your determination. If you paralyze the eye's accommodative power with a cycloplegic drug (like Cyclogyl® or atropine), you need not fear accommodation as a contaminant of measuring the position of the far point plane. With care, however, you *can* do this without cycloplegia if you always move the target *from* a blurred zone *towards* the clear range, and stop at the position of *initial* clarity of a small target letter. That is the position of the far point plane.

A myope will see everything beyond his far point blurredly; but at, or closer than this plane, his eye will create sharp imagery as well as the emmetropic one. At distances nearer than the far point, the myope also has to accommodate to see sharply, but he has a "head start" on the emmetrope and will need to exert less accommodation at a comparable distance. It is as if the myope's eye had a "built-in" plus lens of a magnitude equal to his myopic "error"; so, let's begin now to think of the myopic eye's error as being a *plus* error.

A 5 D myope has an extra + 5 D of power and allows him to see clearly at 20 cm with no accommodative exertion at all; this same surplus + 5 D "built-in" lens provides him with the means to see at a 10 cm distance by using only 5 D of his own accommodation. The myope will have a closer near-point-of-accommodation than an age-comparable emmetrope, and 10 D actual accommodation will allow the myope to see at closer distance than that same 10 D exerted by the emmetrope.

Working a sample problem will help cement some of these points:

PROBLEM:

A 4 D myopic eye has a near point of accommodation of 8 cm. Determine the following:

a) far point
b) amplitude of accommodation
c) range of accommodation
d) How much accommodation must he exert to see detailed print at 10 cm?

ANSWER:

a) Far point $= \frac{1}{-4\text{ D}}$ or 25 cm in front of the eye.

b) The "amplitude of accommodation" is the total number of diopters of accommodation available to an individual — his maximal capacity for focusing — the dioptric difference between his near point and his far point. This myope's near point of accommodation is at 8 cm. This means that with maximal accommodation active, he can see a point which has a vergence of $\frac{1}{-.08} = -12.5$ D.

Since he is 4 D myopic, he will need to exert only 12.5 — 4 or 8.5 D of accommodation to see at a distance of 8 cm. (The emmetrope has to accommodate the full 12.5 D). Thus, the amplitude of accommodation is 8.5 D.

c) The "range of accommodation" is that actual distance through which the eye can see clearly in going from no accommodative effort to a maximal one.

With no accommodation, the far point is at 25 cm.

With maximal accommodation, the near point is at 8 cm.

Therefore, his range is 25 to 8 cm (17 cm long).

d) A point 10 cm away has a vergence of — 10 D, and from an emmetrope would demand 10 D accommodation. Since this eye is 4 D myopic, it has a "head start"; it will need to accommodate only (10 — 4) or 6 D to see this object point clearly.

The handling of problems of this type should become second nature to you.

Now after introducing the concept of the far point, I want you to consider *axial* myopia — myopia due only to the fact that the axial length is too long for the eye. So, assume the dioptric power is "normal" here, at 60 D.

PROBLEM:

How *much* longer is this eye than the "reduced eye" if it presents 5 D of myopic error?

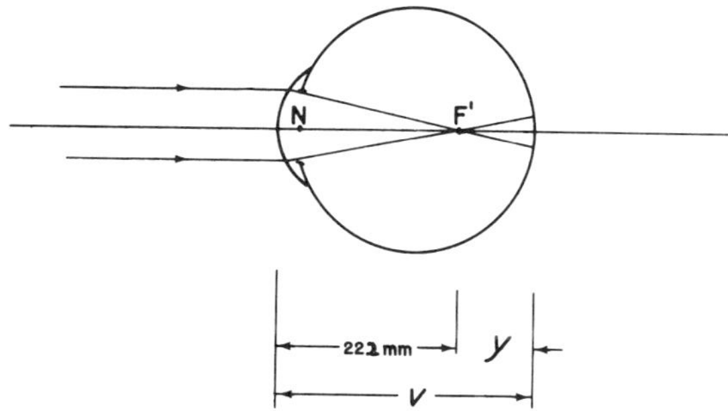


ANSWER:

Let $y$ = number of mm increase in length over the emmetropic eye to yield 5 D of error.

What we must *first* find is the image distance $v$ which corresponds to the object distance $u$ for an object located at the far point; then we can determine $y$, which will be equal to $v - 22.2$.

Since the myopia was given as 5 D, the far point must be located 20 cm in front of the front surface of the eye; and $U = -5$ D

$$U = -5 \text{ D}$$
$$P = 60$$
$$V = ?$$
$$U + P = V$$
$$-5 + 60 = V$$
$$+55 = V$$

Since the image vergence $V = \dfrac{n'}{v}$,

then $$v = \frac{n'}{V}$$

and $$v = \frac{1.33}{55} = 24.2 \text{ mm}$$

since $$y = v - 22.2$$
$$y = 24.2 - 22.2 = 2.0 \text{ mm}$$

This linear distance represents 5 D of axial myopia, so, *each* diopter is equivalent to $\dfrac{2.0}{5 \text{ D}}$ or 0.4 mm of axial elongation.

Look now at an example of *refractive* myopia where only the dioptric power of the eye is at fault; here, it is too strong. Again, use our "reduced" eye as a model; the "corneal" power would be 65 D instead of 60 D if we assume the presence of 5 D of refractive myopia.

PROBLEM:

What is the axial equivalent of 1 D of myopic refractive error?

ANSWER:

The focal plane F′ for images at infinity would be located as follows:

$$P = \frac{n'}{f'}$$

$$f' = \frac{1.33}{65} = 20.5 \text{ mm from the "cornea".}$$

REFRACTIVE MYOPIC EYE

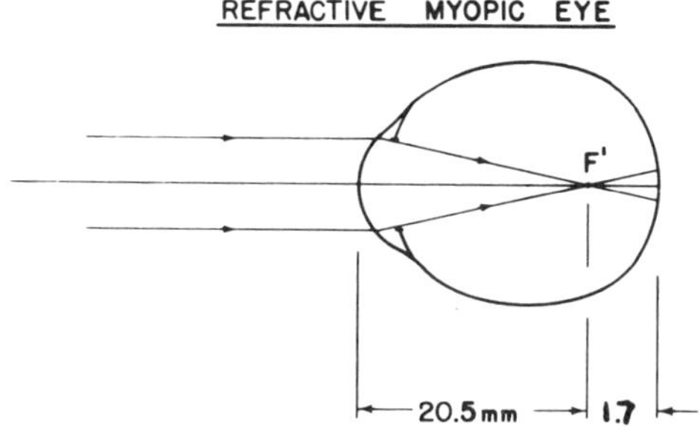


Thus, F′ is short of the retina by (22.2 — 20.5) mm or 1.7 mm.

Each diopter of error would be roughly equivalent to $\frac{1.7 \text{ mm}}{5 \text{ D}} = .34$ mm of axial distance.

To summarize these last two calculations:

With pure *axial* myopia, 0.4 mm of axial elongation is equivalent to 1 D of error.

With pure *refractive* myopia, 0.34 mm of axial distance is equivalent to 1 D of error.

These figures are approximations for a *model* eye with an *assumed* refractive power located exactly at the "corneal" plane. This is not so with an actual human eye which additionally may be of *any* reasonable refractive power. In spite of this, the approximations given above are fair ones; you will not be far off if you assume an average of .37 mm as the axial equivalent of 1 D of "average" myopia. (Besides, clinically, you cannot determine whether a refractive error is axial or refractive anyway!)

CLINICAL POINT:

The typical ametropias tend to arise gradually, probably through some slight aberration or exaggeration of the normal growth processes of the eye. Occasionally, however, something abnormal occurs to cause the refractive error to change rather abruptly.

For example, the eye will be made functionally too short by something which pushes from behind, indenting the eye or shoving the retina towards the vitreous: a retrobulbar mass, a choroidal tumor (melanoma or metastatic lesion — most commonly breast), a retinal elevation (as seen in a pigment epithelial detachment or even central serous choroidopathy). These problems serve to shorten the axial *length,* thereby increasing the refractive error in a *hyperopic* direction.

On the other hand, if the change in error happens to be toward the *myopic* side, a problem will usually be found with the *refractive* components (except in degenerative myopia): keratoconus, spasm of accommodation, incipient nuclear cataract, subluxed or anteriorly dislocated lens, or any condition which causes rapid shifts in location of the body fluids (as pregnancy, diabetes, acidosis, assorted drugs — sulfonamides, osmotic agents, etc.) — all tend to create a myopic error.

Keep alert to all of the above diagnostic possibilities whenever there are sudden (or relatively rapid) changes in the patient's preexisting refractive state.

* * *

Resuming our discussion of the far point, we shall demonstrate that *every* eye has its own far point — that axial point conjugate to the retina when accommodation is inactive. As we have shown, the myopic eye has a far point which is always located between infinity and the anterior corneal surface, but even the emmetropic eye has a far point; it is located at infinity!

## Hypermetropia

We have dealt so far with one type of refractive error — the "over-powered," myopic eye. There is another side to the coin — the eye that has *insufficient* refractive power or is "too short", or both. This eye is the hypermetropic (hyperopic) or "farsighted" eye. It too must have a far point (an object) which is conjugate to the retina, but where? Let's find out by solving for $U$ in our old standby $U + P = V$. Attribute a "weak" refractive power $P$ of 55 D to our reduced eye; this makes it 5 D hyperopic. Its length is still 22.2 mm.

$$U + P = V = \frac{n'}{v}$$

$$U + 55 = \frac{1.33}{22.2}$$

$$U + 55 = 60$$

$$U = +5\text{ D}$$

HYPERMETROPIC EYE

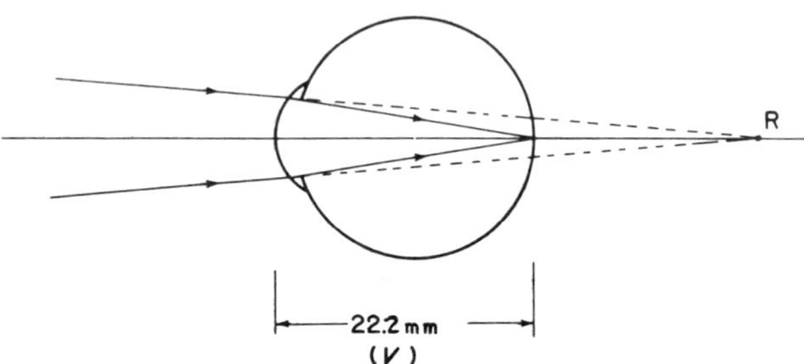


Recall our sign conventions? A $U$ of $+5$ D means that the object which is conjugate to an image on the retina must be located 20 cm *behind* the eye. That is, object rays must be convergent to a point 20 cm behind the eye to be focused sharply on the retina when no accommodation is active. This point, R, signifies the location of the far point plane for this eye since it fulfills our definition.

We already know that no real object emits *convergent* object rays; we can therefore surmise that the hypermetropic eye will never produce sharp point images from real objects, that is, *unless* it can increase its resting eye power (which in this case is too weak) by accommodating.

When this hyperopic eye looks at infinity without accommodating, $U = 0$.

$$U + P = V$$

$$0 + 55 = \frac{n'}{f'}$$

$$55 = \frac{1.33}{f'}$$

$$f' = \frac{1.33}{55} = 24.2 \text{ mm}$$

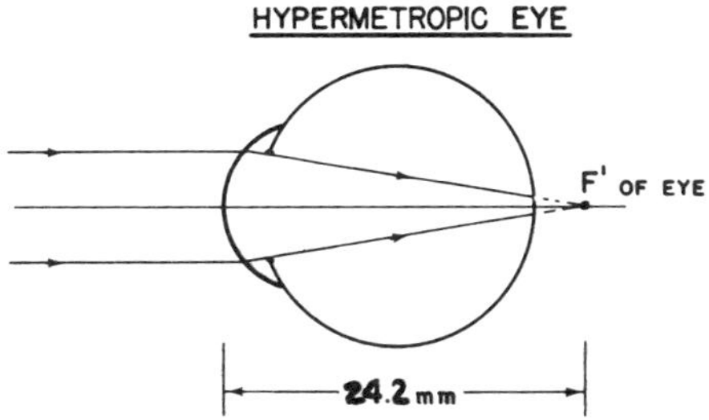


This eyeball is too short by (24.2 — 22.2) or 2.0 mm, and F′ falls *behind* the retina. There is, therefore, a blurred spot on the retina instead of a sharp image point. But, that is exactly what we saw when we investigated the *emmetropic* eye's response to near objects and learned that as an object approached the eye, its image tended to move back off the retina (optically) and blur the retinal image; it was accommodation that "pulled" this image forward onto the retina. So, also here with the *hyperopic* eye; a blurred image on the retina stimulates accommodation which "pulls" the sharp image forward into focus on the retina.

The unaccommodated *emmetropic* eye is "too weak" only in regards to a near object and its associated divergent object rays. For the hyperopic eye, it is *not* object proximity alone that causes the image to recede behind the retina; it is simply that the hyperopic eye is relatively too weak for *all* distances — too weak even to bring *parallel* object rays to a sharp focus on the retina. In emmetropia, accommodation is required for near only, but in hyperopia, accommodation is necessary even for distance and still more so for near.

A hyperope of 2 D will require 2 D of accommodation just to see clearly in the distance. To read material held 20 cm from his eye, he must accommodate as much as the emmetrope ($\frac{1}{.20}$ D) *plus* 2 D more to overcome his hyperopia, for a total of 7 D. Since he must constantly use a higher proportion of his accommodative reserve than his age-matched emmetropic friend, he may well exhibit symptoms of

presbyopia at an earlier age than his confrere.

Go back to the last diagram: the fact that R, the far point, is "behind" the retina and also F′, the eye's secondary focal point, is "behind" the retina leads some to confuse the two. It shouldn't; the distinction between these entities should be crystal clear.

1) F′ is an *image* which is conjugate to an object at infinity, while R is the *object* which is conjugate to its image on the retina.

2) The distance from the retina to F′ is very short (usually in *milli*meters or fractions thereof), while the distance from the retina to R is very long, relatively; it is almost never closer than about 8 or 9 centimeters, but may be "infinitely" long.

Don't confuse F′ and R in diagrams.

This discussion of the far point in myopia and hypermetropia will be of utmost value to you. Of all the subjects you have covered so far and will cover later, this one is probably *the* most important since the far point is the crux of the clinical correction of ametropia. You must be fully comfortable with the far point as a concept, so brand this completely onto your brain: the far point R belongs to the unaccommodated eye, *not* to any corrective lens. Picture R as being rigidly attached to the eye like a barnacle on an oyster; imagine that it protrudes a certain distance from the eye (front or back) as the amount and type of ametropia demands. The action of all lenses must be referred to this far point — it serves as *the* basis for understanding the action of all corrective lenses.

## THE CORRECTION OF AMETROPIA

With your present background, the correction of refractive error becomes a real cinch. In a nutshell, *any* lens that images infinity at the far point of *any* given eye is a "corrective" lens. Since all lenses will, by definition, image infinity at their secondary focal points (F'), all you have to do is place that lens in such a position that its F' coincides with the far point of the eye requiring correction. This will make that particular lens a "corrective" lens for that particular eye. The corollary is just as important; *more* than one lens can "correct" a given eye.

### Hypermetropia Correction

Let's take an example using a 3 D hyperopic eye. As before, our reference plane for the refractive error is at the anterior refracting surface; so, the far point R must be located 33 cm behind that surface. Light rays will have to converge toward R to be imaged on the retina, but light rays from an object at infinity are parallel and will *not* be convergent toward R, that is, without help. So, for *sharp* retinal images, a plus lens will have to be introduced into the light path to add convergence to the light. Let's use a + 2 D lens.

A + 2 lens will "correct" this eye for objects at infinity if (by definition) it is placed so that its F' coincides with the far point R. Since F' of a + 2 lens is located 50 cm behind it, the lens itself must be placed 50 cm from R for it to qualify as a "corrective" lens.

Then, object light rays from infinity will be focused by the + 2 lens at F'; these image rays by the lens become object rays for the eye, and any object rays heading for R will be sharply imaged on the retina.

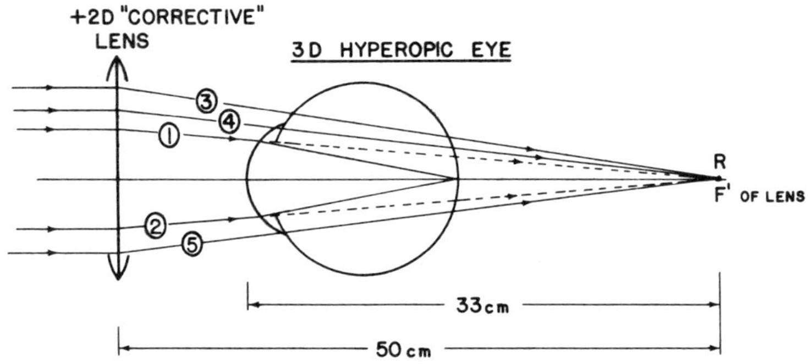


In the figure above, rays 3, 4 and 5 are blocked by the sclera and do not get through the pupil; therefore, they cannot help form the retinal image point. In this figure, only those rays *between* 1 and 2, imaged by the lens at R, actually enter the eye and are imaged by *it* sharply on the retina. 1 and 2 are "limiting" rays — limited by the edge of "the pupil".

If we wanted to place the "corrective" lens smack up against the cornea (like a contact lens), it would also have to be of such power as to have its F' at R of the eye. We know the distance from this surface to R — it is 33 cm since this is a 3 D hyperopic eye. Thus, the corrective "contact" lens must be of + 3 D.

It should now be clear that you *could* use a + 0.5 D lens and have it "correct" this eye's refractive error *if* that lens were held 200 cm from R. The 3 D of hyperopia will still be corrected by it.

So, now you have seen that the lens power which "corrects" a given amount of ametropia is not fixed, and the far point *which belongs to the eye and not to the corrective lens* does stay rigidly in place.

I can hear the cries of anguish now: "Be realistic". "Who would wear a corrective lens at 200 cm for a 3 D hyperopic error?" The question is moot since my point is not that you *would* correct hyperopia with this lens, but that you *could;* you might use *any* plus lens, just as long as its F′ is made to coincide with R. (You will soon learn that there *is* a use for such a peculiar "corrective" lens.)

O.K. Let's be more practical. Corrective lenses are typically worn in spectacle frames which place them in front of the eyes. Although I can here specify a definite location for the spectacles, this should only be considered an approximate guide since the true position depends on many factors — the type of lens frame and the position of its nose rest, the shape of the patient's brow, position of his ears, how deep set the eyes are in their sockets, etc. Wherever the corrective lens is worn, *that* is the "spectacle lens plane". *It* should be the reference plane for prescribing corrective lenses even though it varies somewhat in position from patient to patient. Arbitrarily here we will place it at 15 mm from the corneal surface. (The distance between the cornea and the corrective lens is called the *VERTEX DISTANCE.)* However clinically, especially for a patient with a large refractive error, this distance cannot be assumed; it must be measured. You will soon see why.

(This 15 mm separation between a "practical" corrective lens and the cornea places the lens just about at the anterior focal plane of the eye. Later, we will take up what this means optically, but, I will "spill the beans" now and tell you this is *not* of momentous clinical importance.)

In any case, back to our + 3 D hyperopic eye, the "practical" corrective lens must have a focal length which is 15 mm longer than the distance (33 cm) between the cornea and R. Its corrective power must be $\dfrac{1}{(.33 + .015)\text{ m}} = \dfrac{1}{.345\text{ m}} = +\,2.9\text{ D}.$

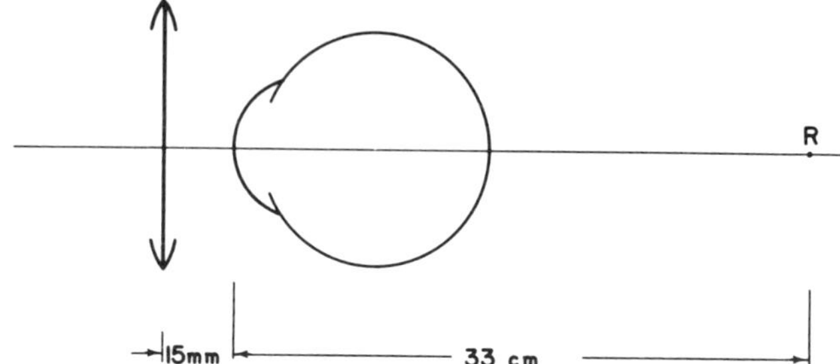


With low powered corrective lenses, it will not make too much difference whether one reckons with the "vertex distance" or not. In this case, there is only 0.1 D difference from the existing 3 D of hyperopic error. (This small amount will just barely be grindable by an optician's lens grinding machine.) However, for an aphakic eye (following lens extraction for cataract), the vertex distance *is* a critical consideration. Aphakic eyes are usually 10 - 12 D hyperopic; a few mm of change in the vertex distance can change the effectiveness of such a spectacle lens greatly.

To demonstrate this point, work the following problem.

PROBLEM:

A + 12 D corrective lens (which was measured at a vertex distance of 15 mm) is prescribed for an aphakic eye. When the patient's optician fits the corrective lenses into a stylish frame of the patient's choice, the lens is placed only 10 mm from his cornea. What is the dioptric error caused by fitting the lens at this closer distance?

ANSWER:

First you must locate R — *that* is the key point no matter what the refractive error is. Since you are given that a 12 D lens "corrects" this aphakic eye, its F′ must fall on R. F′ is $\frac{1}{12\text{ D}}$ or 8.3 cm behind the lens, but the lens itself is 15 mm in front of the cornea. So, R must be (8.3 — 1.5) or 6.8 cm behind the cornea. (See A below.) Now that the position of R is located, we can use any corrective lens we wish depending on its desired position from the eye.

Since this patient's actual vertex distance is only 10 mm, he *should* require a corrective lens with a focal length of (6.8 + 1.0) or 7.8 cm — a + 12.8 D lens neatly fills the bill. (See B below.)

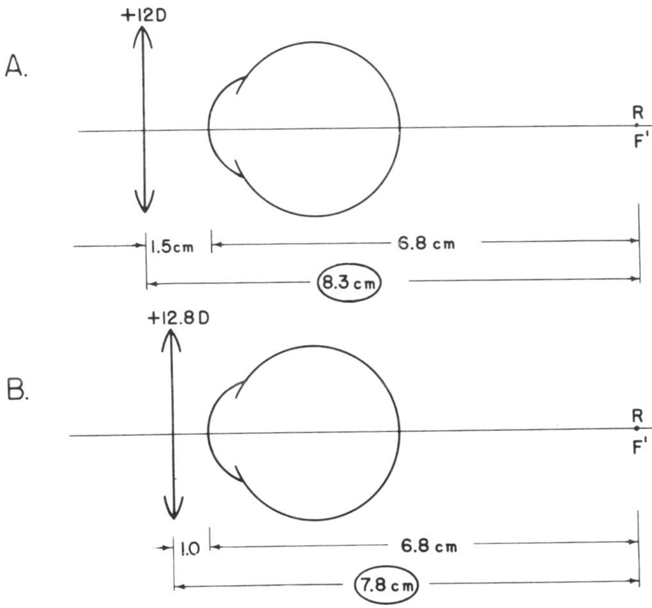


But he is wearing a + 12.0 D lens inadvertently fitted at this 10 mm vertex distance, so he is *under*corrected in plus power by (12.8 — 12.0) or 0.8 D — certainly not an *in*significant amount.

You must pay attention to the vertex distance; it becomes particularly significant with ametropias over 4 Diopters, whether in hyperopia, myopia or astigmatism.

## Myopia Correction

Everything we said about far point correction in hypermetropia also holds true for myopia. For any corrective lens to work, it must image infinity at the far point of the eye. "Optics" takes over from there to assure that a sharp image will fall on the retina.

When an eye is 7 D myopic (with the "error" considered to reside at the corneal surface), it has its far point located 14.3 cm from the cornea.

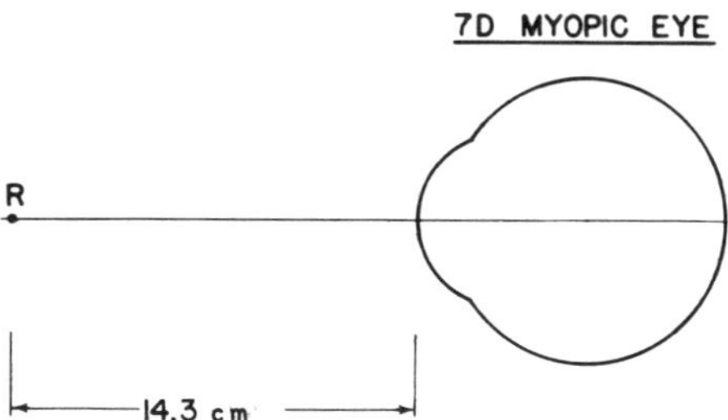


A "corrective" lens is placed so as to take parallel light bundles from infinity and, after refraction by the lens, make them *seem* to the eye as if they arose at R. Such a lens will do so only if its secondary focal point F′ coincides with R.

A number of individual lenses, each of which fulfills this criterion, are shown below:

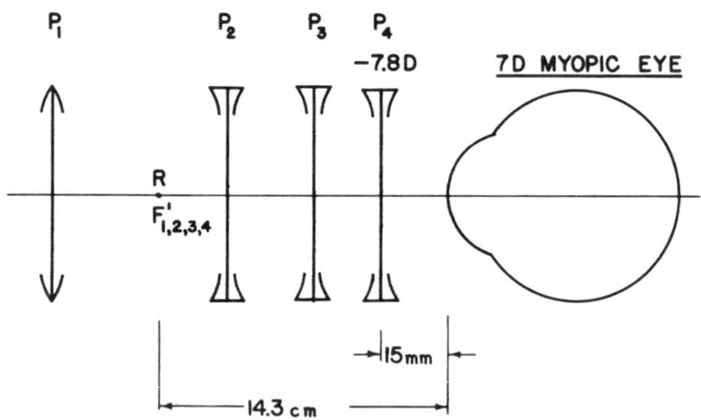

Lens position 4 places the corrective lens in the "spectacle plane" 15 mm from the cornea; its power must be as follows: In this 7 D myopic eye (again with the "error" at the corneal surface), R is 14.3 cm from the cornea. The F′ of the lens must be at that same position, that is, (14.3 − 1.5) cm or 12.8 cm from lens position 4. $P_4$ must then be $\frac{1}{-.128\ \text{m}} = -7.8$ D. $P_4$ must be *minus* since it has to diverge the parallel light from infinity to make it seem to the eye behind it as if the rays came from R.

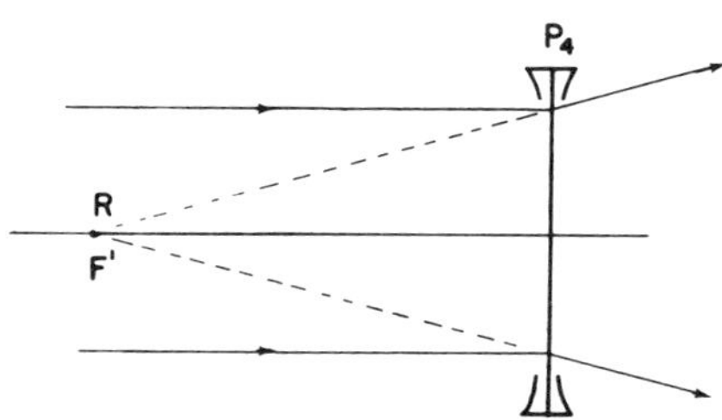


If the corrected lens were located at position 3 above, it would of necessity have a shorter secondary focal length and, therefore, $P_3$ would be greater in minus power than $P_4$. Of all the possible corrective lenses, the one with the *least* minus power would be situated furthest away from R, that is, in contact with the corneal surface — a contact lens of − 7 D power.

Say that the position of corrective lens 2 was only 1.0 cm from R. Its power would have to be a whopping −100 D to "correct" this eye, but correct it optically it surely would, since a sharp image of infinity would then fall on the retina.

Look now at lens position 1 on the *left* of R. Could a lens here also correct this myopic eye? Why, of course it could. However, it would have to be a *plus* lens. If located 20 cm to the left of R, this peculiar "corrective" lens would have to be of + 5 D power (with its F′ at R) — it, like the other corrective lenses, images infinity at the far point, and thereby allows objects to be seen clearly and distinctly by this 7 D myopic eye. (One small point in this case; the image of an object at infinity "corrected" in this manner would be seen *upside down,* but sharp, nonetheless!)

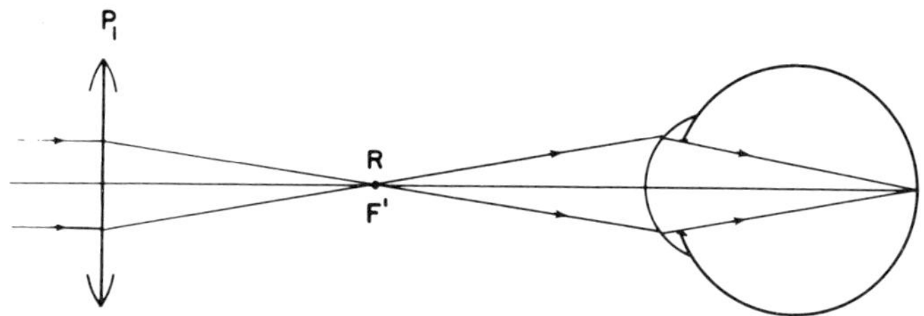


Let's neglect this latter, unusual "correction" except to know it as a possibility.

To summarize for the usual clinical situations:

Plus lenses are used to correct hyperopia; the closer the corrective lens is to the eye, the shorter its focal length must be (greater plus power) since each of these lenses must have their F′ precisely at the eye's far point which is fixed in location *behind* the hyperopic eye.

Minus lenses will correct myopia; and, the closer to the eye the lens is, the longer its focal length must be since R is in *front* of the eye.

As long as you know where the far point is, in myopia and in hyperopia, you can figure out the proper corrective lens very easily.

PROBLEM:

Where would you place a − 8 D lens to "correct" a 3 D myopic eye?

ANSWER:

A 3 D myopic eye has its far point 33 cm in front of the cornea.

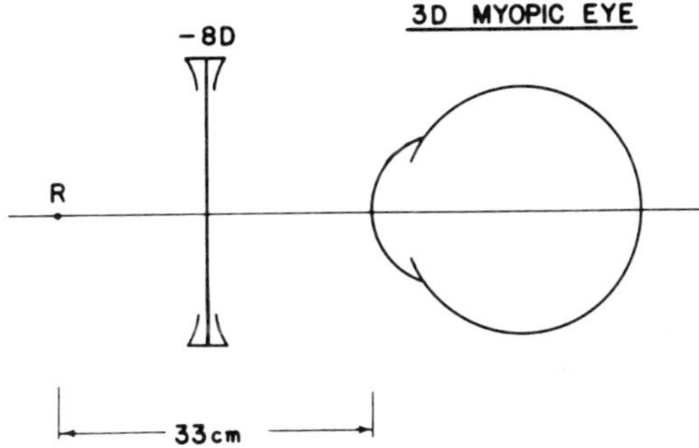


A − 8 D lens has its F′ at $\frac{1}{8\text{ D}}$ or 12.5 cm from it. Therefore, to superimpose F′ on R, the − 8 D lens must be 12.5 cm from R. This places the lens (33 − 12.5) cm or 20.5 cm in front of the eye.

You should now be able to grasp all future permutations and combinations in corrective lens power necessitated by shifts in the vertex distance. Don't bother with the formulas you'll find in other texts. Suffice it to say is that there *are* mathematical descriptions of how much effective power change is induced by each mm of lens shift; however, all you *have* to know is the actual power of any single "corrective" lens. Since this one lens pinpoints the location of R, the far point of the eye, you should be able to determine *any* other one for any vertex distance you want. The beauty of the "far point" and its clinical application to the correction of refractive error should be quite apparent to all.

## Astigmatism Correction

I have purposely steered away from discussing astigmatic refractive error until we had thoroughly digested myopia, hyperopia *and* their corrections. With these in mind, astigmatism "correction" is not tough.

Recall from our session with basic cylindrical lenses and astigmatic imagery that the refractive surfaces in these instances are toroidal; they can be described by noting the meridians of maximum and minimum powers which, in *regular* astigmatism (the type we are concerned with), are always perpendicular to one another. (Re-read that section if you have forgotten since you must understand it to get anything out of what follows.)

Just as a lens may possess astigmatic power, so also may an eye; and just about every eye is afflicted with *some* astigmatism, which usually resides more in the cornea than in the lens. As with the spherical refractive errors, we do not care about the *total* power of the refracting surfaces; it is only the *error* which is important. For astigmatism, the error can be expressed in either spherocylindrical form (plus or minus cylinder) or the "combined" cylinder form, with each power and its associated axis stated. These three expressions, when properly stated, are all equivalent and describe the same basic eye error.

Take an eye with an *error* of $\left[\begin{matrix} +5 \times 90 \\ +3 \times 180 \end{matrix}\right]$ measured from the corneal surface. (Remember, this is *not* the corrective lens but the basic refractive error. It is comparable to saying a myopic eye of 5 D has an error of + 5 D — "too strong" by 5 D.) This error means the power associated with × *90* (focusing the *vertical* astigmatic focal line) is too strong by 5 D; and that power focusing the horizontal line is too strong by 3 D. Thus, the eye has less *error* in focusing the horizontal line. This means that when the eye is looking at an object point at infinity, the horizontal focal line must be closer to the retina (only 3 D in front) compared to the vertical (which is 5 D in front).

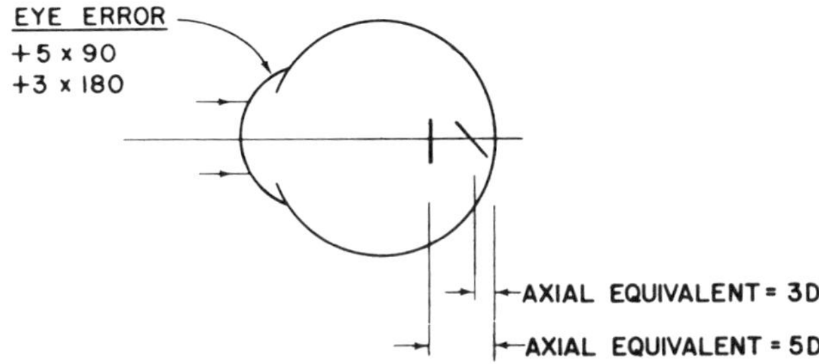

Since the horizontal line-images of each object point are positioned closer to the retina, the *horizontal* lines making up any target or test letter located at infinity will appear clearer than will the vertical strokes. For this eye, a target letter H would look something like that shown below — each object point being represented on the retina as a horizontal ellipse.

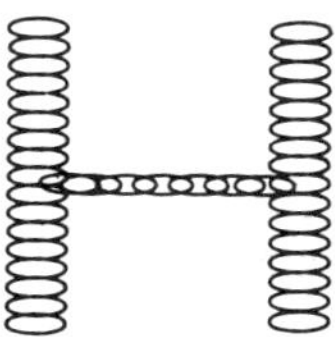

This eye would have two far points — one conjugate to the retina for the horizontal line image and one for the vertical line image. Just as with any *spherical* refractive error, the corresponding far point positions are found by knowing the refractive error of this eye in each of the primary meridians. In this case, the far point for the horizontal line image is 33 cm in front of the cornea and that for the vertical line is 20 cm in front of the cornea.

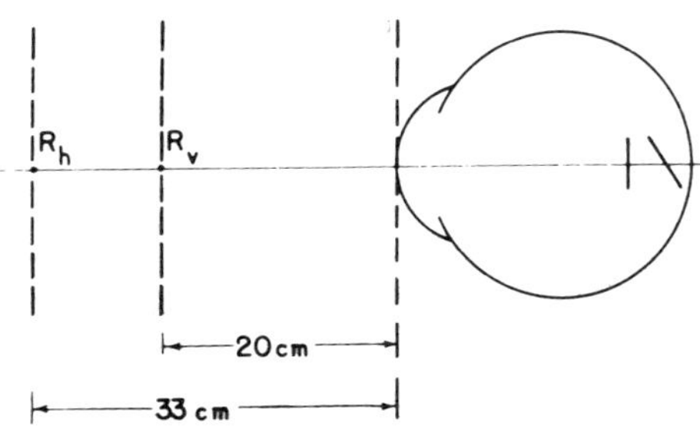

To "correct" this eye, we have to go through the same procedures as for the spherical refractive error — we have to place the corrective lens so as to image infinity at the far point; (the lens can be located at whatever position you choose if its power is appropriately adjusted). However, for the astigmatic eye there are two far points, so this procedure must be done for both primary meridians using a cylindrical lens for each.

For simplicity, let us correct this eye with lenses situated very close to the corneal surface; that is, consider the vertex distance to be zero.

First the horizontal line:

The far point associated with the horizontal line is 33 cm from the cornea, so we will need a — 3 D lens at the cornea to correct it. However, we only want to move the *horizontal* line 3 D backwards to place it on the retina, so we must use a *cylindrical* lens axis *180,* which moves only the horizontal line. Thus, to "correct" the horizontal line, we require a — 3 × 180 lens at the cornea. This lens will take an object point at infinity and create a horizontal line image of it at far point $R_h$. The eye then takes over and will image these rays sharply on the retina as a horizontal line image. Remember, that line image on the retina represents the *original* object *point* at infinity as focused by the *vertical* meridian (which has its axis at 180°).

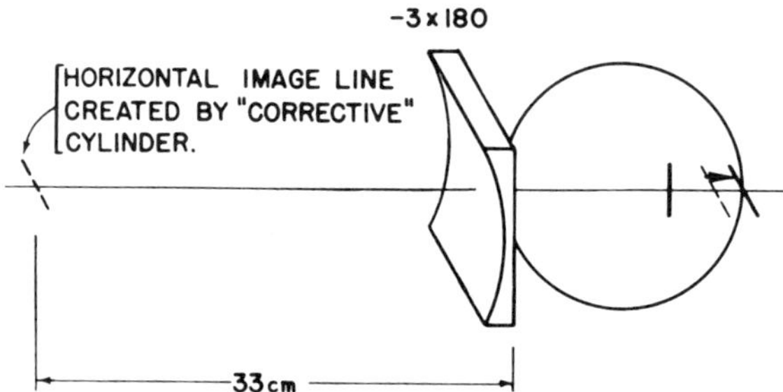


In the figure above, the — 3 × 180 corrective lens moves the horizontal line which was 3 D in the vitreous back onto the retina.

For the vertical line:

Similarly, we need a $-5 \times 90$ lens at the cornea to image the same object point from infinity at the far point ($R_v$) to create a vertical line there. This line will then be imaged by the astigmatic eye back onto the retina, thus "correcting" the power error *associated with* axis 90°.

When both lines are imaged on the retina, there will be no more astigmatism present since Sturm's "interval" will have been eliminated. The eye becomes fully "corrected" (with no line foci present) by the following *corrective* lens: $-3 \times 180°$ combined with $-5 \times 90°$.

So, if you know what the eye *error* is, you not only know where the far points are, but more importantly in astigmatism, you can always *visualize* the line images *within* the eye as it images an object point at infinity.

Get in the practice of visualizing. If I tell you that the eye *error* is

$$\begin{bmatrix} +5 \times 90 \\ -3 \times 180 \end{bmatrix},$$

a vivid picture should pop into your head and should look something like this:

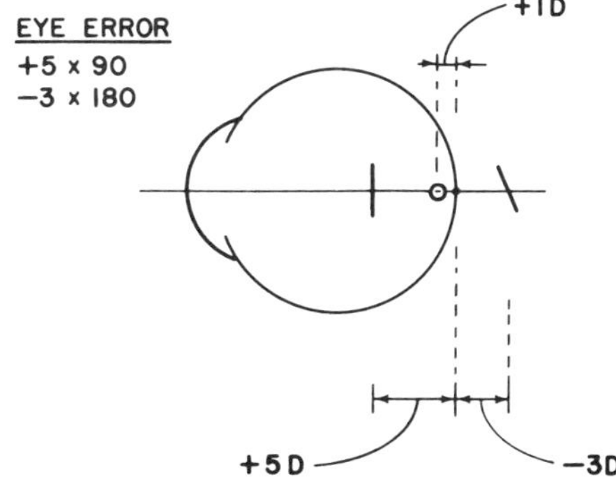


The $+5 \times 90$ means the vertical line is 5 D in front of the retina; the $-3 \times 180$ signifies the horizontal line image is 3 D "behind" the retina. These lines are separated by *8 D,* that is, the astigmatic error is 8 D. The circle of least confusion, which is dioptrically halfway between these lines, must be located 4 D from each; but, that position is 1 D in *front* of the retina.

So, with the eye *error* expressed in the combined cylinder form, you should never have difficulty visualizing the lines. But, the eye *error* in the form given above is not just lying around like a call-girl waiting for a jingle. It is partly hidden, and requires a little effort — some seduction — on your part. The type of information that *is* usually available is the prescription of the *corrective* lens; (again, consider it to be located at the cornea). — 5 + 8 × 180 would be the typical spherocylindrical form of such a lens. Now, you should *still* be able to sketch and locate the positions of the line foci and the circle of least confusion in this eye with that corrective lens *removed.* Your brain should conjure up the *exact same diagram* as above. How do you get there from here?

As you know, we can express this corrective lens in two other ways — you should be able to derive them now, but I'll "spoon-feed" once more.

Since it is much easier to visualize the positions of the line foci when you deal with the combined cylinder form, transpose the spherocylindrical corrective lens using the "cross" diagram.

A — 5 + 8 × 180 lens placed on the cross is as follows:

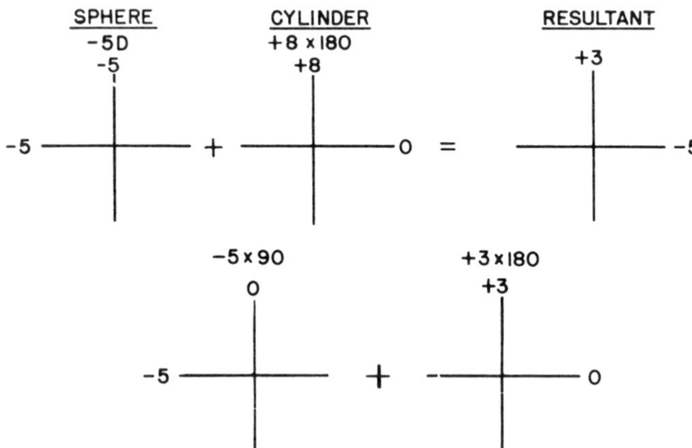

The resultant is shown above, as well as the derived two simple cylinders. Thus, the transposed form for this *corrective* lens is

$$\begin{bmatrix} -5 \times 90 \\ +3 \times 180 \end{bmatrix}.$$

But, a $-5 \times 90$ *corrects* ("neutralizes") an eye *error* of $+5 \times 90$, and a $+3 \times 180$ cylinder corrects an error of $-3 \times 180$. So, the composite eye *error* requiring this corrective lens is

$$\begin{bmatrix} +5 \times 90 \\ -3 \times 180 \end{bmatrix}.$$

We have already seen in our last diagram that this *error* produces the same line foci in the exact positions shown there.

With a little practice — make up some examples yourself — it will not take you long to be able to quickly transpose corrective lenses in either plus or minus spherocylindrical form to eye *error* in the "combined cylinder form". This will allow rapid and easy access to the focal lines created by an astigmatic eye. You will soon appreciate the practical impact of your newfound skill.

We have seen once again that the spherocylindrical lens form is an equivalent, transposed expression to that given by the two separate cylinders. The latter form does allow us to picture more easily the motion of *both* focal lines as each is placed on the retina by its respective cylinder. The spherocylinder lens form can also be pictured to correct the same astigmatic error but will be shown to work somewhat differently though, of course, to the same end result.

The *spherical* part of the *correct* spherocylindrical combination will move *both* astigmatic *lines* an equal amount dioptrically — forward by plus sphere, backward by minus sphere — so that one of the two lines lands on the retina. The *cylindrical* portion moves the remaining focal line (the one parallel to its own axis) back onto the retina and so, "collapses the conoid of Sturm" and eliminates the astigmatism.

Since a plus cylinder can only pull a focal line toward itself, it should be clear that when the spherocylindrical lens is expressed in the *plus* cylinder form, the correct *sphere* moves the entire astigmatic conoid behind the retina and leaves the *anterior* line in contact with the retina; the plus cylinder then pulls the posterior line forward. (See diagram A and B below.)

A minus cylinder can only *push* a focal line back, away from itself; so, in the minus spherocylindrical lens form, the corrective sphere places the conoid entirely within the vitreous with only the *posterior* line in contact with the retina. The minus cylinder then pushes the anterior line backwards to remove the astigmatism (diagrams C and D below).

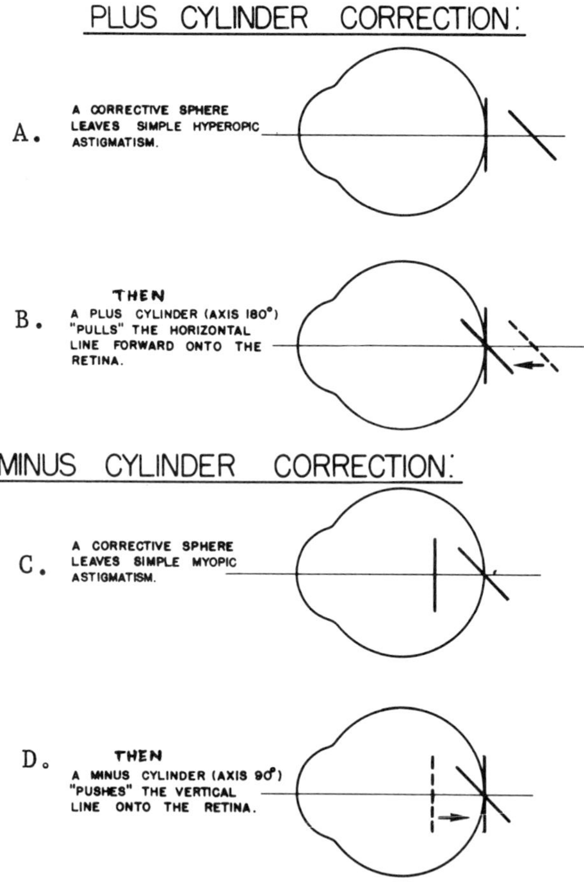

All three corrective lens forms produce the same result on an astigmatic eye; they eliminate the astigmatic error by moving the focal lines into superposition on the retina. Though each of the methods they utilize is different (as you have now seen), they all accomplish the same goal. This is what makes them equivalent.

So far, we have kept our astigmatic correction placed at the cornea. To tidy up a loose end on this subject, let's consider that the astigmatic corrective lens will be worn in the "spectacle lens plane" — a likely assumption. Its power would have to be modified to take the vertex distance into account, just as was necessary for myopia and hyperopia correction, but now separately for each major meridian. I will give one exaggerated example:

PROBLEM:

A corrective lens at the cornea is $\left[\begin{matrix} +6 \times .75 \\ -9 \times 165 \end{matrix}\right]$

The vertex distance is to be 15 mm.
What lens should be placed in the spectacle lens frame to have the equivalent corrective power?

ANSWER:

For the $+6 \times 75$, the far point is $\frac{1}{+6\text{ D}}$ or 16.7 cm *behind* the cornea. A corrective lens at the spectacle plane must have a secondary focal length of $(16.7 + 1.5)$ cm or 18.2 cm. Its power must be $\frac{1}{.182} = +5.50$ D, so, $+5.50 \times 75°$ corrects one meridian.

For the $-9 \times 165$, the far point is $\frac{1}{9\text{ D}}$ or 11.1 cm *in front of* the cornea. The secondary focal length of this corrective lens must be $(11.1 - 1.5$ cm) or 9.6 cm, and its power is $\frac{1}{.096\text{ m}}$ or $-10.4$ D. A $-10.4 \times 165°$ corrects this second meridian.

The complete corrective lens placed at 15 mm from the cornea is

$$\left[\begin{matrix} +5.50 \times 75 \\ -10.40 \times 165 \end{matrix}\right].$$

Notice that the "quantity" of astigmatism present (the difference between the meridional powers) will *seem* to vary with the position of the reference plane; measured at the cornea, it is $[+6 - (-9)] =$

15 D astigmatism, but measured at the spectacle lens plane it is [+ 5.50 — (— 10.40)] = 15.90 D of astigmatism. This astigmatism variation is only *apparent*, not real; what varies is *not* the actual astigmatic error of the eye but the amount of cylindrical *correction* required to eliminate the astigmatic error. We have already shown in our example above that this depends on the distance between the cylindrical corrective lenses and the eye's far points. This is exactly comparable to the spherical situation — the required corrective lens power varies, but this reflects its position in front of the eye and does *not* signify a change in the actual refractive error present in that eye.

## *Astigmatic Terms*

For completeness, I simply want to define a few astigmatic terms with diagrams; they should be intuitively obvious by now anyway, but these are used frequently in conjunction with uncorrected errors or partially corrected errors. (I used them myself in the last diagram.) Actually, the only basic difference between all of them is the superimposed amount of spherical error.

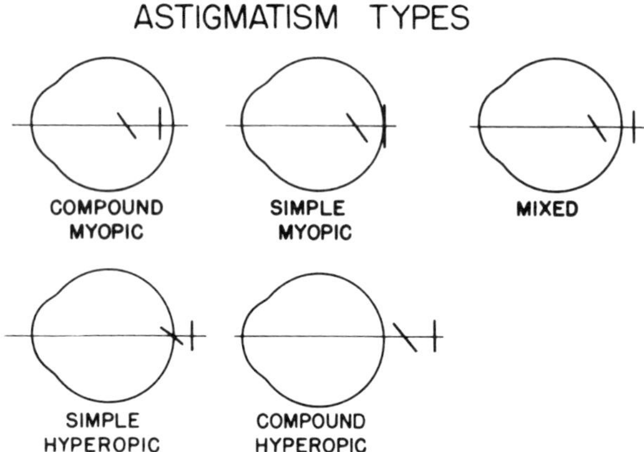


It is clear that a *corrective* lens of $\left[{+4 \times 90 \atop -3 \times 180}\right]$ corrects a *mixed* astigmatism since the vertical line is behind the retina while the horizontal is in front of it. But, a + 3 — 2 × 175 corrects a compound hyperopic astigmatism (both lines are behind) and a $-4 + 4 \times 33$ corrects a simple myopic astigmatism (since one line is on the retina, the other, in front). If you are unable to tell quickly just by looking at the prescription, the simplest way to discover which type you are dealing with is to express the corrective lens (or its converse — the eye error) as a "combination of cylinders" rather than spherocylindrically. In either case, if you can do this easily, it is only a manifestation of the skill you have now gained in dealing with cylinders.

## *"With" or "Against-the-Rule" Astigmatism*

Another set of terms which are found riddled through the ophthalmic literature are "with-the-rule" and "against-the-rule" types of astigmatism. These are not very important terms *per se,* but are used so often, you should be familiar with them.

The only reason astigmatism is "*with*-the-rule" is that it is commoner than "against-the-rule" (at least throughout most of one's life; there is some tendency for advancing age to cause patients to slip over into the "against-the-rule" moiety).

The only way I can keep the terms straight is to *remember* one fact: *"with-the-rule" astigmatism is corrected with a PLUS cylinder whose axis is vertical.* (The axis doesn't have to be exactly at 90°; anywhere from 65 to 115° or so will qualify.) A $+ 4.50 + 1.25 \times 80$ corrects "with-the-rule" astigmatism.

If someone ever asks you about any relationship requiring information about "with" or "against-the-rule", you should be able to figure the problem out from the one fact I have given above.

Now, test your reasoning and your ability to utilize this one fact in answering the following questions:

Which of the following statements are *true?*

STATEMENT 1)

A $+ 3 - 4 \times 180$ corrects "with the rule" astigmatism.

ANSWER:

Expressed in the plus cylinder form, the above lens is transposed to $- 1 + 4 \times 90$. Since this corrective lens has the plus cylinder correction with the axis at 90°, the statement must be CORRECT.

STATEMENT 2)

An *eye error* of $\left[\begin{smallmatrix}+3 \times 90 \\ +5 \times 180\end{smallmatrix}\right]$ is "with the rule".

ANSWER:

This *eye error* would be *corrected* by the following lens:

$$\left[\begin{matrix}-3 \times 90 \\ -5 \times 180\end{matrix}\right].$$

This lens transposed to the plus spherocylindrical form is $-5 + 2 \times 90$. Since the plus cylinder axis is at 90°, it corrects "with-the-rule" astigmatism, so, this statement is also CORRECT.

STATEMENT 3)

"With-the-rule" astigmatism is present in an aphake whose corneal curvature measurements indicate 43 D of power in the vertical meridian and 47 D in the horizontal meridian.

ANSWER:

These measurements show the power of the vertical meridian is too *weak* compared to that in the horizontal. This would be corrected by a lens which adds 4 D of power to the vertical meridian. A lens which does so while leaving the horizontal meridian alone is a $+ 4 \times$ *180* cylinder. (Remember, the power is *not* added in the meridian of the axis but 90° to it.) However, a $+ 4 \times 180$ corrective lens corrects "*against*-the-rule" astigmatism, so Statement 3 is INCORRECT.

STATEMENT 4)

The astigmatic line images in uncorrected "with-the-rule" astigmatism would look somewhat as follows:

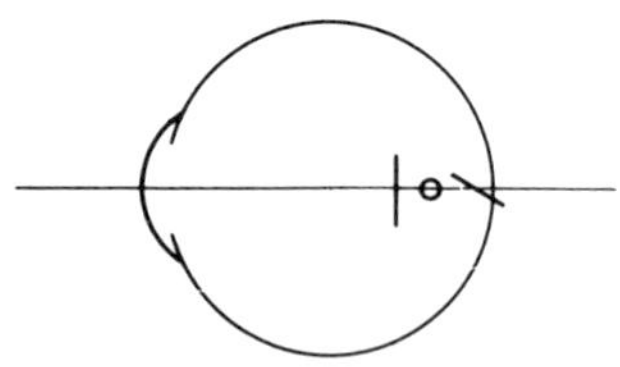

ANSWER:

Here, the vertical focal line is anterior to the horizontal, and so, the plus power associated with the axis at 90° (× 90) is greater; the corrective lens for this eye must be $\begin{bmatrix} -P_1 \times 90 \\ -P_2 \times 180 \end{bmatrix}$ with $-P_1$ in diopters greater than $-P_2$; say, $\begin{bmatrix} -5 \times 90 \\ -1 \times 180 \end{bmatrix}$.

This is "against-the-rule" astigmatism (see reasoning in STATEMENT 2 above) therefore, STATEMENT 4 is INCORRECT. In "with-the-rule" astigmatism, the horizontal focal line is *always* anterior to the vertical.

See how remembering just *one* fact about "with-the-rule" provides you with the capability of working out others?

## *Clinical Tests For Astigmatism*

Following this exploration into the correction of astigmatic ametropia, I feel it is appropriate to slide over into a discussion of a couple of clinical tests for astigmatism. Both of these can now be completely understood. Whenever you use either or both, you should be able to picture for yourself how the astigmatic lines are positioned within the eye and how you are moving those lines with the lenses you are adding during your clinical refraction of that eye. These two tests are as follows: 1) the radial astigmatic charts and 2) the Jackson "cross-cylinder".

### Radial "Sunburst" Dial and/or "Clock" Dial

These targets present a series of heavy lines arranged radially every 30°, either as a "Sunburst" or as a full "clock". (See figures A and B below.) The "Lancaster" dial is a 90° cross (figure C). This "cross" is rotatable to any angular position and so can be used in conjunction with either of the others.

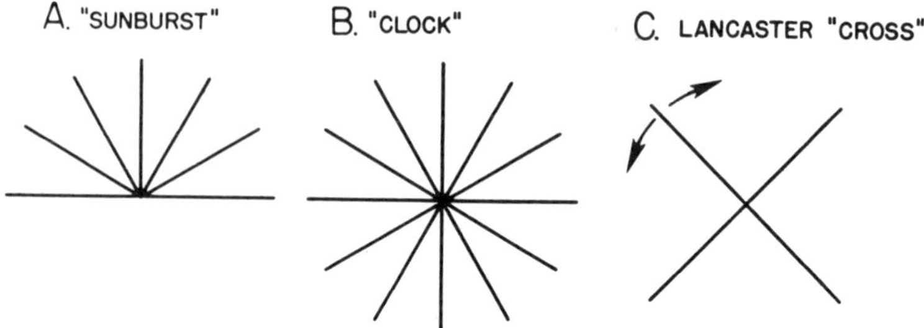


All the lines of the target are printed in equal widths with the same ink density. (The target itself is usually placed at optical infinity.*)

An astigmatic eye looking at this dial will see one of these lines as "blackest", sharpest, or clearest. That line must represent the direction of one astigmatic focal line in the eye — that particular line which is *closest* to the retina. The other line will, of course, be perpendicular to this one; but it can be either in front of or behind the retina — you cannot know which.

Let us assume it is the *vertical* line of the "Sunburst" dial which looks "darkest" to a patient. The two diagrams below show possible positions for the astigmatic line images relative to a number of different retinal positions. (Remember, each pair of astigmatic line images is created by only one of the object points making up this "dial" target.)

* For the human eye, a distance of 6 meters or 20 feet is considered "optical infinity", and for practical purposes does not stimulate any appreciable accommodative response. We should know that light from this distance *does* have a divergence of $-\frac{1}{6}$ Diopter at the eye; this amount, however, is considered negligible and unimportant clinically.

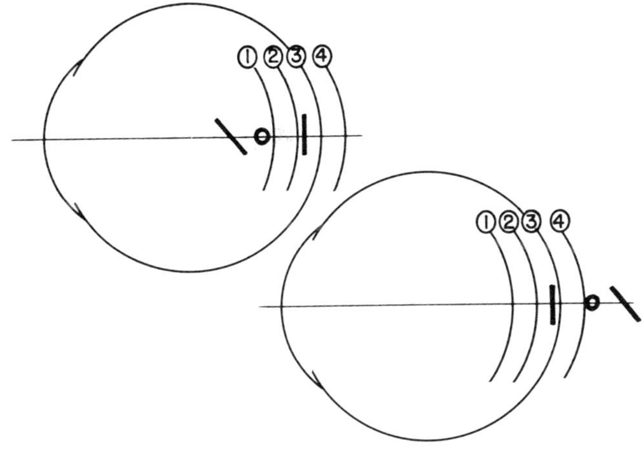


The vertical line will only appear *perfectly* sharp when it is *on* the retina. If it is not *on* but only clos*er* to the retina than the horizontal line, it will appear sharp*er* than the horizontal. With the astigmatic lines as shown above and the retina in *any* of the positions (1 - 4) shown in either of the two diagrams, it is the vertical line that will appear the sharper. Of course, if the circle of least confusion is on the retina, all the lines of the "Sunburst" dial will appear equally blurred.

If one or both of the focal lines falls "behind" the retina, the eye may tend to accommodate, and if it does, you may have trouble determining the true amount of astigmatic error present clinically. This difficulty — the possibility of stimulating accommodation — is gotten around very neatly in the refraction room by a technique known as "fogging".

First you will need to have some idea of the approximate correction. This you can determine as part of your routine refraction by using retinoscopy, or just by looking at the patient's old glasses, etc. (Do keep in mind that this is not a book on refraction technique, but one to allow you to understand principles.) To this approximate correction, add sufficient plus lens power in front of the cornea to make the astigmatism "compound *myopic*", that is, enough plus to bring *both* focal lines into the vitreous. Then, you know it is the posterior-most line that will be closer to the retina, and it will be in front of it. *This* is "fogging" the patient.

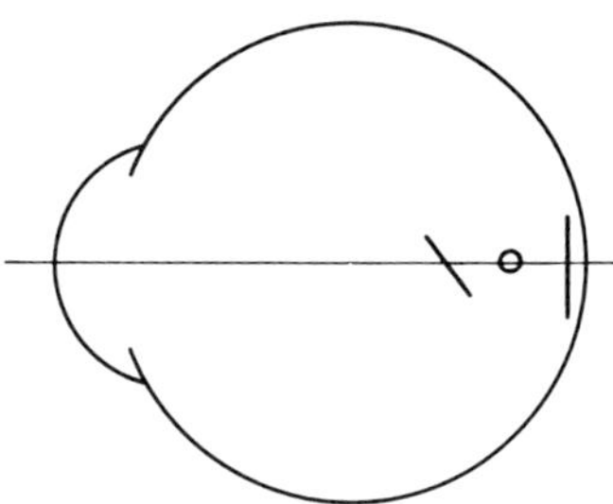

Obviously, the addition of plus sphere to make the eye artificially myopic can be overdone. Typically you should add only enough plus to blur the visual acuity down to 20/40 or 20/50.

Accommodation will be suppressed by the "fog". Now when the patient says that the vertical line on the "Sunburst" target seems to be the "blackest", *you* know that it is the vertical focal line that must be the one closer to the retinal surface but, because of the "fog", you also know it is in *front* of it.

To correct the astigmatism, you must move the horizontal focal line posteriorly. What lens will do this? A *minus cylinder* × *180* (as you should be shouting loud and clear). Enough minus cylinder is added so that the patient claims all the lines look equally dark; then you know that the astigmatic interval has been collapsed, and that any blur remaining is due only to uncorrected myopia. So, at this point you put away the astigmatic dial for this eye and bring out your Snellen acuity chart. Now you can "unfog" by adding minus sphere until you obtain the best subjective visual acuity. Neat, eh wot?

Two clinical points must be made about the "dial" tests:

POINT 1)

What is the relation between the *axis* of the corrective lens you should use and the direction of "sharp lines" seen by the patient under "fog"?

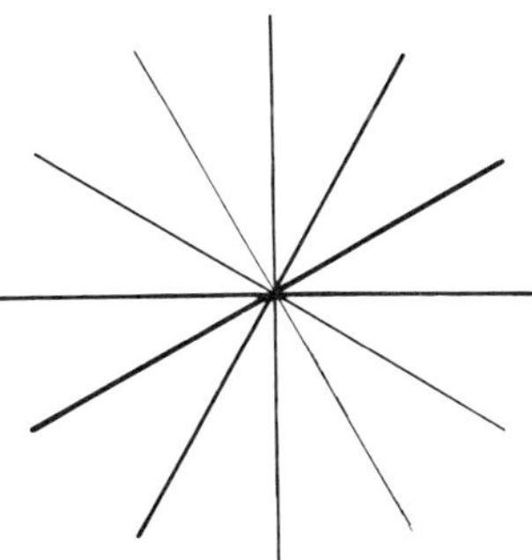

Say, a patient "sees" the dial lines above where those running from 2 to 8 o'clock seem blacker than the others. Thus, *you* know the focal image line in *that* direction is nearer the retina. But, the minus cylinder you plan to use will correct the astigmatism by moving the *more blurred* focal line backwards, towards the clearer one; so, the corrective axis must be positioned parallel to the 5-11 o'clock direction since that's the direction of the line you want to move. The question is, "what *axis* is that?" To help confuse your reasoning, the patient is reading the dial like a clock from *his* position, and you, as a refractionist, are *facing* him! You can, in your mind, visualize the direction of the 5-11 o'clock line as superimposed onto the patient's eye. This direction will correspond to a meridian of 60°. (Remember the *left* ear is the 0 — 180° meridian for both eyes.) So, you should have arrived at the conclusion that the corrective minus cylinder for this patient must be placed with its axis at 60°.

Did you get that answer? Great! Although you may have fumbled trying to convert the patient's "sharp line" response into a proper position for the corrective minus cylinder axis, you managed to come up with the correct answer by reasoning through the required steps. Hooray for you!

Though you should completely understand the above reasoning, you may find it simpler to use a little mnemonic: take the clock hour of the sharpest line (the patient's *subjective* response) and multiply it by *30°* (always 30°); *that* is the axis for the corrective minus cylinder. If the patient says the line from 4 - 10 o'clock is clearest, multiply the 4 (always use the *smaller* hour) by 30°; the corrective *minus* cylinder will be at × 120°, which you *quickly* and knowingly slip into your trial frame or phoropter.

POINT 2)

The astigmatic dial charts are much easier to use if your refracting trial lens set (or phoropter) contain *minus* cylinders. In this way you are always moving the anterior (most blurry) line backward towards the retina which provides an increasing subjective clarity.

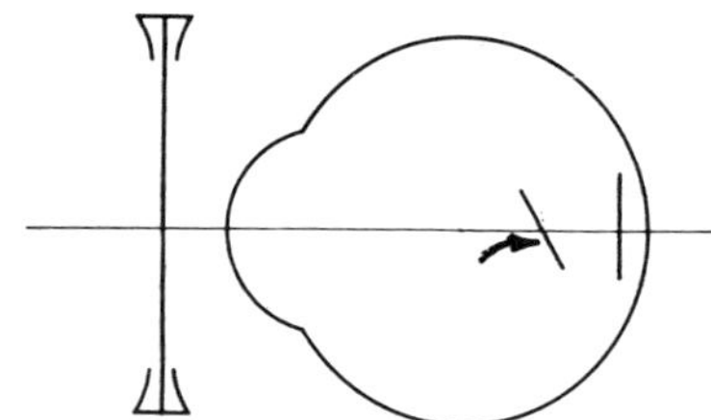


If instead you insist on using *plus* cylinders for this test, you are not complying fully with its intended design. A plus cylinder, which requires that its axis must be placed parallel to the line you wish to move, will "pull" its focal line toward itself, and therefore, more into the vitreous. (The plus cylinder axis will have to be held parallel to the more *posterior* line since "pulling" the anterior line forward would only increase the astigmatism.) By "pulling" the posterior focal line forward towards the anterior one, you do make the chart lines appear of equal intensity (and the end point *is* equality of *all* the lines on the astigmatic dial chart); however, here, they will all simultaneously appear to become *more* blurred. With full plus cylindrical correction, the collapse of the conoid must take place further into the vitreous than with minus cylinder correction.

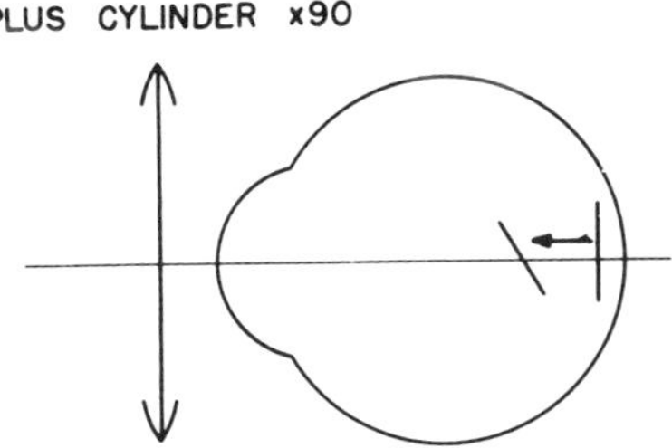


To make the chart lines appear sharper (actually, to make the test work properly), whenever you add plus cylinder, you must compensate by adding *minus sphere* to shove *both* lines back toward the retina. For each 1 D of plus cylinder, you should add about 0.50 D of minus sphere. (This is called "double-clicking".)

In any case, you *can* use the Lancaster dial with plus cylinders, but it is more cumbersome than with minus cylinders. To make things even rougher for the plus cylinder user, he has no mnemonic rule for positioning the corrective plus cylinder axis, and he must either remember which subjective "clock hour" goes with which axis or he must stop to figure it out every time.

* * *

This completes the "dial" method of correcting astigmatism. Before moving to the "cross cylinder", we should expand on a subject already introduced — the *spherical equivalent.*

If you will recall, the spherical equivalent power (that focusing the "circle of least confusion" in the astigmatic image) is always halfway dioptrically between the powers of the two major lens meridians. If a lens is expressed in the "combined cylinder" form, we can find the spherical equivalent — that is, a *spherical* lens that could substitute for a toric one and maintain an image in the plane of the latter's circle of least confusion.

PROBLEM:

What is the spherical equivalent of the following lens?

$$\begin{bmatrix} +0.75 \times 90 \\ -3.25 \times 180 \end{bmatrix}$$

ANSWER:

Find the position which is dioptrically halfway between the two focal lines. This would be $\frac{+0.75-(-3.25)}{2}$ or $\frac{+4.00}{2} = 2.0$ diopters away from each line. So, the *axial* position which is 2 D away from each focal line is the spherical equivalent. In this case, 2 D posterior to the vertical focal line (positioned by $+0.75 \times 90$) is $(+0.75 - 2.00)$ or $-1.25$ D. *This* is the spherical equivalent for the above lens.

If a lens prescription is given in the spherocylindrical form, the spherical equivalent can be found by taking half of the (plus or minus) cylinder power and adding it algebraically to the spherical power.

*EXAMPLE:*

The spherical equivalent of $+1.00 - 2.50 \times 50$ is $\frac{1}{2}(-2.50) + 1 = -0.25$.

Try the following to test your own skill.

PROBLEM:

Find the spherical equivalents of each of the following lenses:

A. $+4.00 + 2.00 \times 90$
B. $-1.00 + 3.00 \times 180$
C. $-6.00 + 1.00 \times 75$
D. $-2.00 - 5.00 \times 29$
E. plano $+ 3.00 \times 50$

ANSWERS:

*Spherical equivalent*

A. + 5.0 D
B. + 0.50 D
C. − 5.50 D
D. − 4.50 D
E. + 1.50 D

You must be able to determine quickly the spherical equivalent of any spherocylindrical lens.

At last, we come to the second of our clinical tools to deal with astigmatism. By now, you should have acquired sufficient ammunition to use this really big gun intelligently to refine the cylindrical *power* and *axis* during clinical refraction. This weapon is the Jackson "cross-cylinder" (after Edward Jackson, the brilliant and innovative, late Denver ophthalmologist).

## The "Cross-Cylinder"

What *is* a "cross cylinder?" It is a specific type of cylindrical (toric) lens that is composed of a plus cylinder and a minus cylinder of equal powers ground onto one lens, with their axes at right angles to each other. Each one of the following three *different* lenses is a "cross cylinder" — each can also be expressed in all three types of cylinder form transpositions we have studied.

$$\left[\begin{matrix} +0.50 \times 90 \\ -0.50 \times 180 \end{matrix}\right]; \qquad +1.00 - 2.00 \times 73; \qquad -0.75 + 1.50 \times 127$$

It is easy to recognize a "cross cylinder" lens when that lens is written in the "combined cylinder" form. (See above.) However, you should not be fooled when the cross cylinder is written in the plus or minus spherocylindrical form either; in both, the strength of the cylinder is always two times and of *opposite* sign to the power of the sphere.

*All "cross cylinder" lenses have a spherical equivalent power of zero* ("plano"). If any one of these lenses is placed in front of an eye with any type of refractive error and with any amount of cylindrical error, the position of the circle of least confusion will *not* be changed. That is one of the key points in the operation of this lens as a clinical test. Now let's see how such a lens is used.

### *Use of the Cross Cylinder for POWER Refinement*

If an eye has the astigmatic error shown below and we place a *plus cylinder* × *180* before it, only the horizontal line will be moved forward — more into the vitreous. A *minus cylinder* × *90°* in that same position would move only the vertical line posteriorly. A *"cross cylinder"* contains *both* a plus and a minus cylinder at right angles to one another, so if it is held in front of this eye with its plus cylinder axis at 180°, the minus axis will automatically be at 90° and both focal lines will be moved simultaneously away from each other, as shown below. The astigmatism will be *increased.*

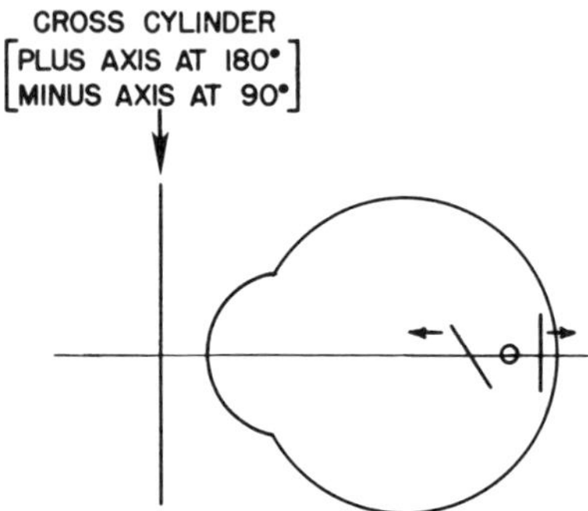


Let's turn the cross cylinder so that the plus cylinder axis is vertical instead of horizontal. With this axis now at 90°, the focal line which will be moved by the plus cylinder is the vertical one, and it will be pulled forward, while the horizontal one (affected simultaneously by the minus cylinder axis now at 180°) will be pushed back. This will *decrease* the amount of the astigmatism.

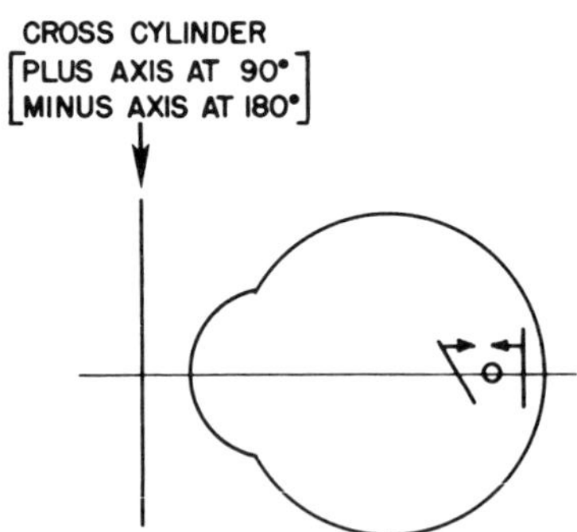

NOTE: The last two figures both show the presence of "compound myopic astigmatism". This is *NOT* where the focal lines *should* be for the proper performance of this clinical test, as will soon be explained. The lines and circle of confusion are shown this way for diagrammatic simplicity only. The circle *should* be on the retina, not in the vitreous.

In any case, the *size* of the circle, which represents a single object point, is dependent (among other factors) on the amount of astigmatism present. When the circle is on the retina where it belongs and the eye is studying a line of Snellen letters on a distant chart, the addition of a cross cylinder with its plus axis at 180° would cause that line to blur subjectively since the *total* astigmatic error is *increased* by it. (Even though the plane of the circle of least confusion stays put, each circle obligatorily increases in size as the astigmatism increases.) But with the plus cylinder axis turned to 90°, the size of the blur circle is decreased; this increases the subjective clarity of a line of Snellen letters located in the distance.

Clinically, you have to ask the patient which of the two cross cylinder positions provides the better, clearer image. When he tells you, follow the directions provided by the cross cylinder itself. Slip the proper *cylindrical* lens into your trial frame or phoropter. An example follows:

The patient tells you that the Snellen letters seem to be clearer when the plus cylinder axis of your cross cylinder is at 90°. Thus you know that the cross cylinder in that position must have reduced the amount of original astigmatism. This directs you to add more plus cylinder with its axis at 90° to your trial lens. (If you happen to be using *minus* cylinder trial lenses, follow the direction given by the *minus* cylinder axis of your cross cylinder, and add or subtract *minus* cylinder from your trial lens appropriately.)

You continue to follow the "instructions" given you by your cross cylinder, adding or subtracting cylindrical lenses as directed, until the patient notices *no difference* between flips of the cross cylinder. As soon as the patient is unable to detect a difference between the two positions of the cross cylinder, you have totally "collapsed" the previously uncorrected astigmatism of the eye. The only astigmatic error now present in the system is that induced by the cross-cylinder lens itself.

The combined figure below shows the optical condition of "equality" between "flips". The two focal lines created by the cross cylinder itself are shown; the circle is *on* the retina. When the patient proclaims that subjective "equality" is present, the focal line positions have simply been interchanged.

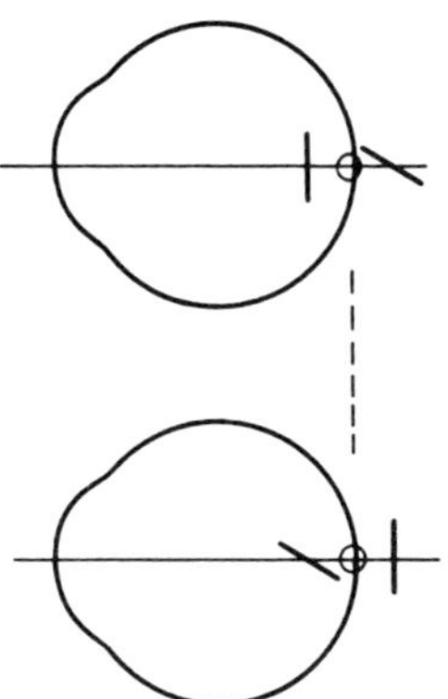

CLINICAL POINTS:

If you continue to add only plus or minus *cylinder* as directed by the cross cylinder, you *will* be collapsing the conoid, *but* you will also be moving the circle of confusion by moving only *one* of the two focal lines. For every 0.50 D of cylinder you do add, you will simultaneously move the circle of least confusion 0.25 forward or back, depending on whether you are adding a plus or minus cylinder. So for practical purposes, you must compensate for the shift in the circle's position by adding a 0.25 Diopter of *sphere* of *opposite* power every time you introduce 0.50 D of new cylinder. Thus, if you add + 0.50 cylinder × 90, you should also add − 0.25 D sphere to maintain the position of the circle of least confusion in its original position — hopefully on the retina.

At the "equality" endpoint, we have shown that there *is* astigmatic error present with the cross-cylindrical lens in place; an alert patient will frequently remind you that he sees better *without* that lens. Because you always would like patients to believe that *you* have the upper hand and know what is going on, it is better to warn them ahead of time; "I realize that the lens I'm about to introduce may blur the target letters somewhat, but all you have to tell me is *which* lens position is the *better* — one or two!" (meaning one of the two flipped positions).

Another refraction hint: Whenever giving patients a choice between any two different lenses but especially when using the cross cylinder, always make the lens switch abruptly and cleanly. This will make the comparative evaluation much easier for the patient. You should shift between the two positions quite rapidly and allow the patient no longer than *one second* at each position to decide.

Typically, clinicians become squirmy and impatient while the examinee sits trying to make up his mind (the patient isn't very comfortable either)! You really simplify his task by forcing him to choose rapidly. Tell him you know you're changing the lenses quickly, but what you want from him is a quick, "first impression" and *not* a carefully thought out, deliberate, judgment decision. So, "push" him and "lean on" your patient for a quick response while reminding him that there is no "wrong" answer.

You may wonder why I dwell so much on this simple point. The reason is that only rarely have I found it performed correctly, and any incurred inaccuracy is usually blamed on the *test!* Performed properly, it will not only speed up your refraction, it will also make it more accurate. But please, temper this instruction with common sense. Older patients tend to be "slower" in general so you will have to adjust your pace somewhat, but even then, keep "pushing". By giving them time "to think", you are only allowing them the opportunity to adapt to the stimulus; this will only make their selection *more* difficult.

The rapid transfer in position of the plus axis and the minus axis is facilitated by the construction of the cross cylindrical test lens. An arm is mounted at 45° to the two principal axes. On the standard cross cylinder, the meridians marked are the *axes;* the plus is indicated with a small, white plus or a white dot; the minus is marked in red.

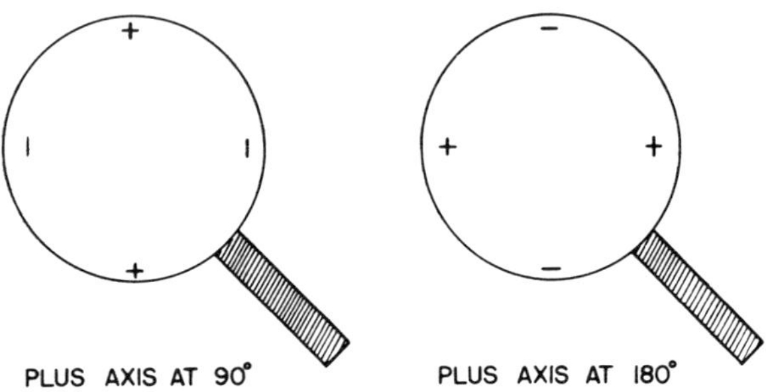


A rapid rotation of the knurled arm between the examiner's fingers quickly interchanges the plus cylinder axis and the minus cylinder axis.

Another tip which you may also consider superfluous but, performed well, will speed up your refraction and reduce frustration all around. You will greatly facilitate your patient's choice of one lens position over another by your giving a "name" to each one — calling it "one or two", "A or B", etc. This immediately identifies for the patient which lens position you are asking him about. I've heard even experienced refractionists say "which is better — *this* lens or *that* one", leaving about 3-4 seconds for a choice. During this prolonged period of time, the patient and the doctor lose communication and neither knows which position *"that* one" is! It will make no difference whatsoever if you rechristen the same lens with a new name during the next sequential presentation. Lens 2 can become lens 3 in the next paired choice and thus will not be confusing for the patient. So, make it easy for both of you; *name* each lens position and make sure you do so *simultaneously* with each lens presentation, *every* time you flip it back and forth.

I would like to demonstrate that when you flip a cross cylinder between its two positions you create an astigmatic *change* of *four* times the power of the cross cylinder; that is, a .25 cross cylinder (— 0.25 + 0.50 × 90) will cause *1.00 D* of astigmatic *difference* between positions "one" and "two" of the cross cylinder, no matter what the eye's refractive error.

First, look at an eye with an existing error of 1.00 D astigmatism:

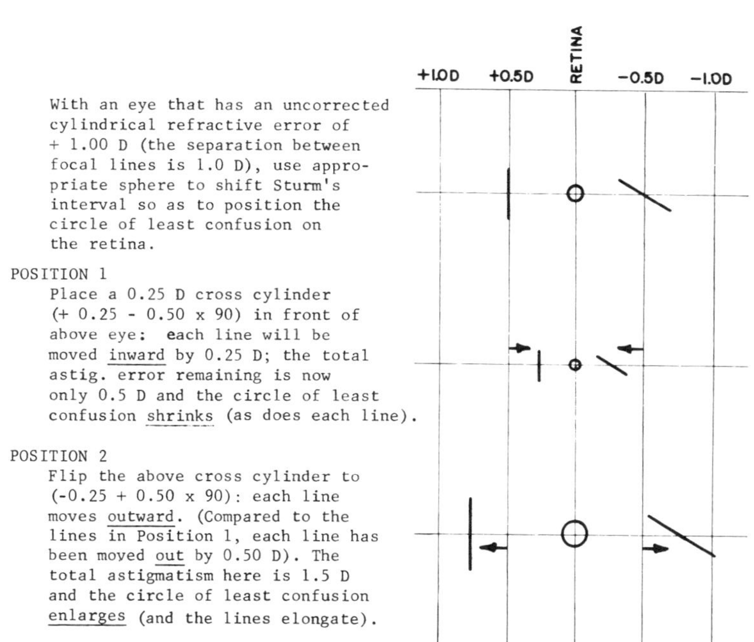


Thus, the *difference* in astigmatic error between POSITION 1 and POSITION 2 is (1.5 D — 0.5 D) = 1.00 D. Q.E.D.

If instead the eye is *emmetropic,* it is a bit harder to "see" that this much shift still occurs, though it really does. The 0.25 cross cylinder introduces only + 0.50 diopter of cylinder in each position. Flipping the cylinder simply switches the positions of the focal lines and so does not "seem to change the *quantity* of astigmatism; there *is* only .50 D present. Realize, however, that since the lines are interchanged, each must have moved by 0.50 diopter, so, there must have been a "change" of *one* full diopter (adding together the two line movements). Look at the diagram sketch below:

Place the 0.25 cross cylinder before an eye which has *no* cylindric eye error:

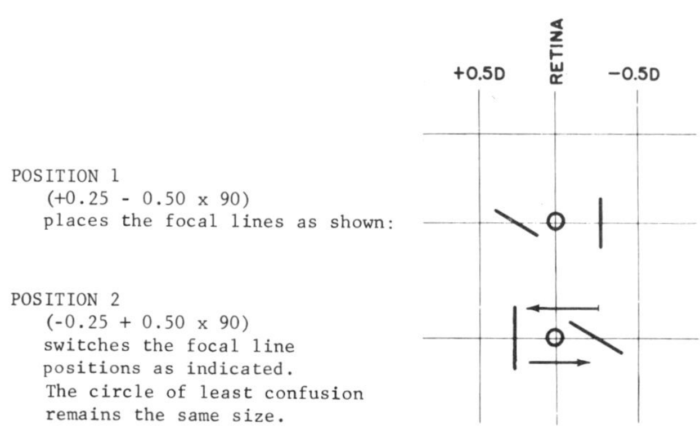


Although the astigmatism present in each of these two positions is only 0.50 D, the algebraic *difference* between them is + 0.50 − (− 0.50) = + 1.00 D. This difference is not as "visible" as in the previous example with a pre-existing astigmatic error; yet, it is there nonetheless. So, *any* cross cylinder will induce a shift of four times "its name" between its two flipped positions.

CLINICAL POINTS — Practical hints for the use of the cross cylinder:

1) The cross cylinder is not used to make a rough, initial approximation of the amount of astigmatic refractive error present; it *is* used to *refine* a close approximation to the existing error.

2) Before starting to use the cross cylinder to refine the cylindric corrective lens *power,* have the patient's *best* visual correction (sphere *and* cylinder) in place. Have him able to read the *smallest* line of the acuity chart that he can — 20/20 if possible, while using the closest you can get to his "full" correction. If you are in doubt as to whether or not you should add another −0.25 D of sphere to sharpen the patient's vision before using the cross cylinder, *do so.* It is better to have him slightly overcorrected in minus than to have him myopic. It is quite *incorrect* to "fog" or blur the vision before proceeding with the cross cylinder refinement. I say this in spite of the fact that "fogging" was proposed for *this* test by none other than Jackson, the inventor, himself. (Even great men do make errors — he *was* wrong here). If you *do* have the patient "fogged", he will have more tendency to vacillate in his responses and switch "back and forth".

3) After making sure the patient's *best* refraction is in place, use the cross-cylinder with the patient looking at a line of Snellen target letters about 1-2 lines *larger* than his threshold acuity line. If the maximum acuity is 20/25, use a 20/30 or 20/40 line for the cross cylinder test. You want to present *almost* threshold letter size, since with this type of challenge the patient's judgment is more critical. But you can't use the *smallest* letter size since the introduction of the cross cylinder itself, even in an emmetropic eye, introduces its own cylinder error (a + .50 − 1.00 x 180 cross cylinder introduces 1.00 D of cylinder) and this astigmatism will blur his retinal image too much. I would suggest that the refractionist use a cross cylinder of the *least* power possible, gauged for the maximal acuity level. (I have already pointed out that the cross cylinder is *named* for its spherical component. The above lens is called a *.50 D* cross cylinder; a −0.12 + 0.25 × 72 is a *.12 D* cross cylinder, etc.) See the following chart for the appropriate cross cylinder lens to use:

| *Maximal corrected visual acuity* | *Size of cross cylinder to use* |
|---|---|
| 20/15 — 20/20 | .12 D |
| 20/25 — 20/30 | .25 D |
| 20/40 — 20/50 | .50 D |
| 20/60 — 20/100 | 1.00 D |

For 20/200 and poorer acuities, cross cylinder refinement is not important and probably a waste of effort.

4) For the cross cylinder to help you refine the astigmatic *power* of a corrective lens, the cross cylinder axes (and the corrective cylinder) *must* be aligned parallel to the primary astigmatic meridians of the eye; you must be "on axis". Only then can the "cross" cylinder move the focal lines predictably as described. We will soon see how you make sure the axes *are* correct.

* * *

You now know how to refine the corrective cylinder *power* with a cross-cylinder. We still have in our bag the other beautiful application for this instrument — the refining of the corrective cylinder *axis.* Though the practical techniques for the two determinations will look different, they are in fact based on optical principles which are really very closely related. These principles pertain to how cylinders will add together when superimposed.

I explained the optics of the *power* check first because this is simpler to understand. But, clinically, as I have already cautioned, the proper use of the *power* check depends on good alignment between the cross cylinder *axes* and the major astigmatic meridians of both the corrective trial lens and the patient's eye. Thus, the check for correct *axis* using the cross cylinder must be done first, prior to checking the *power.*

Now, to complete your understanding of how the cross cylinder can be used to refine the corrective *axis,* you will need to examine more closely what happens optically when two cylindrical lenses are added. Return with me now to more of the "basics".

## *Addition of Two Cylinders*

When two cylinders are superimposed, the resultant optical effect is not *always* obvious. It *is* easy, however, to visualize what happens when their axes are *aligned,* parallel to each other. In this case we have already seen that the principal meridian powers can simply be added algebraically; a $+ 2 \times 90$ added to a $+ 2 \times 90$ combines to give a $+ 4 \times 90$; a $+ 2 \times 180$ adds to a $- 4 \times 180$ to yield a $- 2 \times 180$. It is quite a bit more touchy when the principal meridians are not parallel but misaligned by some angle. First, let's consider the situation when the two cylinders are of *like sign* (when both are plus, or both are minus cylinders).

At first glance it might seem that the cylinders would add "normally", like physical vector forces which act at some angle to each other — that is, a $+ 2 \times 90$ and a $+ 2 \times 80$ combine to yield a $+ 3.8 \times 85$ — the axis being midway between the two when the powers are equal. This is true, but *in addition* something unexpected happens; as if out of nowhere, some plus sphere is also produced!

The greater the angle between the two like cylinders, the greater the amount of sphere induced and the *less* the resultant cylinder (the *axis* always being half-way between when the two cylindric lenses are equal in power). This increase in sphere and decrease in cylindrical power continues as the axes are rotated until they are fully 90° apart. At that time, . . . well, look at our crosses (in the figure below) — we have $+ 2 \times 90$ added to a $+ 2 \times 180$:

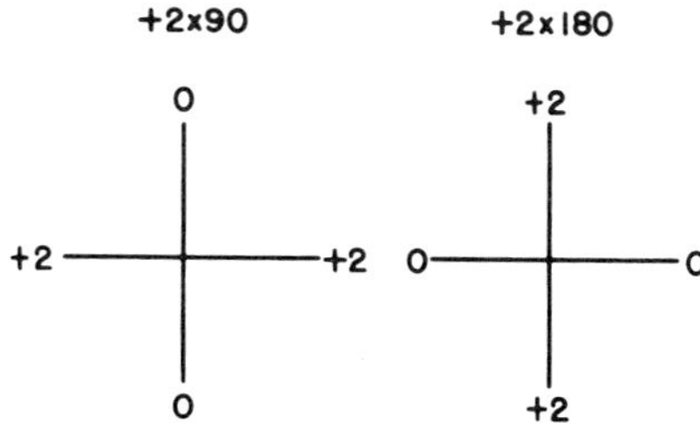

As we can see, when the axes of the like and equal cylinders are crossed at 90°, the resultant is a + *2 sphere*!

So, with the above example, as the angle between the two axes varies from perfect alignment (when the resultant sum is *plano* + *4* × *90*) to where they are 90° apart (yielding a resultant of + 2.0 D sphere), there will be a progressive increase in sphere and decrease in cylinder power. But please note, at "in-between" axes the exact amount of resultant sphere and cylinder is *not* easy to obtain. At 45° separation of axes for example (with one axis at 90°, the other at 45°), the resultant is 0.6 + 2.8 × 67.5°.

As you might guess, it gets more complicated rapidly with more complicated spherocylindrical additions. There *are* complex formulas written which can be used to calculate the resultant obtained when *any* two lenses are superimposed at *any* angle. (An easier, graphical method is also available). However, both the graphical and mathematical methods are so cumbersome that I gleefully state they are relatively unimportant for us — since we *do* have an "out"! Every clinic refraction room has a lovely "computer" in the *lensometer,* the everyday instrument used to determine the power and axes of spherocylindrical lenses. This gadget can instantly determine the resultant power and resultant axes of any two (or more!) lenses in combination. Just slip them together into the lensometer, clamp them in place and read the instrument as if there were only one lens in place — the proper resultant will be given to you in a flash without complicated formulas or graphs. Why learn something you'll never use when you have at hand the means to solve such problems readily? The lensometer provides the easiest, most pragmatic way to obtain a quick and accurate answer, and the simplest way has to be considered the most *scientific* way to do so.

So, we have just looked at combining cylinders of *like* sign when the axes are misaligned. It is even more surprising to find what happens when two cylinders of *unlike* sign are combined "off" axis. Take a practical example:

When you fully "correct" an eye with a built-in astigmatic error of − 2 × 80°, you should use a + 2 × 80° spectacle lens. What if you mistakenly put the corrective lens at axis 90° instead of at 80°

where it belongs? You would *not* obtain a resultant with its axis between the two; you unexpectedly induce a new spherocylindrical combination with its plus cylinder axis located 45° to the "bisector" (85°) axis! That is, the plus cylinder axis of the new resultant is at 130°, (the minus cylinder axis would be at 40°). So you see, the new resultant obtained by misjudging the eye's real axis by only 10° is — .35 + .70 × 130°. (For this eye, this resultant is a new "error" which also would require correction).

What you as a clinician should remember is that the *plus* cylinder axis of the *resultant* error is 45° to the "bisector angle"; but in which direction is that? It is always found on the *same side* of the "built-in" eye cylinder axis that you have placed the axis of the plus cylinder of the "corrective" lens. "Come again?" you say.

Example: An eye's "built-in" minus cylinder axis is at 130°; a "corrective" lens *plus* cylinder axis is *mis*placed at 125°; the resultant *plus* cylinder axis will be at 82.5° (in the same direction as the erroneous placement of the plus cylinder). The minus cylinder axis, of course, will be on the other side — at 172.5°.

With this background on how cylinders act when superimposed, we can finally understand how the cross cylinder can be used to refine the *axis* position of the corrective cylindrical trial lens. (Remember, to allow you to achieve optimal results from the cross-cylinder when refining either power *or* axis, do not have any "fog" present).

### *Use of the Cross Cylinder for AXIS Refinement*

Instead of aligning your cross cylinder axes *parallel* to those of the *corrective* lens cylinder (as you did to check on the "power"), you must align the cross cylinder axes parallel to the induced *resultant* axes. As pointed out above, these resultant axes are approximately at 45° to the axis of the corrective trial lens you are using.

To reiterate: 1) there *will* be a *resultant* cylindrical error present whenever your corrective trial cylinder is *not* perfectly aligned with the eye's refractive error, and 2) the *resultant* astigmatic error will produce focal line images. To move those line images toward or away from each other (as you did in "collapsing the conoid" when you refined the *power* of the corrective cylindrical lens), you must place the cross cylinder axes *parallel* to those resultant focal lines. This will be approximately accomplished if your cross cylinder axes straddle (at 45° to) your correcting cylinder axis.*

Let's take an example: Assume that the eye shown below has an *error of* $-2 \times 80°$. The "correcting" lens is misplaced 10° "off" the proper axis, that is, the $+2$ cylinder is placed with its axis at 90°. The "induced error" due to this 10° misplacement is $-0.35 + .70 \times 130$. (Notice that this error itself is a "cross-cylinder"!) This *resultant* error is equivalent to the following "combination of cylinders":

$$\begin{bmatrix} +.35 \times 130 \\ -.35 \times 40 \end{bmatrix}$$

Therefore, you can easily see that the focal line which is oriented at 130° is in vitreous (since it is pulled forward by the $+0.35$ D power), while the line at 40° is pushed "behind" the retina.

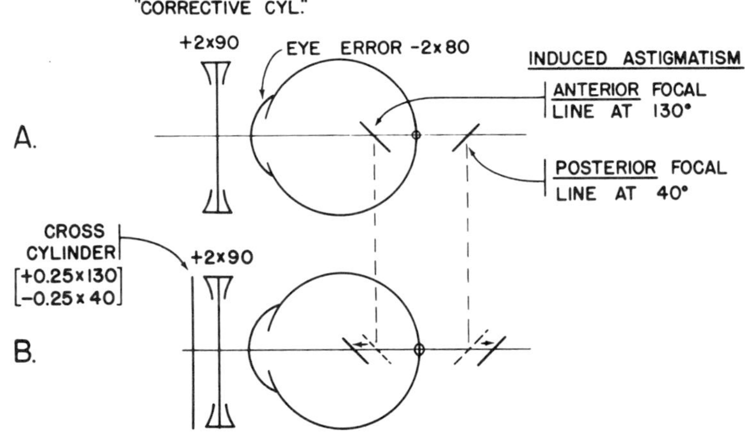


* Theoretically, you should not straddle the axis of the *correcting* cylinder but of another meridian; that latter meridian is located somewhere *between* the correcting cylinder's axis and the axis of the "eye error". However, in clinical practice, this distinction is *not* important. (Actually, it is only the position of the "correcting" cylinder axis that is firmly known anyway.) The diagram on p. 183 shows the typical *clinical* straddling position and not the theoretical one.

If you now place your cross cylinder's plus axis parallel to the *induced* plus cylinder axis, the meridional powers will, naturally, add directly. The cross cylinder $\left[\begin{array}{l}+0.25 \times 130 \\ -0.25 \times 40\end{array}\right]$ will "pull" the 130° line (already in the vitreous) further into the vitreous, and "push" the line at 40° further back; this *increases* the astigmatism present, enlarges the circle of least confusion, and further blurs any target letters the patient sees. (This situation is diagrammed in figure B above.)

When the cross cylinder is "flipped" 90° to that shown in B, the astigmatic focal lines will then approach each other, decrease the astigmatic error, and the patient will exclaim that this second position of the cross cylinder allows clearer vision. This subjective improvement occurs, then, whenever the "cross-cylinder's" *plus* axis falls on the induced *minus* cylinder axis; this tends to neutralize the induced error. Now the examiner should look at his cross cylinder and note the exact position of its *plus* axis, since the latter indicates that the induced *minus* cylinder axis is also at the same orientation (here, at 40°). What the examiner will see when *facing* his patient (who now "sees better") is the cross cylinder in the position shown below; the plus axis is superimposed on the *induced* minus cylinder axis.

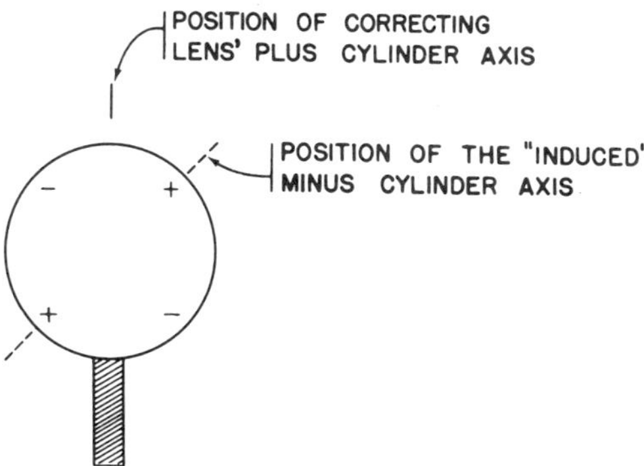


Since we now know that there indeed *is* an induced minus axis we can conclude that there must have been some misalignment of the original corrective lens' *plus* axis to produce it; in addition, we also know that the corrective lens' axis must have been positioned too far on the *other* (counterclockwise) side of the axis of the "built-in" eye error to put the induced minus axis where it is. So, in order to improve matters, we must rotate the *corrective* lens' axis in a *clockwise* direction, that is, *towards* the plus cylinder mark on the cross cylinder lens.

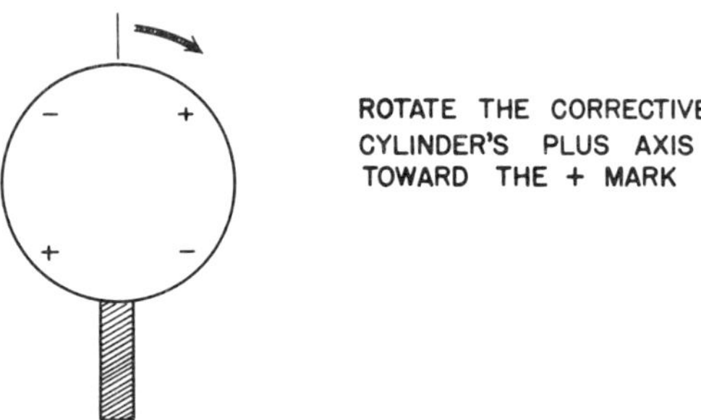


You obviously do not have to reason this out every time you do the test; just rotate the corrective lens' *plus* axis towards the *plus* axis mark on the cross cylinder whenever the patient tells you that particular cross cylinder position is best. If you are using a *minus* cylinder corrective lens, you will have to rotate that lens towards the indicated *minus* cross-cylinder axis.

Continue to refine the position of the corrective cylinder axis by flipping your cross-cylinder back and forth, while straddling each *new* axis, as the patient tells you how the clarity between the two flipped positions compares. You should continue "flipping" until the patient notices "no difference" between the two positions of the cross cylinder. When this occurs, the correcting lens cylinder axis will be properly aligned with the built-in eye cylinder, and the only cylindrical error now present will be that created by the cross cylinder lens itself.

In using the cross-cylinder, remember to check for *axis* refinement first, then check for *power;* and then, if you really wish to be accurate, recheck the axis again.

I have devoted quite a bit of space on the cross cylinder and its use. You should be able to judge the commensurate esteem in which I hold these magnificent tests!

I would like to tie together the subjects of astigmatism, the "stenopeic slit" and the "pinhole" — a clinical screening test for refractive error. Try as I might, I am unable to explain it any more clearly than I did in a previous article*; I am therefore reprinting it here.

## An Elongated Pinhole — The "Stenopeic Slit"

On beginning an optics conference last year with a prominent residency group, I mentioned casually that among the subjects to be covered during that conference would be the principles behind the use of the stenopeic slit. "The what?" was the general reaction. What really surprised me was that some of the more highly trained residents had been taught neither how to manipulate this useful tool nor even what it *looked* like, even though most of the residents had at times wondered what *was* "that peculiar-looking disç" lying around in a trial set. Because of this experience, I felt that it might be instructive to review here the use of the *pinhole* and then go on to study the *elongated pinhole* and its usefulness in arriving at an estimate of cylindrical refractive error.

### *Pinhole*

In emmetropia, every point in an object of regard is brought to a point focus on the retina. Thus, if one neglects the effect of diffraction, there is a *point-to-point* correspondence between object and image. The *sum* of all point foci yields a uniformly sharp retinal image of the object being scrutinized. If a refractive error exists (any kind of error), a "blur circle" is formed on the retina instead of a "point." The size of that blur circle is directly proportional to the size of the subject's pupil.

---

* Rubin, M. L.: "The Elongated Pinhole" Survey of Ophth. 13:355-359.

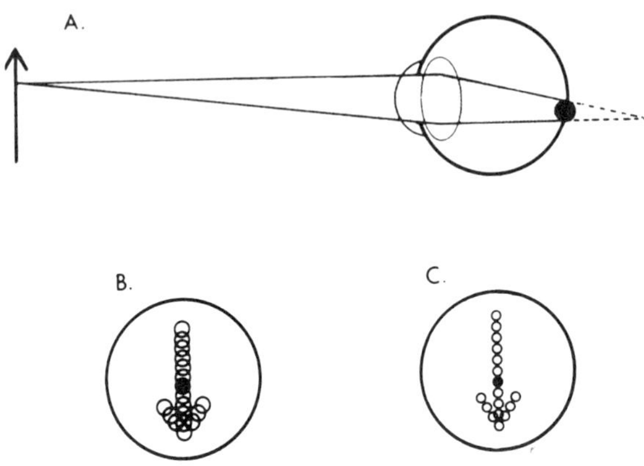


FIG. 1. *A*, light rays from each and every point on an object form a blur circle on the retina of a hypermetropic eye. *B*, the retinal image in this eye is the composite of all blur circles, the size of each being proportional to the diameter of the pupil. *C*, if a pinhole is held in front of this eye, the blur circles are decreased in size, thus sharpening the over-all retinal image.

In Figure *1A*, light from every point on the object is brought to a focus behind the hypermetropic eye, since its refractive power is too weak. What *is* on the retina is a blur circle, or a series of blur circles when the entire image is considered (Fig. 1*B*). If the pupil were made smaller, the diameter of that blur circle would decrease, and the retina would "see" an image more like the correct "point image" (Fig. 1*C*). If a pinhole aperture were used immediately in front of this eye, it would act as an artificial pupil and the size of the blur circles would be correspondingly reduced. This is the reason that many ametropic individuals (especially myopes) squeeze their eyelids together to see more sharply, since in this manner they narrow their pupillary apertures.

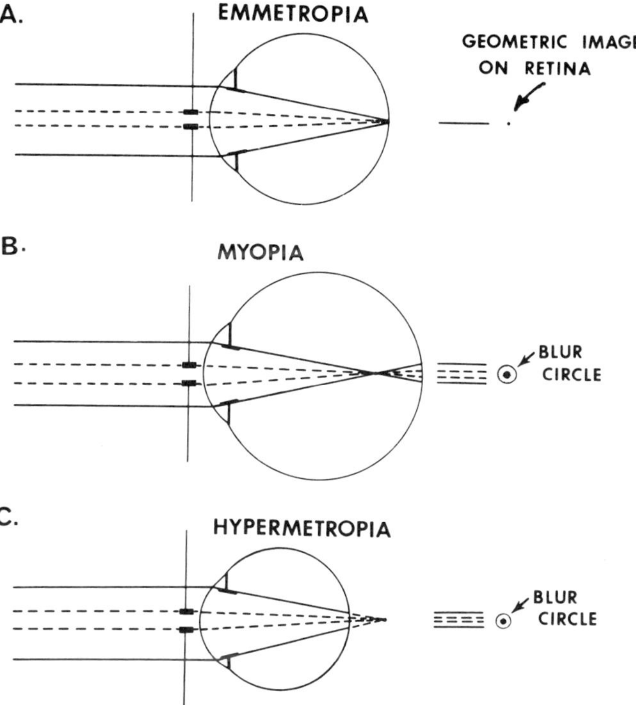


FIG. 2. The effect of a pinhole introduced in front of an emmetropic eye, a myopic eye and a hypermetropic eye, respectively (neglecting the effect of diffraction). The pinhole has no effect on the emmetropic eye *A*, but in both *B* and *C* the blur-circle size on the retina is shown to decrease with the pinhole aperture.

Figure 2 demonstrates the effect of the pin hole in an emmetropic eye, a myopic eye and a hypermetropic eye. This figure shows the comparable sizes of the retinal image of a point source with and without a pinhole held in front of the cornea.

Although the pinhole actually decreases the total amount of light falling on the retina, it does create a smaller blur circle when there is any ametropia existing, and thereby improves the sharpness of the retinal image. However, the pinhole itself does induce a diffraction effect also, and thus tends to blur slightly the sharp point imagery normally present in emmetropia. (The *best* acuity in an emmetrope *with* a pinhole is rarely better than 20/25 even though, without the pinhole, vision may be 20/15). In ametropia, however, the beneficial effect on blur circle size certainly outweighs the diffraction effect. In summary, the pinhole "pinholes" the blur circle and makes it more like a "point."

* * *

Here, I would like to interject with two other phenomena which were not mentioned in the original article; these are so interesting that you as clinical students of optics ought to be made aware of their existence.

The pinhole described above will not only improve acuity in any type of refractive error but can also tell the examiner whether the refractive error is *hyperopic* or *myopic!* If the pinhole is moved side-to-side in the frontal plane while the patient is observing some target through the hole, the ametropic patient will note that the *target* also appears to move — the same direction as the pinhole in myopia or the opposite way in hyperopia. *No* movement will be evident if emmetropia is present. The optical basis for this is shown in the following diagram:

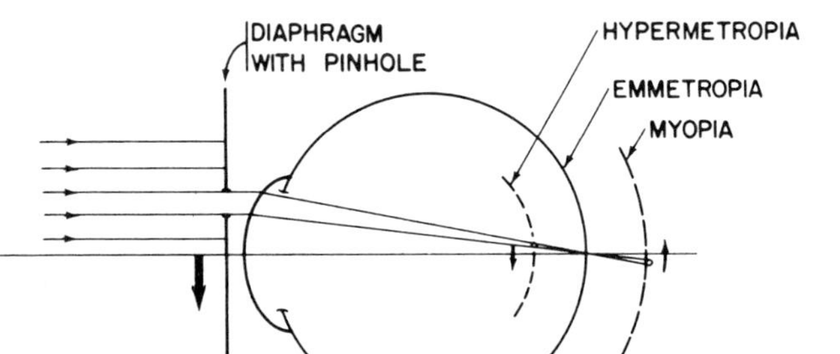

The *sharpest* image point created by the pinhole will be in the position where the emmetropic retina would be located, that is, in the "emmetropic" focal plane. As the pinhole is moved *down* across the pupillary opening, the small "blur circle" which would be found on a hypermetrope's retina will move in this *same* direction; this is visualized by the patient as motion opposite to (or "against") the direction of the pinhole motion — vice versa for the myope. (The arrows in the above diagram point to the directions of motion of the pinhole and the *retinal* images.)

What I have shown here in ametropia occurs with the tiny, circular retinal image of a single object *point;* it is clear that the *entire* image — a composite of many image "circles" — will shift in the same direction as the individual "circles" do.

This subjective test can also be used to locate and measure the near point of accommodation. The patient holds a test card up close and tries to read small print while looking through the "moving" pinhole aperture. If he can indeed accommodate for a given distance, there should be no subjective appreciation of any motion of the target when the pinhole moves. If there is, the target is too close, and should be moved back until no "motion" is noted; in that position, the near point will have been located.

These are actually quite sensitive subjective tests and can detect from 0.25 D to 0.50 D of error.

Another type of motion induced by the pinhole in an ametropic patient creates one more subjective test for myopia vs. hyperopia. This one is not quite as sensitive as the former but is still quite vivid.

Here, the pinhole is moved towards-and-away from the patient while he continues to view some target. On moving it away, if the error is uncorrected myopia, the target will visibly *shrink* in size; if it is undercorrected hyperopia, it will seem to expand. There will be no size change if a full "correction" is in place.

The optical explanation of this revolves about the fact that the pinhole functions as a *universal* corrective lens, and we will soon see that corrective lenses induce magnification or minification, the respective effect being greater as the corrective lens is positioned further from the eye. (In hyperopia, retinal images are magnified as the corrective lens is positioned further from the eye; in myopia, the retinal image is minified.) The pinhole mimics this "lens effect" exactly (and for the same reason) as it moves to and from the eye. This finding as it pertains to lenses will be expanded upon later.

Another interesting phenomenon in a similar vein can be demonstrated using the pinhole. This also is based optically on the fact that the pinhole serves as a *universal* lens:

Hold a pinhole close to your eye. Also hold up any plus lens, say a + 4 D sphere at about 25 cm, and look through both. You will have created a small Galilean telescope of about 4 × magnification power with the pinhole as the "eyepiece". Now, if you watch closely while you move the + 4 D lens further away, you will observe a continuously increasing magnification, like a zoom telescope! The pinhole serves as a continuously variable-minus eyepiece lens. (The complete "lowdown" on telescopes later.)

CLINICAL POINT:

For any particular patient the pinhole makes for a good, quick check on whether or not it is refractive error *per se* that is the cause of a decreased acuity; and, as already explained, because of the blurring effect of *diffraction,* an accurate lens-correction of the refractive error will almost certainly yield a better acuity measurement than the pinhole itself provides without the corrective lens.

Although this is the general rule, the alert student should be aware of exceptions. Sometimes, vision through a pinhole is *much better* than can ever be obtained with corrective spectacles or contact lenses. This finding might even be *expected* in some patients with keratoconus, or those with discrete, small opacities of the cornea or lens, or in cases of irregular astigmatism (when the major meridians are not at right angles to one another) as in a distorted but clear cornea warped by a stromal scar. In such patients *without* the pinhole, the refraction occurring at the surfaces may be irregular and non-predictable, and if any opacities are present, they can cause scattering — all these effects destroy the possibility of sharp retinal imagery. However, with a properly positioned pinhole, a patient can select a small clear area through which light can pass, unimpeded by any ocular obstacles, and thereby create a good quality of retinal image. Thus, despite diffraction, in these patients, vision through the pinhole will be better than without it. (Do keep in mind that this appliance will *not* offer acuity improvement to a patient with a *diffuse* opacity of the transparent media, such as in corneal epithelial edema, mature cataract, vitreous haze, etc.)

Now, back to the reprinted article.

## *The Stenopeic Slit*

To understand the basics of the stenopeic slit, the fundamentals of astigmatism must be kept clearly in mind. Study the diagram in figure 3*A* which shows a cylindrical lens grasping a bundle of incident light and bringing it to an a-stigmatic (nonpoint) focus, that is, two line foci with Sturm's interval between.

The length of the astigmatic focal line images is also directly proportional to the size of the aperture as well as to the *amount* of astigmatism, and inversely proportional to the total spherocylindrical power influencing any given line. Let us, however, restrict our attention to the influence of the aperture size. If we now cover up part of the lens so that we narrow its vertical dimension (Fig. 3*B*), the vertical line is decreased in length while the horizontal is not affected. If we further narrow the vertical to a horizontal slit-like opening (called the *stenopeic slit*), the vertical line will be "pinholed" as shown in Figure 3*C;* the length of the horizontal line still is uninfluenced.

With the slit in the horizontal position, then, there is a position of the image plane (retina) where there is practically a point-for-point representation for every point in the object, and hence a position of sharpest imagery. A patient who has his retina located at this position will have a "sharper" image than one who has his retina located anywhere else behind this lens. If the slit is turned vertically (Fig. 3*D*), it is the horizontal line which is "pinholed," but it does not move forward or back; the position of the "clearest imagery" has now moved from $x$ to $y$.

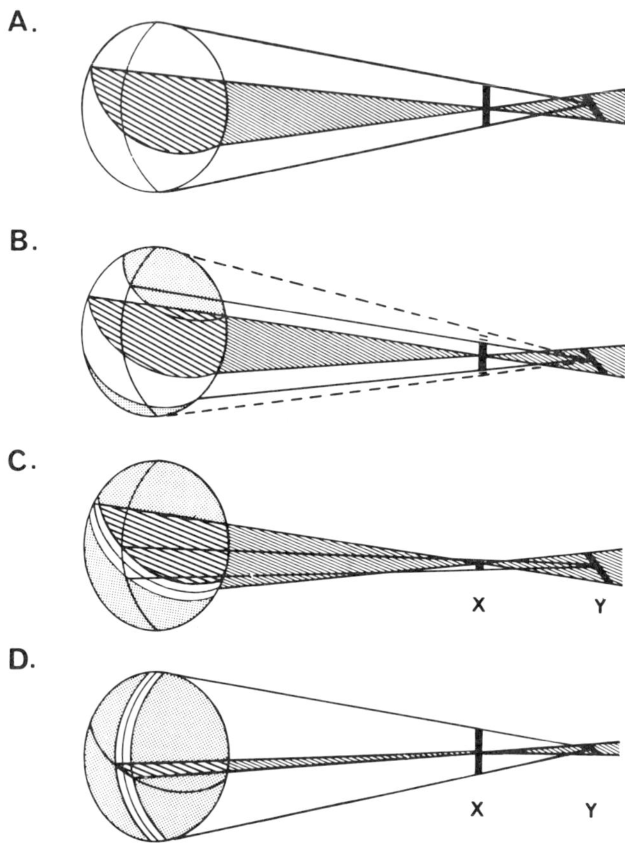


FIG. 3. *A,* the optics of an astigmatic lens forming two-line foci. *B,* the vertical aperture is narrowed slightly. This shortens the vertical focal line, without affecting the horizontal. *C,* the vertical aperture is narrowed to a horizontal slit; the vertical line is now point-like — still there is no effect on the horizontal line. The horizontal slit-like aperture is comparable to the stenopeic slit. *D,* with the slit vertical, the *horizontal* focal line becomes point-like; now the vertical focal line is unaffected.

We should be able now to surmise how to use the stenopeic slit clinically. It should be clearly understood, however, that the slit is helpful only as a *guide* to the subjective refractive error. Retinoscopic findings, if they can be obtained, are vastly superior to stenopeic approximations. On the other hand, if you have no old correction to go by and the retinoscopic reflex is of poor quality (secondary either to opaque membranes or lens capsule remnants, or affected by a pupillary opening too small to yield a recognizable reflex), the stenopeic slit can be very helpful.

Spherical lenses should be used by trial and error (by diopter or even two-diopter steps) to position the conoid somewhere in the "ball park" of the retina. Do not try to get the "best" acuity with spherical lenses, since in doing so you will probably bring the circle of least confusion on the retina; this will only complicate the cylindrical error determination to follow. When a reasonable acuity, say 20/200 or better, has been arrived at, bring up a stenopeic slit into the plane of the trial lens frame and rotate it to the angle that provides the best vision. If the patient claims that there is no "best" position, there are some obvious possibilities: (1) no astigmatism is present, (2) his astigmatic interval is equally mixed (equally spaced around the circle of least confusion which now is on his retina) or (3) he is physically incapable of recognizing any change in the image "quality" caused by the rotation of the slit. To test the possible existence of (2), throw in a + 1 or — 1 diopter sphere on top of whatever sphere you have in place, and try rotating again. If the patient still cannot see any difference, forget the whole thing and prescribe the best *sphere* determined by trial and error.

If, on the other hand, the patient is able to notice a difference in acuity with various positions of the slit (this is the most likely situation), place it in the position of best acuity and add plus or minus sphere on top of the slit to give the patient the sharpest possible vision. (You might need to refine the slit position slightly.) What you are now doing is "pinholing" the focal line positioned perpendicular to the slit opening, and moving that "pinholed" line onto the retina with sphere. (With the slit vertical, the horizontal line is pinholed as shown in Figure 3*D*). The *total* amount of sphere present (say, + 11 D to start with, plus 2 D more to clear up the target with the slit in whatever position you find — say at 90°) gives you the dioptric error of the focal line perpendicular to the slit (here, + 13 D for the horizontal line).

Next, rotate the slit exactly 90° to the first position. This blurs the existing retinal image and transfers the "pinholing" effect to the other focal line. You will now have to add *either* plus or minus sphere, whichever improves the acuity. (In using this method, you have no way of *knowing* which direction to go; you will have to try both, by trial and error, and determine whether the addition is plus or minus. Actually, that is the beauty of this technique. You "automatically" come out with the correct cylinder sign.) If you have had to add a plus sphere to clear the retinal image, the newly "pinholed" line must, before your lens addition, have been behind the retina, the plus sphere serving to pull it forward onto the retina. If you had to use minus sphere, the newly "pinholed" focal line was of necessity in front of the retina and your minus sphere pushed the line back onto it. It should be clear that whatever amount of sphere addition was necessary *that* is the amount of astigmatism present. If a plus sphere was necessary, then a corrective *plus* cylinder is required for a final corrective lens; if a minus sphere was necessary, then a corrective *minus* cylinder is required in the final prescription. In either case, the corrective axis is perpendicular to the slit in its second position; however, it is much easier to think of prescribing the corrective cylinder axis *parallel* to the *first* position of the slit. (Say you require − 3 D more sphere with the slit horizontal; the cylindrical prescription is − 3 × 90°. The *spherical* correction is that amount of sphere found to give the clearest vision with the slit in its initial position, + 13.00; final prescription, then, is + 13.00 − 3.00 × 90°.)

SUMMARY OF STEPS

1. Approximate sphere (without slit) to measurable, but not best, acuity.
2. Rotate slit to best angular position — "position 1."
3. Add plus or minus sphere to get the best acuity at slit-position

1. Refine angle slightly, if necessary: the *total* sphere in position 1 is the corrective sphere of the prescription.

4. Rotate slit 90° beyond position 1; this is "position 2."

5. Add plus or minus sphere to maximum acuity; the additional sphere is the cylindrical power which must be added to the corrective sphere of step 3. The cylinder is plus if the added sphere is plus; the cylinder is minus if the added sphere is minus. The corrective lens cylinder axis is parallel to the slit in position 1.

With the optics and the limitations of the technique described, you should be able to use the slit intelligently and "refract" patients who may have been "impossible" previously. At the very least, you will be able to recognize what a stenopeic slit looks like!

(Article *finis.)*

Another article, *"The Little Point That Isn't There,"* deals further with the effect of the pupil's size on visual acuity. It is reprinted in Appendix ***C***.

## NEUTRALIZATION

There are a number of ways to skin the cat of ametropia correction: we have already considered one — far point correction. However, I would like to expound on another concept of correction to introduce a new subject — that of "neutralization."

In all ametropias there is a mismatch of refractive power and axial length; since the latter cannot easily be tampered with, we must resign ourselves to dealing with the refractive power. What we would like to do is to make the eye of "proper" power to correspond to whatever its axial length is — to add dioptric power to an eye which is too weak (hyperopic) or reduce dioptric power (by adding minus lenses) to an overpowered eye (myopic).

If we place a minus lens in direct *contact* with a plus lens, the two powers add algebraically. When both are of identical powers but of opposite signs they "neutralize" each other and leave no resultant refractive power. We can do the same thing with an eye. However, we do not try to "neutralize" its entire refractive power, but only its refractive *error.* An eye which is myopic by 5 D is too strong by 5 D and this can be neutralized by a lens which is a — 5 D lens. This lens must, however, be in contact with the anterior surface of the eye. This same concept holds true for the hyperopic, "weak" eye in which we must supply dioptric power in the form of a plus lens in contact with the cornea. Now why do I stress *contact* between the corrective lens and the anterior refractive surface?

Neutralization involves the "balancing out" of any dioptric power between two lenses leaving an *afocal* system; that is, if the incoming vergence is zero, the outgoing (image) rays for the system of lenses will also have zero vergence. When the lenses are in contact, a minus lens is *exactly* neutralized by a plus lens of equal power.

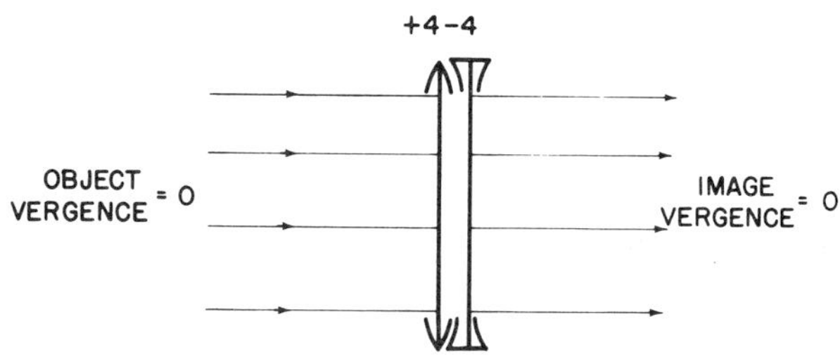


When we separate the two lenses, and we still want to neutralize the — 4 D lens, the plus lens will have to be of less power dioptrically to produce an afocal system. Let's see how much plus is required of the "neutralizing lens". For a — 4 D lens to produce zero *image* vergence, we must have light rays in its object space directed to its primary focal point F. But F of a minus lens is on the *right* of the lens. Thus, light rays in the object space must be converging toward F to enable them to leave in parallel bundles.

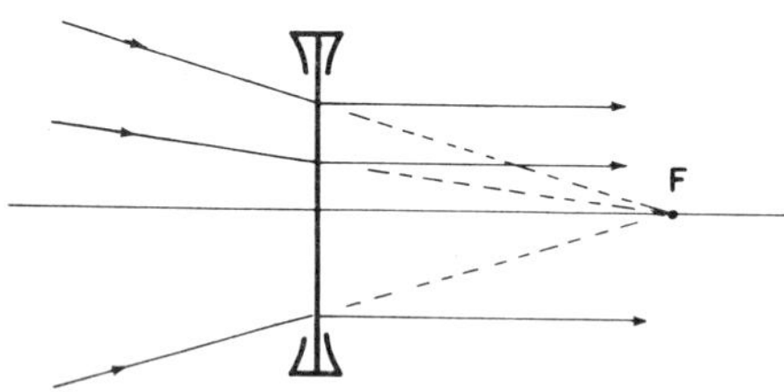

To create that convergent beam, we require a plus lens. We know that this lens can take parallel entering rays and converge them towards *its* secondary focal point F′. So, if this F′ is made to coincide with F of the minus lens, our conditions for a "neutralized", afocal system will be satisfied — the original object vergence is zero and the final image vergence is also zero.

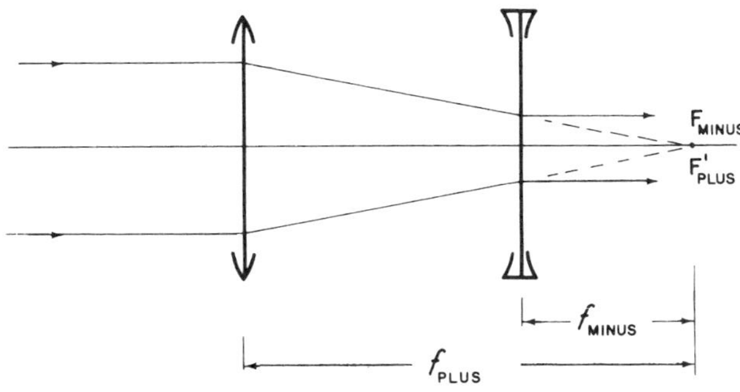


It should now be clear that we can neutralize a minus lens with a plus lens. When the lenses are in contact, the secondary focal length *f′* of the plus lens will be the same length as the primary focal length *f* of the minus lens — thus, both are of the same power. However, when there is any space between the two lenses and we desire a coincidence of the focal points to occur — the condition for neutralization — *f′* of the plus lens must be the longer focal length. The greater the distance between the lenses, the longer *f′* of the plus lens would have to be and, therefore, the *weaker* its plus power. So now we've shown that for the neutralization to occur, the plus lens power will always be equal to (when in contact) or *less than* (with separation) the power of the minus lens.

Though we have been talking about "hand-held" plus and minus lenses, the identical principle holds for "eye error" and "corrective lenses", that is, when we try to "neutralize" any refractive error. A hypermetropic eye has too little power for its axial length and can be considered as having a small extra *minus* lens "built" into the eye. It is corrected (neutralized) by a plus corrective lens which must be greater in power the closer it is to the "minus" lens within the eye. The *"corrective"* lens would necessarily have to decrease in plus power if it were placed further from the eye error. (This is just another way of looking at the identical point which I made when dealing with far-points and corrective lenses!)

So in two different ways I have emphasized that the "proper" corrective lens power can vary yet still be "corrective". Until now, I have avoided the question of what the *patient* notices when one changes the corrective lens power. Let us examine what happens during such power manipulation by closely scrutinizing our "neutralized" afocal system (also called a Galilean Optical System).

## Galilean Optical System

Our previous figure showed that light rays with zero vergence from an object point at infinity also left the system with an image vergence of zero. Though this is true for all object points at infinity, that diagram shows only the rays from that particular point at infinity which is *on the axis.* What about a point at infinity which is off the axis — say one 10° above the horizon?

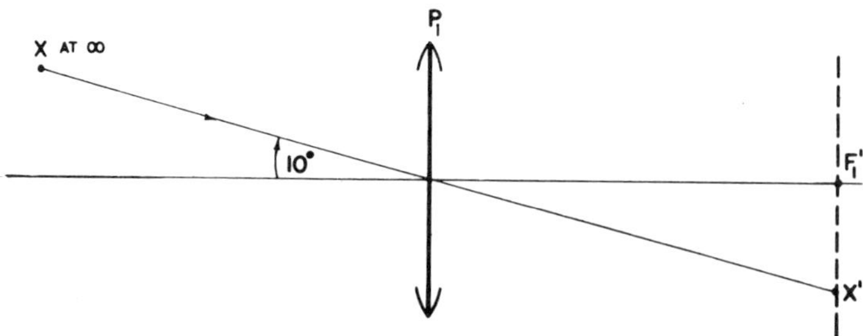

Our first lens $P_1$ would image this point in its secondary focal plane. Since the object is at infinity but off the axis, the image point is determined by the ray from X through the $P_1$ nodal point.

We can draw a few more rays from X.

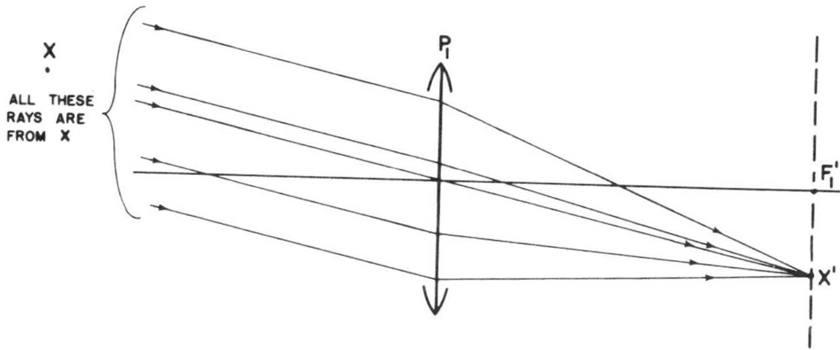


Since X is at infinity, those rays which reach lens $P_1$ will all be parallel and all will have the same 10° inclination to the axis, and after being refracted by $P_1$, all rays will be bent toward the image point X′ in the secondary focal plane (as we saw way back in the early chapters). However, in our Galilean neutralized system, these rays in the image space of $P_1$ cannot actually reach X′; they are intercepted first by our minus lens $P_2$, which is situated in front of the secondary focal plane of $P_1$.

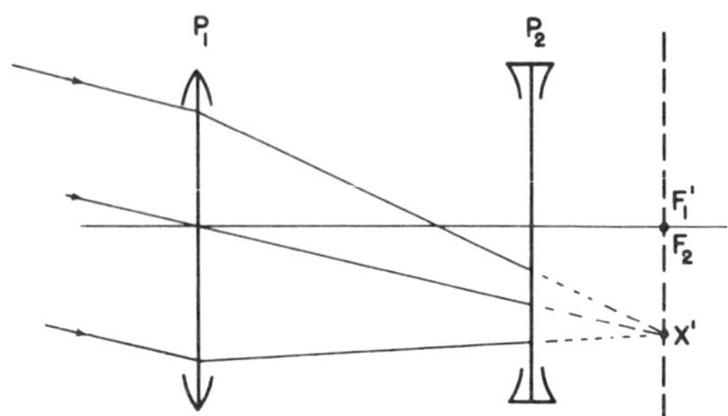


$P_2$ "sees" X' as an "object" point and images it — the question is where? We know that X' is in the *object* space of $P_2$ and is located in the primary focal plane of $P_2$. (Remember our coincidence of F' of $P_1$ and F of $P_2$?) The image of X' by $P_2$ then must be at infinity, and all the rays originally heading for X' must leave $P_2$ in parallel bundles.

O.K., so far, but how do we know at what angle these rays leave? They *could* be bent in any direction (two possible ones are shown in A and B below) and still they could remain in parallel bundles.

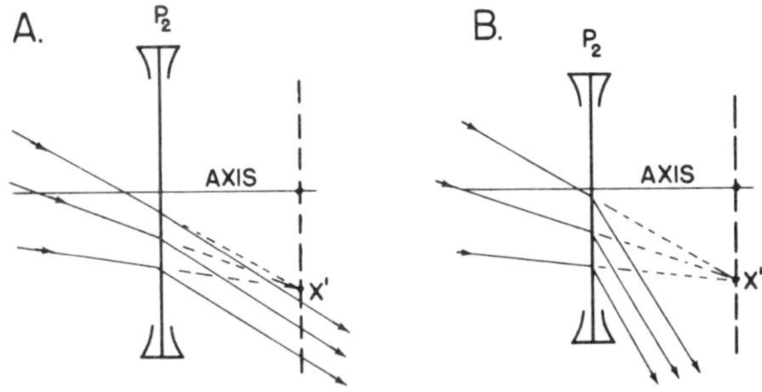

Can you see how we can find the final direction imparted to these parallel bundles of rays? You should. We know that *all* rays from the *original* object point X had to be imaged by $P_1$ at X'. Therefore, we could trace backwards *any* ray originally directed to X' (such as 1, 2 and 3 in the figure below) and they would all have to be parallel to all those original *object* rays with an inclination to the axis of 10° — the same as our "key" ray through the $P_1$ nodal point.

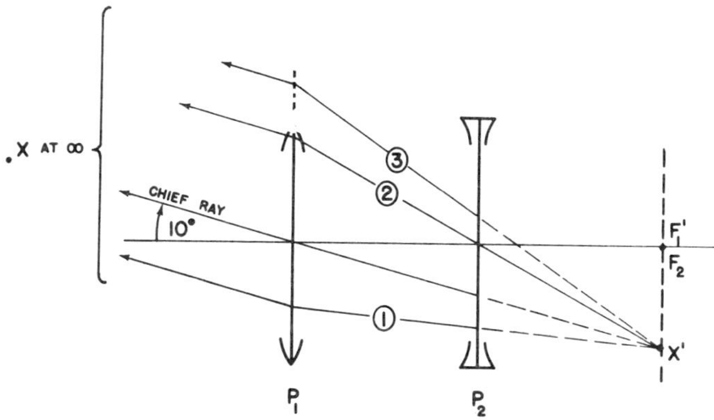


(In order to draw ray 3 back from X' to $P_1$, we had to artificially extend the lens plane $P_1$ upward so that 3 would "intersect" it and change its direction — that ray, in the object space of lens $P_1$, also makes an angle of 10° with the axis.)

In any case, we know that there will be one ray which passes back from X' through the nodal point of $P_2$ (ray 2) and in doing so will not be deviated. AHA! That ray 2 will be the one that determines the *final* direction of *all* image rays from $P_2$ for which the object was X'. That angular direction is shown in the diagram below as $\theta'$.

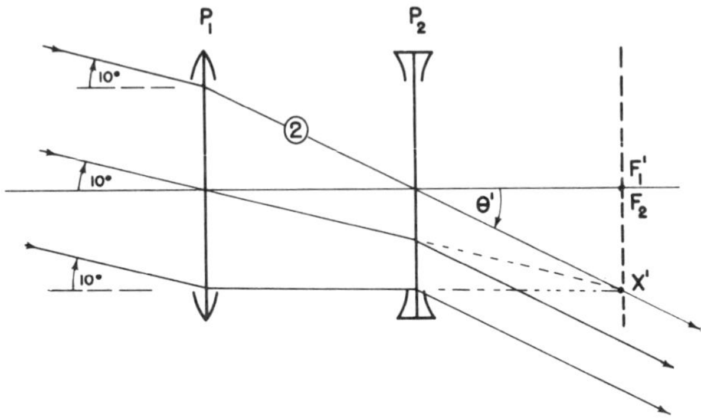


In summary, ray 2, originally from object X, hits $P_1$ at an inclination of 10° to the axis. $P_1$ refracts that ray toward X'. Though this ray is intercepted by $P_2$ (and will thus participate in the *final* image of X' which will be at infinity), it is *not* deviated further since it is the *one* ray which passes through the $P_2$ nodal point. Since we now know *its* direction, we also know the final direction of all the final image rays.

(By the way, our last diagram shows that the final image rays look as if they are "packed" closer together than their corresponding object rays. Many are led to the *erroneous* conclusion that image *mini*fication must have occurred to cause this "packing". You should not be led astray. All those image rays stem from the *same point* as the object rays. Therefore, their apparent density cannot signify magnification or minification. It is only their angular or directional *change* that matters in the determination of magnification.)

We can simplify the above diagram and show only *one* object ray inclined at 10° and *one* (the undeviated) image ray inclined at $\theta'$ which is obviously *greater* than 10°. The final image of our original object point X will be at infinity (since there are parallel light bundles in the final image space), but that image will appear to have a *greater* inclination to the axis than the 10° inclination of our original object point.

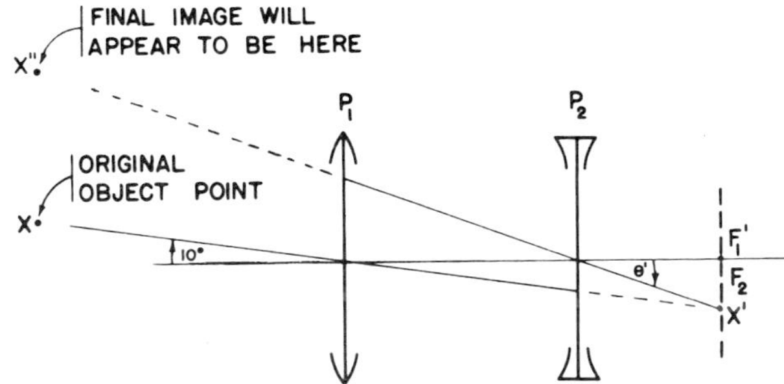


It should be obvious what an eye will see when placed behind $P_2$ if it looks at the final image point X'' through this afocal, "neutralized" lens system. Since X'' is at a greater angle above the horizon than X, the entire object must appear larger — magnified in its angular size. This system of lenses is a Galilean or terrestrial telescope. The image seen will be virtual, upright and *magnified* whenever the *minus* lens is closer to the eye.

You should draw a diagram for yourself demonstrating that if it is the plus lens that is nearer your eye, the image seen will be minified. (Even in this latter, afocal *mini*fication system the plus lens power will always turn out to be less than the minus lens power.)

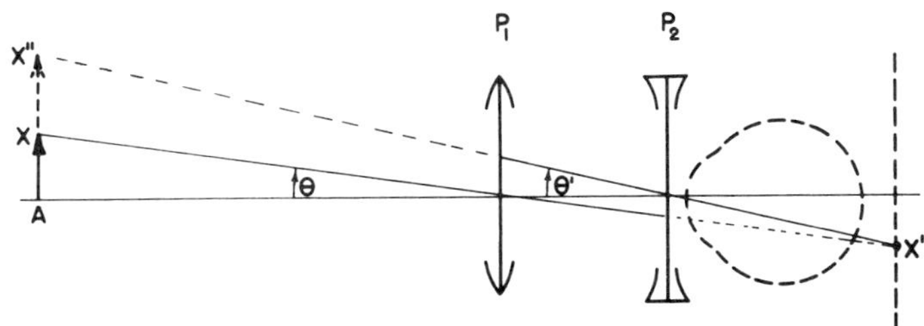


Without the Galilean magnification system shown above, the eye would see object AX; *with* this telescope, it sees image AX'' which would occupy a greater extent of its retinal surface area. The actual magnification, of course, depends on the relationship of $\theta$ (the angular size of the object) and $\theta'$ (the angular size of the image). This angular magnification M is $\frac{\theta'}{\theta}$.

If $\theta'$ is two times as large as $\theta$, the angular magnification would be 2 or 2X — signifying a *100%* increase in the apparent size.

Now let's look and see how we can denote the angular magnification of a telescope if we know the lens *powers* we are "neutralizing". Again to simplify, only one object ray and another image ray are drawn to X'. Image X' is located a distance *h* from the axis:

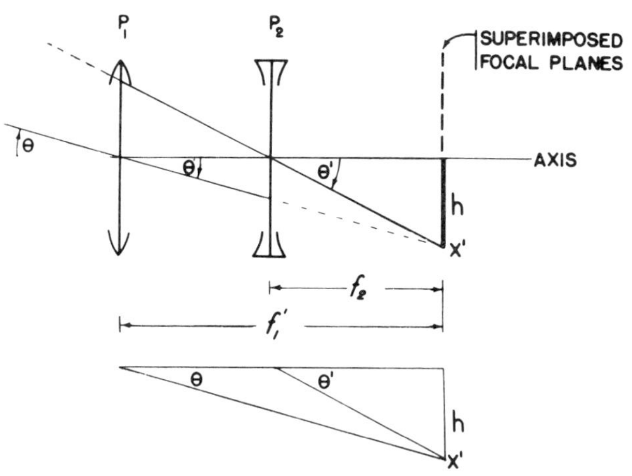


$$\tan \theta = \frac{h}{f_1'}$$

$$\tan \theta' = \frac{h}{f_2}$$

Since for small angles $\tan \theta = \theta$ and $\tan \theta' = \theta'$,

$$\theta = \frac{h}{f_1'} \text{ and } \theta' = \frac{h}{f_2}$$

$$M = \frac{\theta'}{\theta} = \frac{\frac{h}{f_2}}{\frac{h}{f_1'}} = \frac{f_1'}{f_2}$$

In contrast to what you will find later when you study the *astronomical* telescopic magnifying system, in this Galilean system both the object and its images are upright so both $\theta$ and $\theta'$ are measured in the same (here, clockwise) direction; therefore they must be of *like* sign. But recall that to attain this, we had to make $P_2$ a *minus* lens (with its minus $f_2$), so our *general* expression will have to be as follows: $M = \frac{\theta'}{\theta} = -\frac{f_1'}{f_2} = -\frac{P_2}{P_1}$. This is the basic magnification relationship between *any* two lenses which "neutralize" each other. Now if $P_2$ happens to be a minus lens as it is here, M will turn out to be plus, signifying an upright image — *that* was our originally agreed upon convention.

## Magnification Following Optical Correction

At the beginning of this section, we stated we could consider the hyperopic eye as being "too minus", with a built-in minus lens inside — this "lens" is $P_2$. This "error" could be "corrected" by any lens $P_1$ outside the eye; the greater its distance from the eye (that is, the greater the "vertex" distance), the less plus is required in the corrective lens. Finally now, we can see the consequences of varying the position and power of the corrective lens. The further from the eye the corrective lens is, the less its plus dioptric power; but, the *greater* is the magnifying power it presents to the eye! So, as $P_1$ (the "corrective" lens power) decreases, the magnification $(\frac{P_2}{P_1})$ increases.

If $P_1$ is a plus corrective *contact* lens, the magnification $\frac{P_2}{P_1}$ induced by that corrective lens for *that* eye will be minimal — that is, less than that created by a corrective lens in *any* other position (aside from an *intra*ocular* lens!). But keep in mind that even with a contact lens, there will still be *some* magnification present. This is because the eye "error" is not *at* the cornea but somewhere within the eye; that is, some physical separation will still exist between the "error" and its corrective, "neutralizing" lens.

* Such as a Binkhorst lens for an aphakic eye.

CLINICAL POINT:

These optical fundamentals provide the basis for a handy gimmick for aphakes. We have already seen that an aphake of 12 D (with a refractive *error* of $-12$ D measured at the cornea) can be "corrected" by a $+3$ D sphere if it is held about 25 cm from the cornea. (You should be able to diagram *exactly* where it must be held!) Now it is clear that such a "correction" will provide him with $-(\frac{-12}{3})$ or 4X magnification (that is a 300% enlargement) of the retinal image of some object! This simple do-it-yourself telescope with a $+3$ D sphere is especially useful to that aphake who has a reduced visual acuity.

* * *

Magnification is nice when it's necessary but can be a handicap too. Take for example the unilateral aphake (the other eye being emmetropic) — here, even a contact lens correction will create a 7% image magnification to impede a sensory fusion with the image given by the emmetropic eye. I have always been amazed that most individuals corrected this way do *not* have difficulty!

Typically, the average aphakic eye has its corrective lens worn in the spectacle lens plane, 15 mm from the cornea. This affords a magnification factor of about 125% (usually written as "25% magnification and implying an image which appears 25% larger than the object itself). Oops! That's putting the cart before the horse. You yourself should be able to diagram the comparison between the magnified retinal image seen by a corrected aphakic eye with that by an emmetropic eye; but anyway, let's go through it together. First the emmetropic eye:

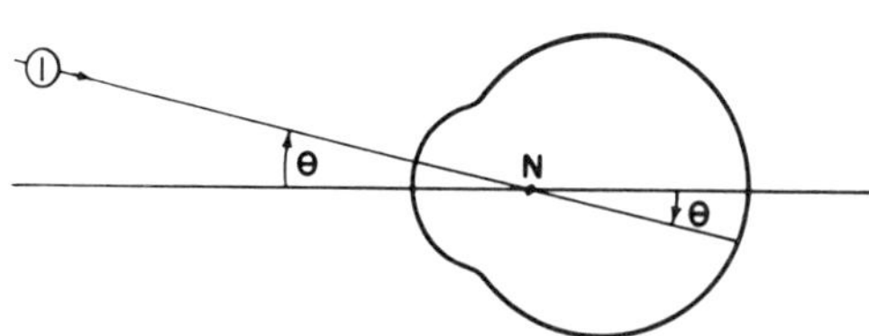

The angular size of the object is given by ray 1 which goes from the tip of the object to the eye's nodal point. This ray subtends angle $\theta$ with the axis. Since ray 1 is undeviated (through N), the angular size of the retinal image is also $\theta$. (Remember that the nodal point ray in *any* eye provides a quick reference to the angular size of the retinal image.)

Now to the corrected aphakic eye:

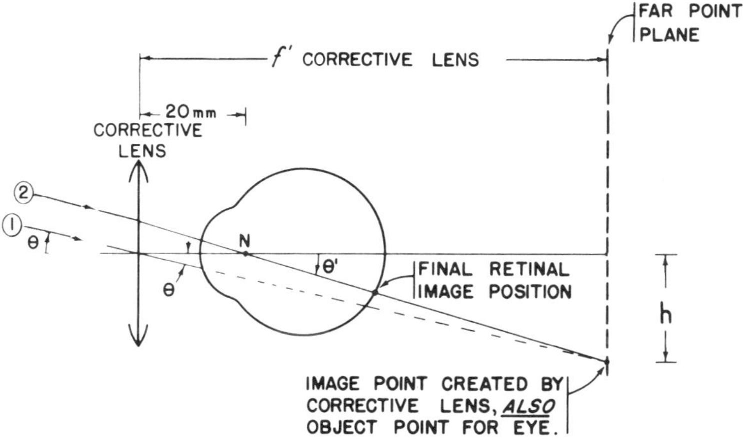


Follow this reasoning closely. In contrast to the emmetropic eye, the aphakic eye requires a corrective lens which here is positioned about 15 mm from the cornea or about 20 mm from the nodal point of the eye. Of course, the secondary focal plane of any "corrective" lens will coincide with the aphakic eye's far point plane. This corrective lens will image every object point from infinity somewhere in its focal plane. Since that image point also happens to be in the eye's far point plane, the aphakic eye will see it clearly.

The original object point (just as in the emmetropic eye) subtends object angle $\theta$ at the lens. The corresponding image point is located in the secondary focal plane of the lens by ray 1 through the lens' axial (nodal) point. That is, if this actual ray were *not* intercepted by the eye, it would proceed undeviated and intersect the secondary focal plane (and far point plane) at a distance $h$ from the axis. This intersection, then, is where the corrective lens images the original object point.

Now get this; once this image is formed by the corrective lens, the eye couldn't care less about that lens; it is only this image point which assumes importance as it becomes the *object* point for the eye. This object point subtends an angle of $\theta'$ at the nodal point of the eye. With reference to the eye then, this "object" point (in the far point plane) and the final retinal image point must both lie along the undeviated ray 2 which passes through the eye's nodal point. (There should be no question about that.) If you now want to see where this ray 2 *originally* came from, you must trace its path back through the corrective lens; you will then find that it must have been refracted by that lens so that in the *lens'* object space it was parallel to ray 1 (as shown in the diagram above).

Since $\theta'$ is angular size of the *retinal* image, we can compare it to $\theta$, the angular size of the original object and arrive at an expression for the angular magnification given by an aphakic spectacle lens. Go back to the last diagram for a little geometry:

$$\tan \theta = \frac{h}{f'}; \ \tan \theta' = \frac{h}{f' - 20 \text{ mm}}$$

Again with small angles, $\tan \theta = \theta$ and $\tan \theta' = \theta'$

$$\frac{\theta'}{\theta} = \frac{f'}{f' - 2.0 \text{ cm}}$$

Since this sample aphakic corrective lens can be considered to have a focal length of about 10 cm, substitute 10 cm for $f'$.

Therefore, $\dfrac{\theta'}{\theta} = \dfrac{10}{8} = 1.25$

or $\theta' = 125\% \ \theta$

Thus, $\theta'$ for an aphakic eye is 25% larger than $\theta$, which is the image size for an emmetropic eye.*

Study the next diagrams for a moment. (Remember, the far point plane stays in the same fixed position relative to the eye and is independent of where *any* "corrective" lens is placed.)

Figure A (hyperopic eye): for a given angular object size $\theta$, in the far point plane there will be a corresponding image size $h_1$ created by that corrective lens. As the corrective lens $P_1$ moves *away* from the eye towards $P_2$, $h_1$ must move towards $h_2$. So, the size of this image will get longer and, as it does, the angular size $\theta'$ of the *retinal* image will also necessarily *increase.* Hence, as the vertex distance of a plus corrective lens increases, the Magnification $\frac{\theta'}{\theta}$ increases too, as we have cried so often.

Figure B (myopic eye): An object of angular size $\theta$ will be imaged by lens $P_1$ in the myopic eye's far point plane at $h_1$. So for the eye, the original object will appear to be at $h_1$. The angular size of the retinal image of $h_1$ as subtended at the eye's nodal point will be $\theta_1'$. If now you move the corrective lens toward $P_2$, $h_1$ will move toward $h_2$ and the angular size $\theta_2'$ of the retinal image will shrink. So here, the Magnification $\frac{\theta'}{\theta}$ *decreases* as the vertex distance of a minus corrective lens *increases.*

In the diagrams below, then, picture that in both hyperopia and myopia, as $P_1$ moves away from the eye towards $P_2$, $h_1$ (in the far point plane) will move toward $h_2$ and the retinal image size $\theta_1'$ shifts towards $\theta_2'$.

---

* Thus, any corrected aphakic eye should be able to read smaller-sized print (and hence demonstrate a "better acuity") on a Snellen chart than would an emmetrope.

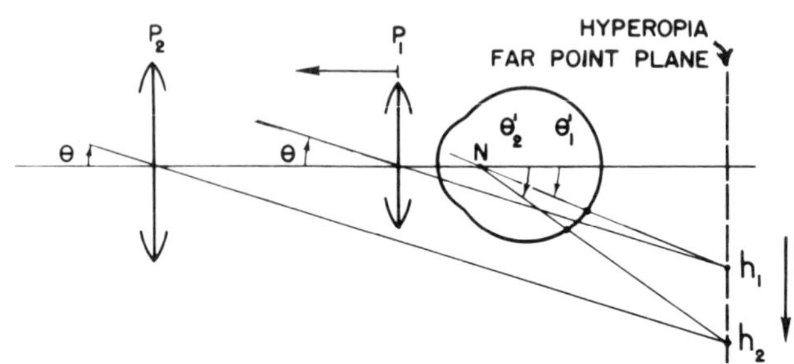


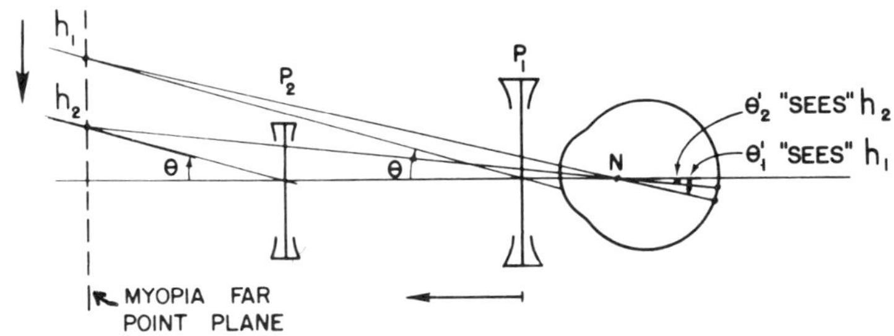


In both A and B above,
as lens $P_1$ moves towards $P_2$, $h_1$ moves towards $h_2$.

The corrected myope has magnification working in reverse for him, as if he looked through the telescope the wrong way. This introduces minification — a *reverse* Galilean effect. The further away from his eye he has his corrective lens mounted, the greater the *minus* power must be, and the more *mini*fication he obtains. His most magnified view of the world will come with a contact lens. No wonder then, the great affinity that myopes have for contact lenses. A 10 D myope will obtain about a 15% increase in retinal image size with a contact lens compared to that afforded by a corrective spectacle lens.

The magnification and minification effects obtained by varying the vertex distances account for many everyday-observed phenomena. Not only will a myope see spatial objects as minified, but, if another observer looks at this myope's fundus through an ophthalmoscope, he will see it *magnified* over that visible in emmetropes. This is because he must use a minus lens in his ophthalmoscope head to compensate for the "built-in" plus error of the myope's eye; and since there will always be a "vertex" separation, he will obtain the same optical effect as when looking through a small Galilean telescope. The *minus* lens is closer to the examiner's eye, and therefore, the image seen is magnified. Vice versa for the image of a hyperopic fundus.

CLINICAL POINT:

A spectacle-corrected aphake has a large retinal image. This means that the image of a given area of space now falls on a greater area of the retina than it did prior to the aphakia. A spot in that retina (say 10° nasal to the macula as subtended at the eye's nodal point) will, through the aphakic corrective lens, project outward onto a tangent screen to a position that is *closer* to the fixation point than would the comparable retinal spot of an emmetropic eye.

Look at a sketch:

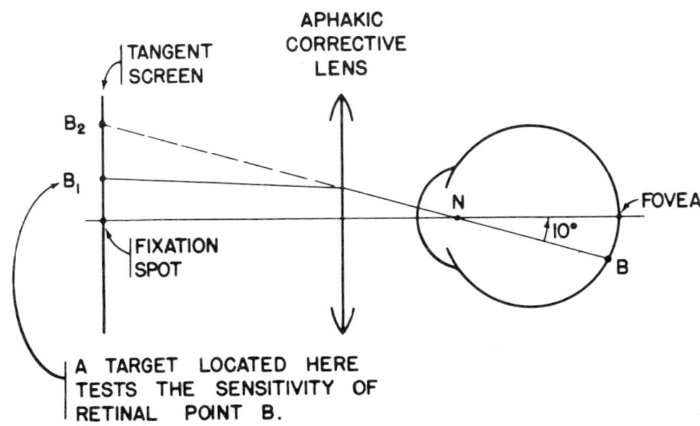

The fovea is aligned with the fixation spot on a tangent screen. In an emmetropic eye, retinal point B, 10° below the fovea, would normally project out onto the screen to point $B_2$. Because of the presence of the aphakic corrective lens, B projects instead to $B_1$. Now look at this same situation from the point of view of an examiner at a tangent screen: If a given sized test object is to be just barely detected by (at the sensitivity threshold of) retinal area B, it would have to be placed at $B_1$, which is closer to fixation than $B_2$. If it were placed at $B_2$, its corresponding image on the retina would necessarily be *below* B and therefore would become a subthreshold stimulus for that retinal area.

This same reasoning can also be applied to the corrected aphake's visual field isopter lines as mapped on a tangent screen; they will be closer to the fixation point than in the non-aphake. Also, the blind spot will be smaller and closer to fixation than will that of a patient who is not aphakic. (Obviously, the reverse is true for a high myope.) You must be aware of this when you plot the field and attempt to search for the blind spot in highly ametropic, "corrected" patients.

* * *

As stated, the magnification effects described so far are those due to vertex distance or the *separation* between elements which are being "neutralized". There *is* a particular vertex distance which has unique properties. That distance is one which places the corrective lens in the anterior focal plane of the eye, roughly 15 mm from the cornea.

## Knapp's Rule

Let us construct the retinal image in an eye which is *hypermetropic*, but so by virtue of its being too short (that is, an *axial* error). We will consider simultaneously *three* such eyes, with the retinas in three separate positions. The anterior segment will be identical for all three eyes since the "error" is in the length, not in the power.

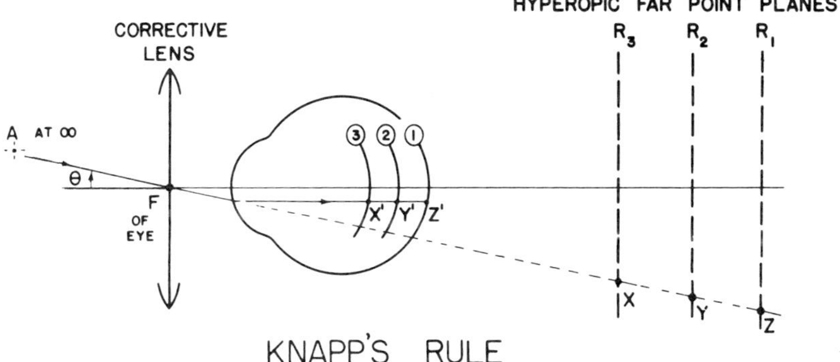


KNAPP'S RULE

Each eye will have a corresponding far point plane. Eye 1 corresponds to $R_1$, etc. Since we have assumed that all of these eyes have the same refractive power, F (the anterior focal point) is in the same position for all these eyes.

If any of these three eyes is to see point A (at infinity) clearly, a corrective lens is necessary to image it in the corresponding far point plane. Let us place the appropriate corrective lens in the anterior focal plane of the eye so that the lens' own nodal point coincides with F of the eye.

Draw a ray from object point A that crosses the axis at the lens nodal point. For all three corrective lenses, that ray will continue on undeviated. Extended beyond the eye, this ray indicates the position in each of the corresponding far point planes where the image of A will be — at X, Y and Z, respectively. (The actual sizes of the images in each of these planes will differ, the largest image being present in the most remote plane). However, since this same ray passes through F of the eye, it will, after refraction by the eye, emerge (in the vitreous) parallel to the axis. You can see that the position of the retina (which would vary with the degree of axial ametropia) will not influence the *absolute size* of the retinal image of A; images X' Y' and Z' are all the same distance from their respective maculas.

(The foregoing analysis also covers myopic correction with the myopic far point plane in front of the axially myopic eye. This will not be diagrammed here.)

Thus, a proper corrective lens located at F in any *axial* ametropia will produce retinal images of the same actual size, no matter what the degree or direction (myopia or hyperopia) of the ametropia. This retinal image size will be the same as that given by any eye with the identical *refractive* power, emmetropic or not. This is Knapp's rule. Know what it is, but realize the obvious difficulty in using it in the clinical situation; you can never know whether an existing ametropia is axial or not (though sometimes, you may have a fair idea — in a monocular high myope, for example, it is likely that the ametropia is axial). In general then, I would like to state my heretical opinion very clearly. For practical purposes, it doesn't make any *real* difference whether an ametropia is axial or refractive; so, Knapp's rule has no *useful* application though it theoretically seems like it should!

## Aniseikonia

When a patient *perceives* a difference in the image sizes seen by his two eyes, no matter what the cause, the defect is labeled *aniseikonia.* The signs and symptoms of this malady are well discussed in other clinical texts. Suffice it to say, size *differences* of less than 5% can induce symptoms. Clinically the size difference perceived by a patient can be measured on an instrument called an *eikonometer.*

Lenses which compensate for this difference can be ground and are called "size" lenses. These are really miniature Galilean magnifiers (or minifiers) with a very small (in millimeters) separation between elements — this separation, however, is one factor which helps create the few percent change in image size which may be desired. (In the "size" lens, the space separating the front and rear dioptric elements is not usually filled with air but is composed of the lens material itself.)

The second factor helping to provide the magnification given by a "size" lens stems from its front and back surface powers which can be varied to change the "shape" of the lens. Thus, even when *no* "corrective" dioptric power is necessary, a simple "size lens" can be made. One such lens is diagrammed below.

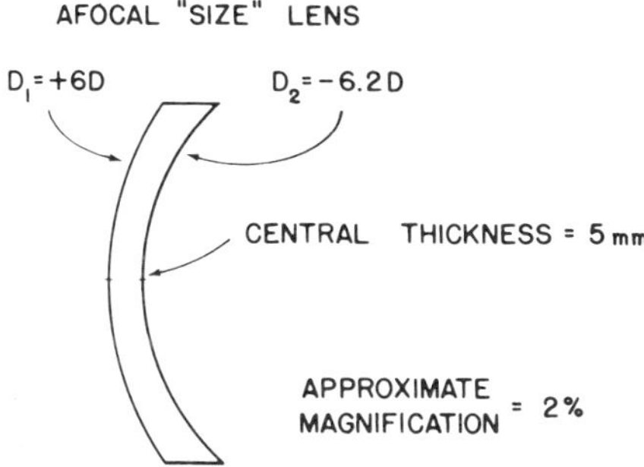


To give you some idea of the range of magnifications allowed by such a lens, I have constructed the small table below. From this table you can determine that a lens of 5 mm thickness and with a front surface power of + 6 D (the rear surface power would have to be about − 6.2 D for an *afocal* lens), would yield about 2% magnification.

*TABLE IV*

*MAGNIFICATION IN %*

| Lens Thickness (mm) | $D_1$ (Front curve in diopters) + 3 D | + 6 D | + 9 D |
|---|---|---|---|
| 1 | 0.20 | 0.40 | 0.60 |
| 3 | 0.60 | 1.20 | 1.80 |
| 5 | 1.00 | 2.01 | 3.00 |

*(For the mathematical buff only* — The M in % $= D_1 \times \frac{t}{n}$

$D_1$ = diopters of front surface power

$t$ = thickness in *cm*

$n$ = index of refraction of the glass)

This table shows us that the *shape* of a lens (as governed by the steepness of its front surface) and its *thickness* influence its magnifying power, just as does the *vertex* distance between the lens and the eye; the latter, however, is more significant in everyday ophthalmic optics. The shape and thickness effects are simply a manifestation of "thick-lens" optics on which I do not wish to elaborate further.

We already mentioned the 4 X magnification an aphake might obtain with a + 3 D lens held about 25 cm from his cornea; however, he must give up something to obtain that magnification with this "corrective lens" and that something is "field of view".

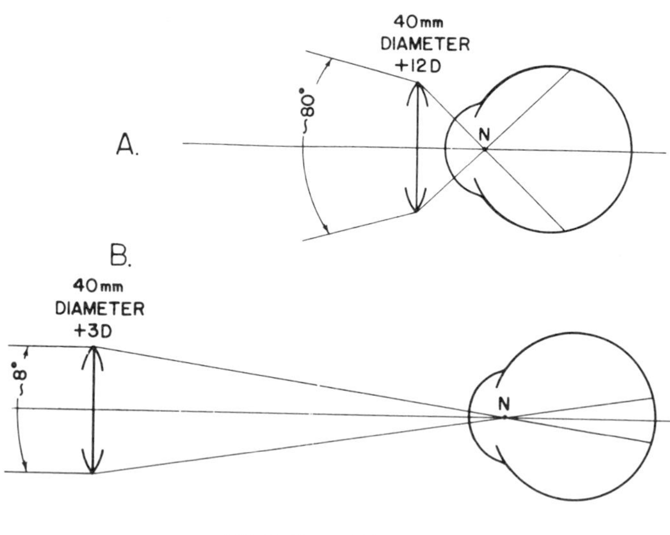


FIELD OF VIEW

As can be seen from the above diagrams, when the corrective lens is in the spectacle lens plane, the field of view (that area of space that is visible through the lens) is much greater than when the lens is situated at a further distance. In the first instance, the field of view is about 80°, whereas in the situation which provides a magnification of 4 X, the field of view shrinks to about 8°.

Our study of "neutralization" of lenses and Galilean optics has led us through a consideration of the retinal image size changes occurring in the correction of myopia and hyperopia. Astigmatic correction with the lens situated in the spectacle lens plane presents the same type of problems; the presence of a vertex distance here also leads to magnification or minification. The difficulty is that this size change is not even — there is a differential, meridional magnification which occurs with the size extremes being in the primary meridians.

The *meridian* with more plus *corrective* lens power creates a greater amount of magnification than does the lesser powered meridian. To visualize this, recall the cross-section of a Galilean system composed of a *plus* corrective lens and a *minus* eye "error"; say the plane of that section was taken in the lens meridian of *greatest* plus power which must also be the meridian of greatest minus error. You should see that this arrangement will yield a *magnified* retinal image in this meridian which must be larger than that in *any* other meridian of this same eye. So, if the corrective lens power in the *vertical* meridian happens to be greater in plus than any of the other lens meridians, there will be *vertical* elongation of the image seen by the patient.

Take another example of a corrective lens of plano $+ 4 \times 90$. Here the maximum corrective plus power is in the 180° *meridian.* Therefore, the horizontal dimensions of the image will be elongated. If the object is a square, the retinal image will look like a horizontal rectangle.

I will not explain this further here but you can prove it to yourself. Place a $-5 \times 90$ cylinder close to your eye to simulate an astigmatic eye's "error". "Correct" this error with a $+ 4.00 \times 90$ held at a "vertex distance" of about 5 cm. You should see that the *horizontal* dimension of any object (say a square) is elongated. The greater the distance the "corrective" lens is held, the more this horizontal magnification is.

Whenever the primary meridians are at 90° and 180°, rectilinear square objects may appear elongated sideways or up and down, depending on the corresponding powers. However, this type of apparent distortion is usually *not* very bothersome to a patient. Even when the corrective lenses have oblique axes (which may cause right angled corners to look somewhat acute or obtuse) patients become very tolerant within a short adapting period of a few days or weeks.

CLINICAL POINT:

A problem may arise, however, when the primary meridians are oblique *and* different in the two eyes, thereby giving a different "tilt" to an image line for each eye; this would necessitate a cyclofusional eye movement to keep from seeing diplopically. You can help reduce this "tilting" type of distortion by *under*correcting the cylindrical power somewhat; this technique is probably better than attempting to "straighten" the distortion by rotating the corrective axis toward the horizontal or vertical meridians (a common practice), since the latter method tends to introduce not only a resultant cylindrical error but also adds a spherical one, as we have seen. (Remember what happens when you combine unlike cylinders off-axis to one another?) However, both methods do work.

## Astronomical Telescope

The Galilean telescope and the principles of its operation are closely related to ophthalmic optics and the correction of refractive error; however, this instrument is not the only *afocal* system providing angular magnification. There is another, extremely useful one called the "astronomical telescope" which produces an *inverted,* though magnified image; otherwise, the optics are very similar to the Galilean system's.

## ASTRONOMICAL TELESCOPE

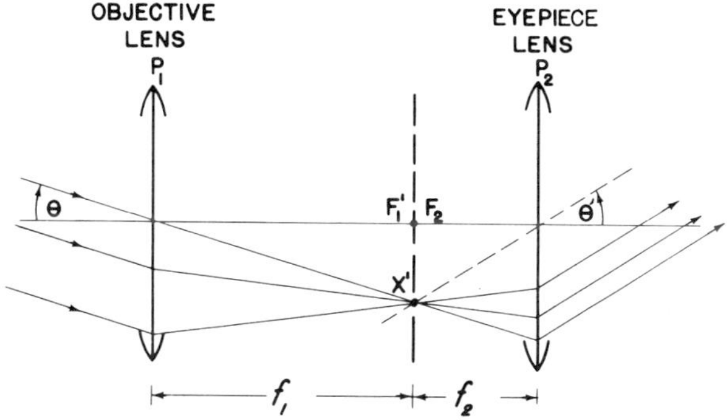


The front (*objective*) plus lens $P_1$ of this magnification device creates a real image in its secondary focal plane ($F_1'$) of an object at infinity, just as it did in the Galilean system. But here the second lens $P_2$ (called the *eyepiece* lens) is *plus*; it is placed so that *its* primary focal plane $F_2$ coincides with the $F_1'$ plane of the objective lens. The image point X′ formed by $P_1$ will now act as an object point for $P_2$. Since X′ is in the primary focal plane of $P_2$, its final image will be at infinity. As in the Galilean telescope, the direction (angle $\theta'$) that the final parallel bundles of light make with the axis is given by a line drawn from X′ through the nodal point of $P_2$ (which is at its axial vertex). At the nodal point of $P_1$, $\theta$ is the angular size of the object.

The angular magnification power of this telescope is $\frac{\theta'}{\theta}$; but as can be seen from the diagram $\theta'$ and $\theta$ are in the opposite directions; the object denoted by angle $\theta$ is above the axis, while the image angle $\theta'$ points to the corresponding image X′ located *below* the axis — this is in contrast to the Galilean system where both the object and image were in the same direction. So, $M = -\frac{\theta'}{\theta}$.

To simplify the above diagram (still considering small angles) look at the figure below:

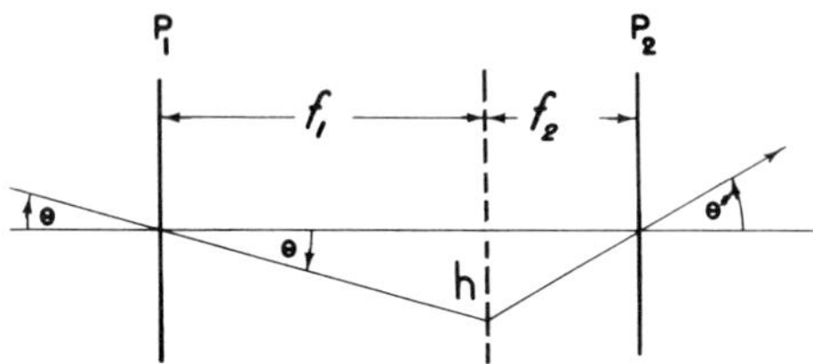


$$\theta = \frac{h}{f_1}$$

$$\theta' = \frac{h}{f_2}$$

Since $\theta'$ is counterclockwise here,

$$M = -\frac{\theta'}{\theta} = -\frac{f_1}{f_2} = -\frac{P_2}{P_1}$$

Notice this is the same expression as that given already for the power of the Galilean telescope! Thus, the Magnification of *any* telescope is simply given by the relationship $-\frac{\text{power of the eyepiece}}{\text{power of the objective}}$.

From this, it should be clear that the *same* magnification power can be provided by *any* pair of lenses that yield a constant ratio of powers; a 3 X telescope can be made up of a $P_2$ of + 9 D and a $P_1$ of + 3 D, or a $P_2$ of + 21 and $P_1$ of + 7. (The distance between the lenses will necessarily be shorter in the latter pair.)

The astronomical telescope principle is utilized clinically in the indirect ophthalmoscope. Also, this same principle (modified somewhat to allow the real object to be very close to $P_1$) is the basis for the standard light microscope.

# The Effect of a Telescopic Magnifier on Accommodation

Both the astronomical and Galilean telescopes are afocal systems, meaning light from a distant object emerges from the instrument in parallel bundles. Since this image vergence is zero, there is no accommodative demand for an eye looking at infinity through such a telescope. What happens to the image seen by an eye if the *object* approaches a Galilean system?

PROBLEM:

Assume an object is located 20 cm from the front (objective) lens of a 2 X Galilean telescope where $P_1 = +10$ D, $P_2 = -20$ D, and the separation is 5 cm. What is the accommodation required to see the object through this telescope compared with the accommodation demanded by the same object located at the same distance *from the eye* (25 cm) without any telescope?

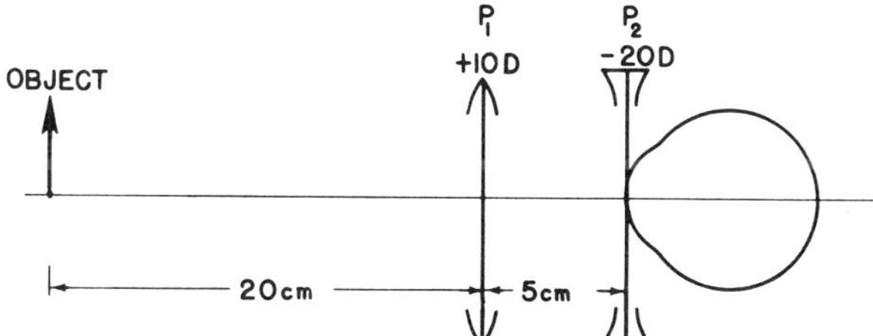


ANSWER:

The object vergence at $P_1$ is $-5$ D.

$$U_1 + P_1 = V_1$$
$$-5 + 10 = +5$$

The image will be 20 cm to the right of $P_1$.

Since $P_2$ is located 5 cm to the right of $P_1$, the object *distance* for $P_2$ is 15 cm to the right (convergent).

$$U_2 = +\frac{1}{.15} = +6.67 \text{ D}$$
$$U_2 + P_2 = V_2$$
$$+6.67 - 20 = -13.33 \text{ D}$$

The final image, then, is located $\frac{1}{-13.33\ D}$ or 7.5 cm in front of $P_2$. So, if an eye is situated directly behind $P_2$, it sees a magnified image of the original object, but it must exert *13.3 D* of accommodation to see it clearly! (*Without* the telescope it would require only *4 D* accommodation since the object is 25 cm from the eye.) Through the telescope, a magnification at near has been gained, but only at the expense of a tremendous stress on the accommodation required to see it.

For practice try another example, taking the same object 25 cm from the eye but making the 2 X telescope *shorter.* This is done by using an objective lens of + 25 D and an eyepiece of — 50 D with a 2 cm separation.

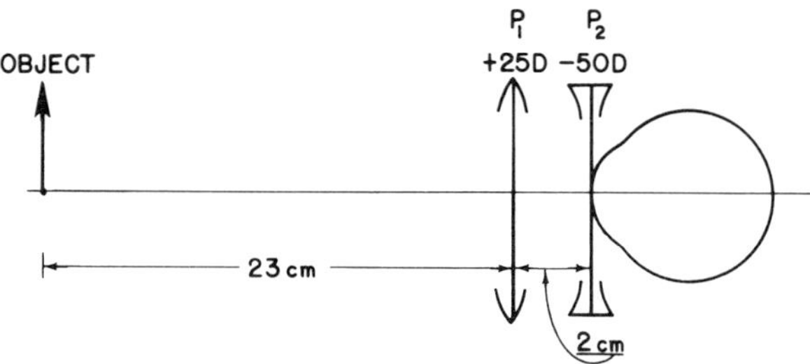


Now,

$$u_1 = 23 \text{ cm}$$
$$U_1 = -4.35 \text{ D}$$
$$U_1 + P_1 = V_1$$
$$-4.35 + 25 = +20.65 = V_1$$
$$v_1 = +4.85 \text{ cm}$$

The distance between $P_1$ and $P_2$ is 2 cm, so $U_2 = (4.85 - 2.00)$ cm = + 2.85 cm.

Therefore,

$$U_2 = \frac{1}{.0285} = +35 \text{ D}$$
$$U_2 + P_2 = V_2$$
$$+35 - 50 = V_2$$
$$-15 = V_2$$

The final image will require 15 D of accommodation for an eye located directly behind $P_2$.

As the telescopic lenses are moved closer together (this, of course, necessitates a change in the respective lens powers so as to maintain the 2 X telescopic power for infinity), you can visualize that eventually they will form an "infinitely thin" 2 X magnifier; in that instance, the accommodation required through it to see an object at 25 cm will be *16 D.*

Though I will not derive it, we have come to another key relationship: the accommodation required to see any point through a Galilean telescope is proportional to the *square* of its angular magnification power. But that's not all; the *total* accommodative demand depends not only on the power of the magnifier but also, naturally, on the position of the object. In our example, we were looking at a point which normally required 4 D of accommodation. So the total accommodation necessitated by a 2 X telescope is $4 \times (2)^2 = 16\ D$ (as given above). The general relationship then is as follows: total accommodation through telescope = (Normal Accommodation required) $\times$ $(\text{magnification})^2$. For viewing at a 33 cm distance through a 1.5 X telescope, you would have to supply $3 \times (1.5)^2$ or 6.75 D of accommodation.

CLINICAL POINT:

What you have just been shown regarding the accommodative demand applies to telescopes which are focused for objects at infinity. A telescopic system *can* be built which allows adjustment for *proximal* objects (as by focusing of the eyepiece, which can also compensate for any ametropia existing in the viewer's eye). This can lessen the strain on accommodation. Specifically for use at near, there are special telescopes available where the integral lenses are set for a close focus; through them, no accommodation will be required to see a near object located in the proper (pre-set) plane. This is the method utilized in the binocular magnifying loupe you probably use in the operating room. It is a small 1.5-2.5 X telescope (only 1-2 cm thick, like that shown in the last diagram) for use at a relatively fixed near working distance. If coupled with some accommodation supplied by the surgeon, it allows a useful *range* of clear vision for eye surgery.

Fact 1: Any afocal Galilean system which produces *magnification* will demand that a greater amount of accommodation be exerted to see a given object than would be required to see that same object sans magnification. (The opposite is true for a minifying Galilean system).

Fact 2: A "corrected" hyperope always has a magnified retinal image present, and the magnification *increases* as the appropriate corrective lens is placed further from the eye.

We should be able to make some predictions by tying these facts together. Since a hyperope who wears a corrective spectacle lens will have a greater image magnification than if he uses a contact lens, it stands to reason that through the corrective lens, *more* accommodation will be required to see a near object than through the contact lens. For low refractive errors, this increase in accommodative demand is not significant. However, when the hyperopic error is greater than 4 or 5 Diopters, the difference can be important, and simply switching the correction from contact lenses *to* spectacles can cause a borderline presbyope to tip over and encounter full-fledged presbyopic symptoms because of the increased amount of accommodation required.

With the myope, the converse is true. A corrected myope tends to have some minification of his retinal image. The further away his corrective lens is from his eye, the greater the minification and the less the demand on his accommodation. So, if a pre-presbyopic myope (for some peculiar reason) decides to switch to a contact lens correction, he will probably find that he cannot read at 20 cm with the ease that he could prior to wearing the contact lenses, especially if his error is greater than 4 D. The higher myope may even require reading glasses to supplement his contacts whereas he did not need them before.

The increased stress placed on accommodation by any magnifying system is the basis of both these clinical experiences.

We should now examine a concrete, reasonable example to observe the *magnitude* of the stress on the accommodation of a corrected myope who switches to contact lenses.

PROBLEM:

A fully corrected myòpe wears a — 10 D spectacle lens 15 mm from his cornea. How much accommodation is required through his spectacle lens compared to the amount required through a corrective contact lens?

ANSWER:

The — 10 D corrective lens tells you that the far point is 10 cm away *from the lens.* The proper corrective contact lens located on the cornea would have a focal length of (10 cm + 1.5 cm) or 11.5 cm; its dioptric equivalent is — 8.7 D.

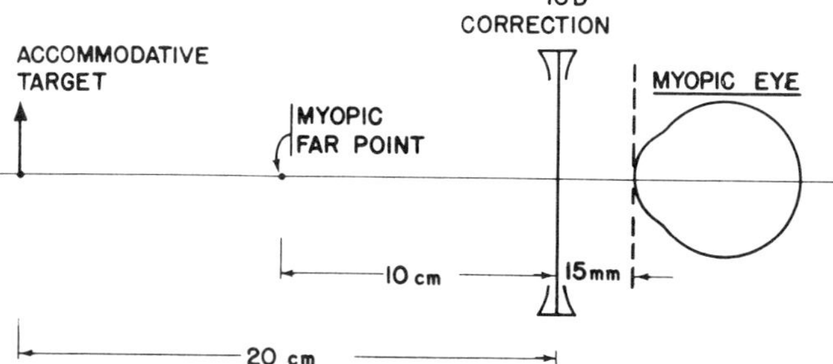


What we are asked for is where the *final* image of an object located 20 cm from the *spectacle* lens plane is placed by each of the two "corrective" lenses — the — 10 D and the — 8.7 D.

---

With a *spectacle* correction:

$$U_1 + P_1 = V_1$$

$$-5 - 10 = V_1 = -15 \text{ D}$$

$$v_1 = -6.67 \text{ cm}$$

$$v_1 + 15 \text{ mm} = u_2 = 8.17 \text{ cm}$$

Thus, as far as the eye is concerned, it must accommodate on an image which is 8.17 cm away; so, the final image presented to the eye has a vergence of — 12.25 D.

With a *contact lens* correction:

The distance from the object to the contact lens is $u_1$.

$$u_1 = 20 + 1.5 = 21.5 \text{ cm}$$

$$U_1 = -4.65 \text{ D}$$

$$U_1 + P_1 = V_1$$

$$-4.65 - 8.7 = V_1$$

$$-13.35 = V_1$$

Thus, the final image presented to the eye has a vergence of — 13.35 D.

---

Since, in both instances, the eye is myopic 8.7 D (referred to the cornea) it is relieved of the responsibility of accommodating to the extent of the first 8.7 D. So, with the — 10 D *corrective* lens, this eye must accommodate (12.25 — 8.7) or 3.55 D to see the object located 20 cm from the spectacle plane; whereas the eye corrected by a *contact* lens must accommodate (13.35 — 8.7) or 4.65 D — that is, 1.10 D more!

(Notice from this example we assume that the contact lens-corrected myope must exert the same amount of accommodation as the emmetrope would. However, even *with* a contact lens, the myope will likely have to exert somewhat *less.* This discrepancy arises because, in our model eye, we neglected the distance behind the cornea that the myopic eye error truly "resides"; because of this oversimplification then, we find that the contact lens corrected eye is equivalent to the emmetropic one.)

CLINICAL POINT:

You are performing the finishing touches on a subjective refraction of a patient. If, after you arrive at his full correction, you still push on and continue to add minus lens power, the patient will tell you that the target letters seem to become smaller. Why?

The *extra* minus power you have added stimulates the patient's accommodation. That *increase* in accommodation can be looked upon as an increase in the plus power "built-in" his eye. These two dioptric powers (minus outside, plus inside) *must* "neutralize" each other exactly if vision is to remain clear; and so, we have created an additional "neutralized" or afocal telescopic system with the two elements separated by the existing *vertex* distance (plus some *more* distance within the eye to the site of the built-in plus). Since here, the "built-in" lens (which corresponds to the telescope's "eyepiece") is plus, this Galilean system is a *minifying* one for this eye.

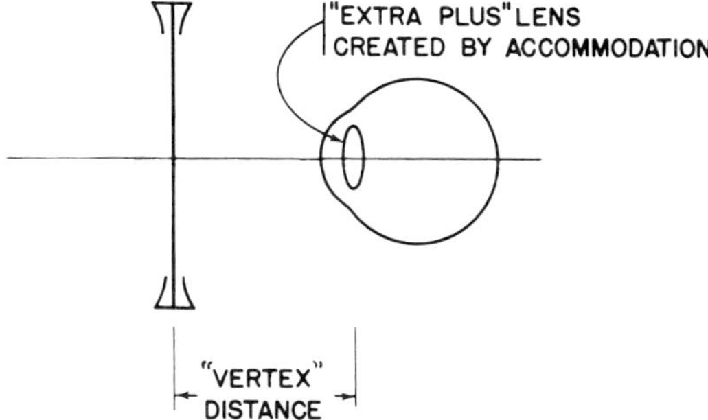


So, placing extra minus in front of an eye forces a patient to peer through a small, reverse telescope. The greater in minus the overcorrection is, the greater the accommodation required to keep the letters clear, and therefore the more the minifying effect of that telescope.

## MAGNIFICATION

So far, we have glibly tossed around the various *types* of magnification. We have mentioned *linear* (lateral) magnification in connection with basic lens imagery and introduced *angular* magnification when dealing with Galilean and astronomical telescopes. We have even implied the presence of a third type — called *axial.* All are often confusing, so now I want to describe each type separately and discuss it briefly so you could get the proper "feel" for all of them.

### Linear Magnification

Linear magnification has to do only with the sizes of images relative to their corresponding objects — not with how big images *look* to an eye, but how big they *are.* The image formed by a lens system can be larger, smaller or the same size, and it can be erect or inverted when compared to the original object. As long as we deal only in the relationships between the actual sizes (meters, inches, or microns) of objects and images, *linear* magnification is the term that applies.

If we know the linear magnification is $\frac{1}{2}$ X, we know the image is half as long and half as wide as the object (the measurements taking place in the planes perpendicular to the axis of the optical system).

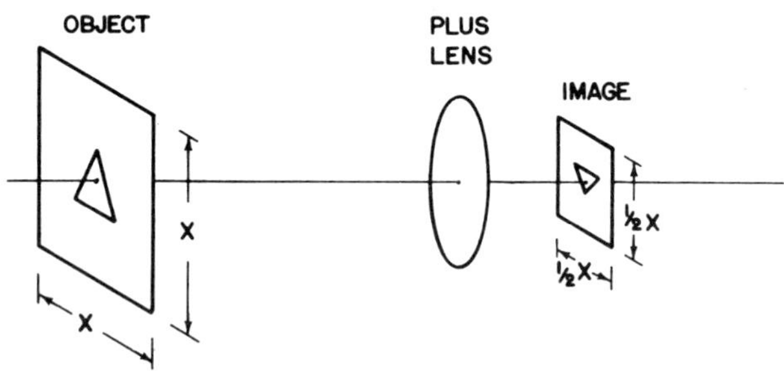


LINEAR (LATERAL) MAGNIFICATION

But *linear* magnification has no real meaning when dealing with objects at *infinite* distances. Since the linear magnification = $\frac{\text{image size}}{\text{object size}} = \frac{\text{image distance}}{\text{object distance}}$, if the *object* distance is infinite, the linear magnification must be zero. When the *image* distance is infinite (as when an object is placed in the focal plane of a simple plus lens), it makes no difference what the *actual* object size is, the image's size (since it is located at infinity) must be infinitely large and, therefore, the *linear* magnification is infinitely large also. In both these instances, the term linear (also called *transverse* as well as *lateral*) magnification has no usefulness.

In optics and ophthalmology we need a way to deal *not* with absolute sizes, but with apparent sizes; how big or small an object or image *looks* to the eye. Angular size and angular magnification provide us with this means.

## Angular Magnification

Angular size has to do with how big an image *looks* to an eye. The same fixed size of object or image will appear larger if you approach it and smaller as you recede. The image on your retina is magnified angularly (that is, it subtends a larger angle at the eye's nodal point) as an object is brought closer. But, it is only *magnified* angularly in relation to some other angular size — that given by that same object at some fixed reference distance. For example, say an object located at an arbitrary reference distance (say 20 ft.) subtends 5° on my retina; when it approaches to half its previous distance (10 ft.), its angular size on my retina has increased to 10° — a 2 X angular magnification has occurred. If instead, the object recedes to twice its original distance (to 40 ft.), its angular size on my retina has shrunk to 2.5°, and the angular magnification has become only ½ X. But this is only with reference to its angular size at the 20 ft. distance. So, it should be obvious that in order to state a magnification figure, you *must* give (or assume) a reference distance; the *change* in appearance from that reference angular size *is* the angular magnification provided by any optical device. This point is so important and yet it is often misunderstood.

When the object is at *infinity* and an angularly magnified image is supplied by a telescope, there is really no need to specify any other fixed reference distance to calculate the angular magnification since the angular size of the *object* at infinity can serve as *the* reference. However, for magnifying objects close-by, we *do* need a reference distance, as you will see. Let's examine how a simple plus lens works as a magnifier.

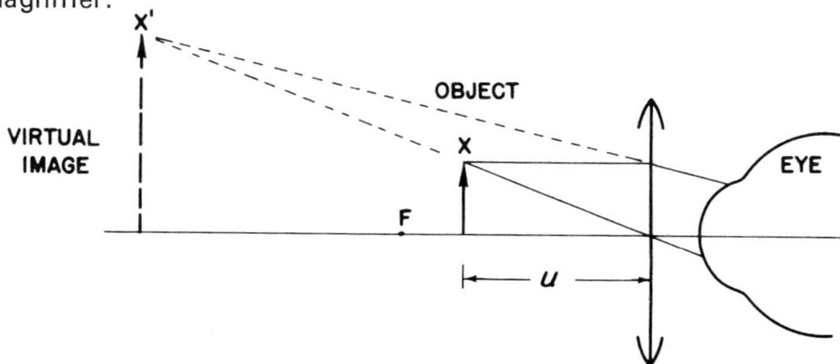

Whenever any object is placed just *inside* the primary focal point of a simple plus lens, a magnified virtual, upright image will be formed. An eye situated behind that lens will see the image. If we fix the object distance $u$ from the lens, the image size and position will also be fixed. Any eye movement relative to the lens cannot influence the *actual* image size; however, its apparent (angular) size *may* change, just as moving closer or farther from any real life object would change its retinal image size. (As the object in the figure below moves closer to the eye, its *angular* size increases on the retina ($\theta' > \theta$) though its *actual* size stays the same.)

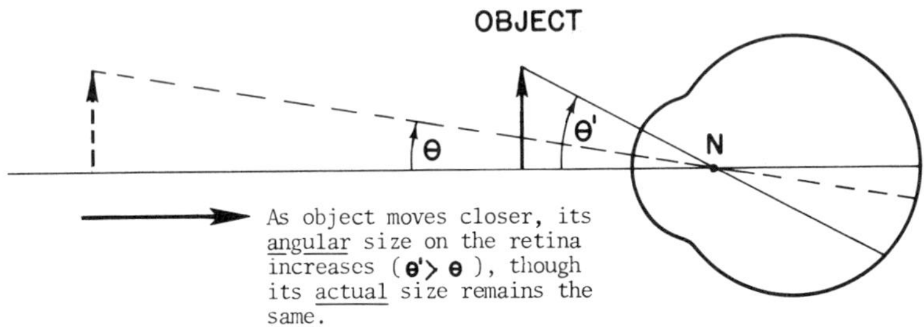


I want particularly to emphasize the word *may* (influence the apparent size) because it *may not* also! Say the image is located at a great distance away, that is, if object distance $u$ is equal to the focal length of the simple plus lens; now if the eye moves closer or farther from the lens (mm or yards), the *angular* size of that image on the retina will *not change.* It would only change if your eye moved an *appreciable* distance away from the lens as compared to the image distance, just as, for example, if you were five miles away from a mountain and moved a few meters closer to it; this would not really increase the visual size of that mountain. To accomplish that, you'd have to move say ¼ to ½ mile closer. Only then might you notice any increase in its angular size on your retina. Similarly here, the image is at infinity and you would not observe any shrinkage in its angular size as you receded from it while keeping it in view through the lens.

As shown in the diagram below, $\theta'$ is the angular size of the *object* subtended at the lens; $\theta'$ is also always the angular size of the *image* subtended at the lens.

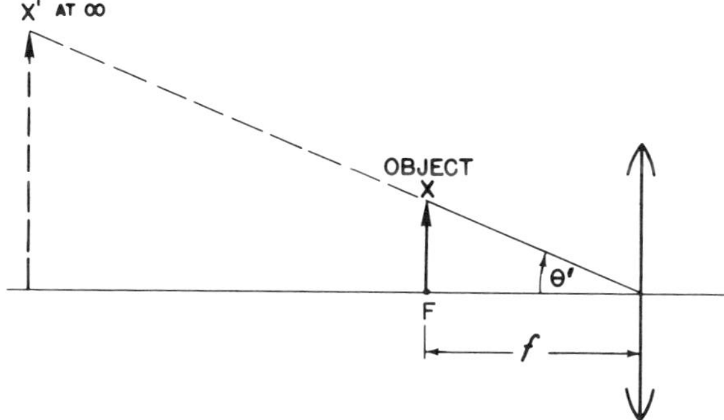


When the object is at F of the plus lens magnifier, the angular size of the *retinal* image created will also be $\theta'$. How do we know this?

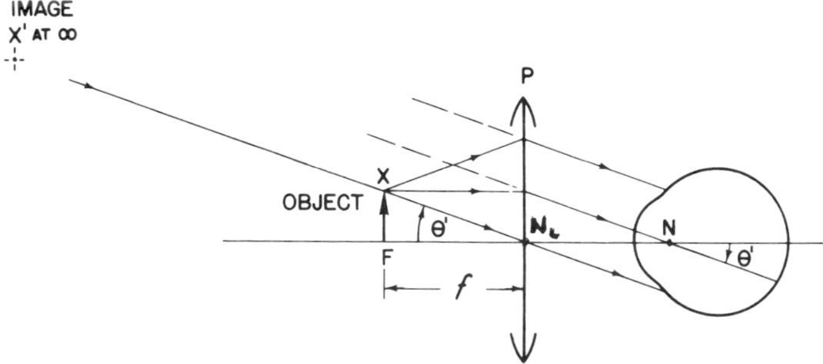


Well, you should be able to see that *all* rays from the object tip X will, in the image space of the lens, be parallel and must all cross the axis at the same angle as that ray which goes through the *lens'* nodal point;

that angle is $\theta'$. Since *one* of those image rays will pass through the *eye's* nodal point, the angular size of the retinal image will be $\theta'$ too! And, it will always be $\theta'$ unless the eye is so far back that light from X which passes through lens P cannot enter the pupil — a situation which depends on the *diameter* of the lens and not its power.

The angular size of $\theta'$ does depend on the size of the object and on the focal length $f_1$ of the lens. Study the last diagram; as $f_1$ gets shorter, the retinal image size $\theta'$ must get larger; that is, $\theta'$ is inversely proportional to *f*, or stated more clearly, the greater the dioptric power *P* of the lens, the greater $\theta'$ will be. Remember that we are still considering only small angles, so tan $\theta' = \theta'$; therefore,

$$\theta' = \frac{\text{XF}}{f} = \frac{\text{object size}}{f} = \text{(object size)} \cdot \text{(Power of the lens)}.$$

But to determine how much "magnification" lens P offers us, we must be able to compare *this* retinal image size $\theta'$ with some *other* one having to do with the same object. Remember, magnification is a relative term which states that a retinal image is bigger or smaller than *something.* So our task is to find a reference distance at which to place that same object, determine its angular size in that instance, and compare that angular size with the $\theta'$ created by lens P. You *could* put that object anywhere you wanted and obtain some magnification "number". *Where* to put that object is what must be agreed upon (not calculated, but arbitrarily picked) as a standard.

What that standard distance should be has been argued about for years, but for most purposes 25 cm has been agreed upon and is used here and in most other textbooks. However, you should know that there are certain situations where 40 cm is a better standard. The specific letter sizes of Louise Sloan's reduced-vision acuity charts, for example, are calibrated in M (magnification) units for a 40 cm reading distance. At that position, a patient with poor acuity is asked to read. If a given sized print can just barely be read, the M unit label for that sized print "tells" an examiner directly the power of a low-vision aid which might be a useful one to try first for that patient.

But as stated, the *standard* reference distance is 25 cm. An object on the visual axis would subtend angle $\theta$ when it was placed 25 cm from an eye. (See next figure.)

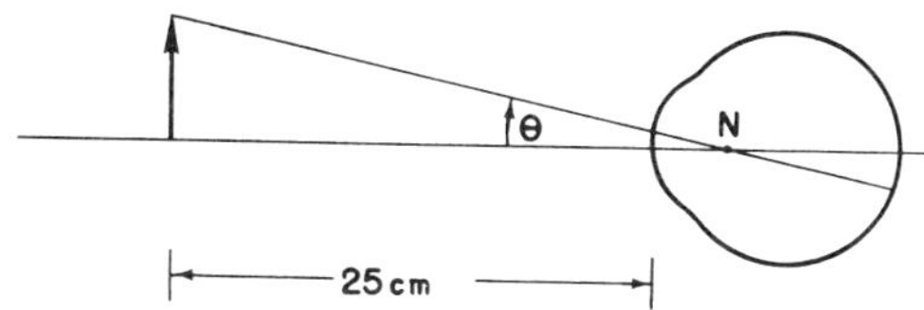


We can return now to the angular magnification provided by lens P — the simple plus lens magnifier. With small angles:

$$\theta' = (\text{object size}) \cdot (\text{Power of the lens})$$

and
$$\theta = \frac{\text{object size}}{25\text{ cm}} = \frac{\text{object size}}{.25\text{ meters}}$$

$$M = \frac{\theta'}{\theta} = \frac{\text{object size} \cdot (P)}{\frac{\text{object size}}{.25}} = \left(\frac{1}{4}\right) P$$

and the magnification is seen to be independent of what the object size actually is.

The ¼ represents our chosen standard distance; if *another* distance *d* is used as a reference, the magnification then would be $P \times d$ (in meters).

So we have found that the angular magnification of a simple plus lens magnifier $= \frac{P}{4}$; so, a + 8 D sphere is called a 2 X magnifier, and a + 20 D sphere is called a 5 X magnifier.

The same reasoning applies to the eye itself. If an eye, *without* any magnifiers, looks at any object located at our reference distance of 25 cm, it also would obtain a *unitary* (1 X) angular magnification. If that same object is now brought closer, to 12.5 cm, and if that eye can accommodate the 8 D required to see it clearly, the angular size on the retina will be doubled and the angular magnification due to this proximity and to the required accommodation will become 2 X. Thus, it is clear that an increase in the accommodative response of an eye can add to the magnification supplied by a simple plus magnifier.

With a 25 cm reference distance, the magnification given by the eye's accommodation is in a ratio of $\frac{4\text{ Diopters}}{1\text{ Mag unit}}$, the same as that provided by a lens; so, the magnification due to accommodation *adds directly* to that due to the lens itself. To determine the *full* magnification possible with a simple lens, we use the *general* expression which takes into account the eye's accommodative ability; that expression is $\frac{P + A}{4}$. In our previous example, then, a + 8 D (2 X magnifier) lens, when coupled with an eye which has, say, 8 D of accommodative amplitude, will provide a total possible magnifying effect of $\frac{8 + 8}{4}$ or 4 X.

CLINICAL POINT:

The important clinical implication of the above is that a 10 year old child with poor (say 20/100) vision at 20 feet will rarely require much in the way of a high plus simple magnifier (as that given by a high bifocal add) for reading. Accommodative amplitude is usually ample to allow him to bring reading material close enough to provide a sufficiently enlarged retinal image without requiring any further help from a lens. Kestenbaum pointed out that a good approximation of the near add required by a patient with reduced vision is the reciprocal of his visual acuity at distance. An acuity of $\frac{20}{100}$ would require $\frac{100}{20}$ or a + 5 D add to aid in near work.* For the child in the above example, with sufficient accommodative power, this add would probably be superfluous, as explained.

* * *

Previously we showed that moving the eye back from the plus lens magnifier will *not* change the angular magnification power of the lens and will not change the image size on the retina (assuming the object is in the anterior focal plane of the magnifier). But something does change as the eye moves back, and that is the field of view. The extent of the image which is visible to the eye is maximal when the eye is in contact with the lens but decreases as the eye recedes.

---

* That is, to read that size of print which is about the size of pica typewriter type — the 14/42 letter size on the 14" near-acuity card.

CLINICAL POINT:

The direct ophthalmoscope, our handy-dandy gadget for examining the interiors of the eye, makes use of the eye's dioptric components as a simple magnifier. The ophthalmoscope itself is only a flashlight, arranged so that you can sight just alongside the light beam. All those lenses on its dial are only to compensate for your refractive error, the patient's refractive error, any amount of accommodation you exert (but shouldn't) and the height or depth of any fundus lesion you are examining. It is very likely all these will sum to zero and you will need no supplementary lens in your ophthalmoscope dial to see the fundus clearly. Then, you are using the patient's cornea and lens powers combined as a plus magnifier.

We learned that the dioptric power of an emmetropic eye is roughly 60 D. Thus the angular magnification of fundus detail is $\frac{P}{4} = \frac{60}{4} = 15$ X; that is, the optic nerve head (which is in reality about 1.5 mm in diameter) appears 15 times larger through the ophthalmoscope than it would look if it were situated 25 cm from your eye with the lens and cornea removed.

* * *

Our psychological apparatus is such that we are enabled to substitute a change in an object's angular *size* for a change in its *distance* from ourselves, and, we can do so rather freely, but within limits. Thus, when the image size of an object which *can* exist in various sizes is increased on the retina, we can perceive either that the object itself has enlarged or that it came closer to us; usually, we "feel" the latter has occurred. But, if we *know* that its distance couldn't change, the increase in angular size will be interpreted as the object enlarging.

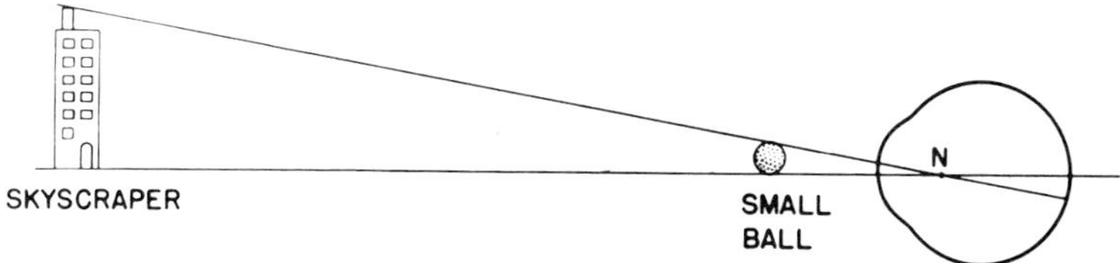

As shown in the above diagram, the angular image height on the retina of both the ball and the skyscraper is the same. Since a ball can come in any size, its retinal image size does not psychologically fix its distance away and the ball may be "seen" either as a small one up close or a large one farther away. A building, however, is *known* to be of a certain size, therefore, it is the *distance* that becomes the variable.

With both the skyscraper and the ball, I am assuming that no other cues are present which help set the distance more accurately — such as shadow, color, overlap, etc.* Any of these cues (as well as your "mental set") can tip the scales towards your making a clear choice as to whether it was distance or size that *really* changed. Typically, telescopes and binoculars seem to bring objects "closer" since the objects visualized are familiar ones whose size is known. Microscopes on the other hand make objects appear "bigger" since there is no such familiarity with the object's true size. All these instruments, of course, can only enlarge the size of the retinal image of the object.

So, angular magnification (in contrast to linear magnification) involves the eye or a camera, and the true size of the image matters only indirectly, since it is only *one* of the variables, the other being its *distance.* The term "angular magnification" takes both variables into account in one datum which hides the *actual* value of either. Your personal judgment of the apparent magnitude of either will depend on the balance of many psychological factors, but for our subject "angular magnification", it is only the *angular size* that counts.

## Axial Magnification

There is yet another type of magnification which spills over into both the linear and angular types; this is *axial* magnification. It is not usually discussed in the standard optics text. I am mentioning it here so that you can be aware of its presence and of its particular significance, especially in relation to indirect ophthalmoscopy.

You have been shown that with a telescope the stress on an observer's accommodation is proportional to the *square* of the *angular* magnification provided by the instrument. This is a manifestation of

*Rubin and Walls: *FUNDAMENTALS OF VISUAL SCIENCE, pp. 359-403.*

how a small change in the object's axial (fore and aft) distance (that is, as it comes closer) causes a *magnified* axial shift in the position of the final image; thus the final image appears to be *much* closer and thereby demands much more accommodation. *This* is axial magnification.

Even when dealing with *linear* magnification (in the figure below) ($M = \frac{\text{image size}}{\text{object size}} = \frac{y}{x}$), *axial* shifts ($\Delta x$) of the object will create *axial* shifts ($\Delta y$) in the image.

AXIAL MAGNIFICATION

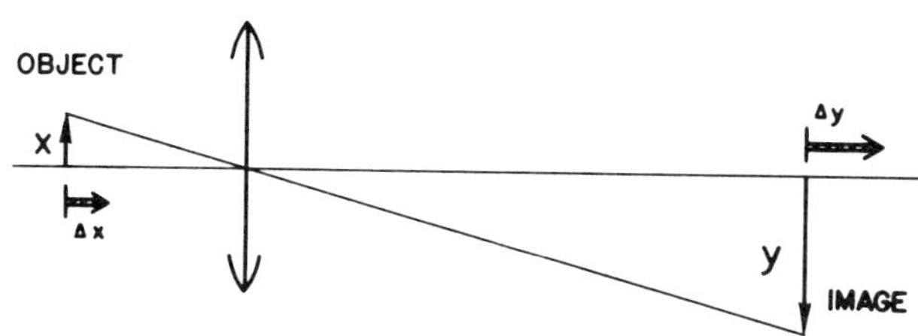


These shifts ($\frac{\text{axial image shift}}{\text{axial object shift}}$) are proportional to $(\frac{y}{x})^2$, that is, to the square of the linear magnification. (This is detailed in one of my previous articles and will be summarized later in the section on the indirect ophthalmoscope, page 292.) So, the *axial* magnification is related to the square of the *linear* magnification* just as it is to the square of the *angular* magnification.

Now you shouldn't be caught unaware when you see the term *axial* magnification again.

---

* This is not quite true when one deals with a refracting *surface,* which, of course, has different indices of refraction for the object and image spaces. For the refracting surface, the axial magnification $= (\frac{n}{n'} \times \frac{v}{u}) \times \frac{v}{u}$. (Call this relationship "Formula One.") If you will recall, I showed (in Appendix A) that, with a refractive surface, the linear (lateral) magnification *M* was equal to $(\frac{n}{n'} \times \frac{v}{u})$. So, by substitution in "Formula One," $M_{axial} = (M_{linear}) \times \frac{v}{u}$. But, with a lens in air, $M_{linear} = \frac{v}{u}$. So, for the *lens,* $M_{axial} = (M_{linear}) \times (M_{linear}) = (M_{linear})^2$. This is what is stated in the text above.

# THE PRISMATIC EFFECT OF LENSES

It's been some time since we studied something about prisms and their ability to change the direction of light beams. It should come as no surprise to learn that lenses also exhibit *prism* power because they also change the direction of light rays, but their ability to do is not constant. Any light ray that falls on the surface of a lens is subject to a constant *vergence* effect by that lens; however, the *prismatic* effect by that lens on any incoming ray varies, depending on specifically where on that surface the ray strikes.

Let's take a concrete example: The light coming from an axial point at infinity falls onto a lens of + 2 D power. The rays will be brought to a focus 50 cm behind the lens at its secondary focal point F'.

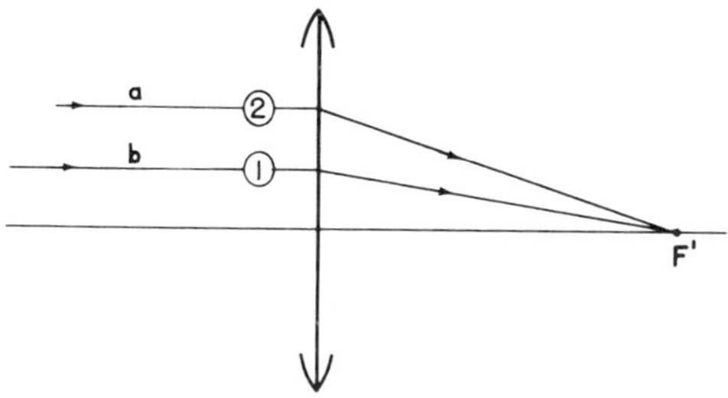


Since all parallel incident rays will focus at F', each ray must have been bent a different amount by the lens; those rays striking the lens close to its axis will be deviated only slightly while those hitting the lens nearer to its periphery will, of necessity, be bent a greater amount if they are to cross the axis at F'. The degree of bending (or deviation) exerted by the lens *at* the point that an incident ray touches the lens is its "prism power". So, the prism power of the lens at point 2 affecting ray (a) must be greater than that affecting ray (b) which strikes the lens at point 1.

To calculate how much prism power is exerted by the lens on these 2 rays, assume that point 2 is 2 cm from the lens axis and point 1 is 1 cm from the axis. Since F′ is 50 cm from the lens, ray 2 is deviated from the straight-ahead direction by such an angle that it is displaced 2 cm at 50 cm as shown below:

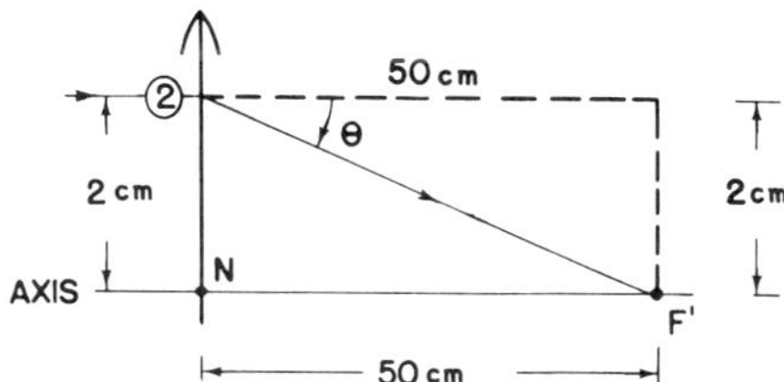


The tangent of the angle of deviation $\theta$ must be $\frac{2}{50}$ or .04. Back when we described how to measure angles, we noted that 100 times the tangent of an angle is equivalent to that angle expressed in prism diopters. So, angle $\theta$ must be 4 prism diopters.

The next diagram analyzes what happens at point 1. Here, $\theta$ is equal to $\frac{1}{50}$ or .02, and this is equivalent to 2 prism diopters:

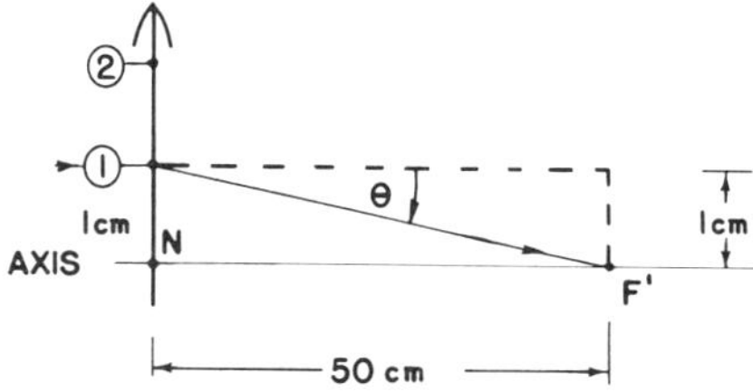


So, the prism power exerted by the lens at position 2 is $4^{\Delta}$ while that at position 1 is $2^{\Delta}$. This was a specific example; let's generalize:

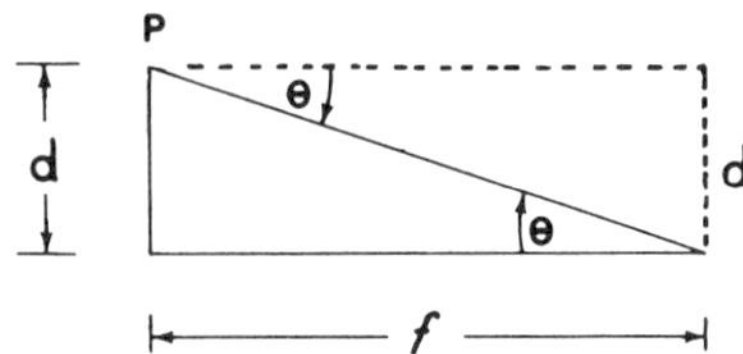


Let $d$ = the distance from the lens axis of the point we wish to examine, and $f$ = the focal length of the lens in question. Both these distances were expressed in centimeters in our example, so let's continue with the same units. We have seen that the prism power (in prism diopters) of the point in question was 100 tan $\theta$. Tan $\theta$ = $\frac{d}{f}$ so, 100 tan $\theta$ = 100 $\frac{d}{f}$. If the lens power is expressed in diopters of vergence, $\frac{1}{P}$ = $f$, when $f$ is given in meters; if instead, $f$ is in *cm*, then $f = \frac{100}{P}$;

$$100 \tan \theta = 100 \frac{d}{f} = 100 \frac{d}{\frac{100}{P}} = d \cdot P$$

Thus, the angle $\theta$ in prism diopters = $d$ (cm) × $P$ (Diopters). This is known as Prentice's rule and is a most useful one for the clinician. This rule is very valuable in arriving at the prism power induced by *any* point in *any* corrective lens. Just remember that $d$ is measured in *cm* to the *optical center* (at the lens axis).

PROBLEM:

What is the amount of prism induced 7 mm from the axial point of a minus spherical lens with a focal length of 8.5 cm?

ANSWER:

$$P = \frac{1}{f}; \qquad P = -\frac{1}{.085} = -12 \text{ D}$$

The prismatic power = $d \times P$ = (.7 cm) (− 12 D) = 8.4 prism diopters.

To determine the optical *effect* of this prism induction, not only must we know the *amount* of prism, we must also know which way the prism base or apex is oriented, as this tells us the *direction* of deviation of the rays caused by that particular lens point. The base of the prism is always at that lens position that is "thickest," so the base tends to be central in a plus lens and peripheral in a minus lens; but not always, such as in the following extreme example:

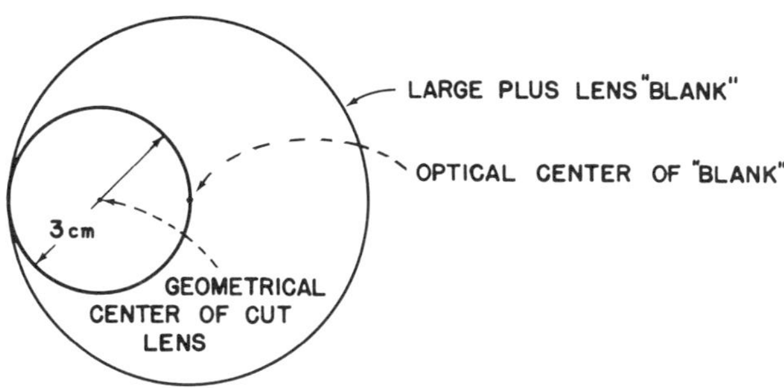


A 3 cm diameter circular lens is cut out from a much larger + 3 D "lens blank". The diagram shows how the smaller 3 cm *plus* lens can be fashioned with its optical center at an *edge* and not at its geometric center; thus the base here is *not* central as it usually is in a plus lens.

For the measurement of the induced prism, you have to know the location of the optical center (which is concomitantly the prism *base* with *plus* lenses). To determine the prism power at the *geometrical center* (1.5 cm from any edge of the small lens above), we refer to $d \times P$. The distance $d$ in cm must be from the *optical* center which is 1.5 cm away at its right edge (as shown).

The lens power P is + 3 Diopters. So, the *prism* power is $1.5 \times (+3) = 4.5^{\Delta}$ *Base* to the right (as you face it). If this is a "corrective" spectacle lens and is worn so the nose is on the left of our drawing, this $4.5^{\Delta}$ is directed *Base Out;* if the nose is on the *right,* this same prism is directed *Base In.* If this lens is rotated appropriately in front of the patient's pupil, the prism base *can* also be straight up or straight down. However, if there *is* vertical prism induced, you must also name the eye to designate the direction; Base *Up* OD is exactly equivalent (as far as the eye muscles are concerned) to an equal amount of Base *Down* OS.

Moreover, a point on the lens can exhibit both vertical *and* horizontal base effect *simultaneously*, if its direction from the lens axis is off at some angle which is not a straight vertical or horizontal one.

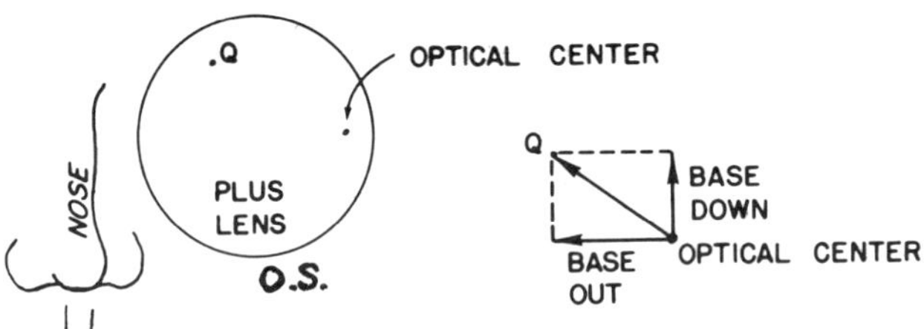


If the above lens is a plus lens and if the nose is on the left, Point Q introduces vectors of Base Out *and* Base Down prism effect.

If the visual axis of an eye looks through the *optical center* of any spectacle lens, there will be *no* prismatic effect induced. But if the eye rotates to look up, down, or through any positions other than the optical center, prism *is* induced, and any object viewed through that portion of the lens will appear to be displaced (*toward* the apex of the induced prism). For example, study the figure below which shows an eye peering through the lower periphery of a minus lens:

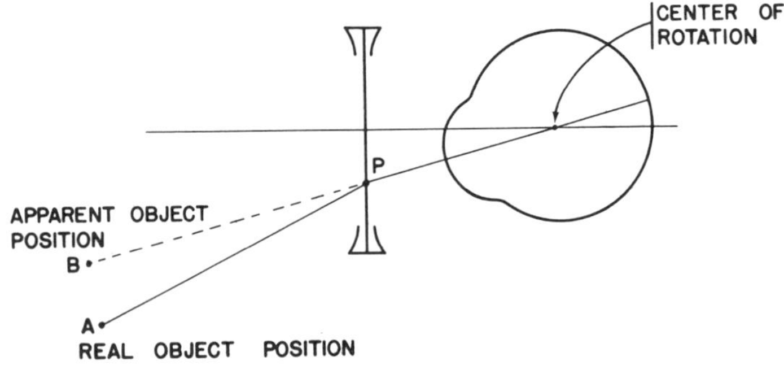

In this figure Point P in this minus lens introduces Base Down prism and makes the object at A appear to be at B. So, the object will appear to be displaced *upward.* Therefore, to align the macula with the displaced image, the eye has to rotate down a lesser amount than it would if the lens were not there.

Let's now look at a clinical example:

PROBLEM:

What is the prism induced when the two visual axes intersect a pair of spectacle lenses at points 1 cm directly below their optical centers? (OS: — 12 D; OD: — 10 + 2 × 180°).

ANSWER:

O.S.: −12 D

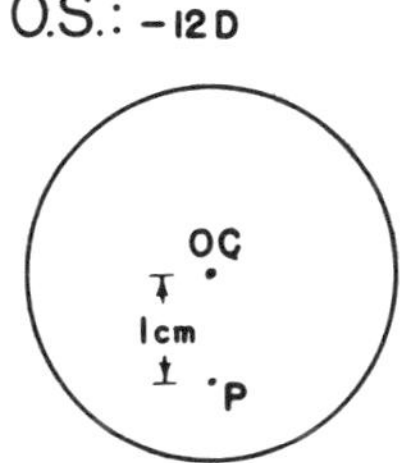


O.D.: −10 +2 x 180

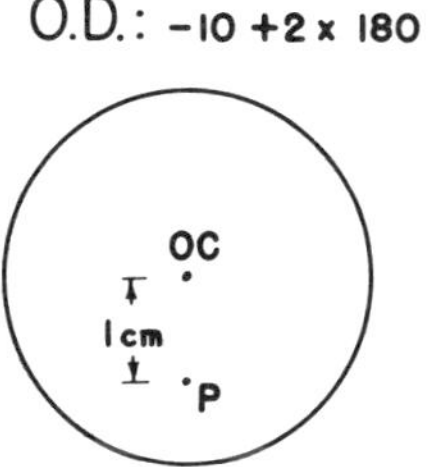


Prism induced is $d \times P$.
(1 cm) × (12 D) = $12^{\Delta}$

Since this is a minus lens, the base of the prism (thickest part of the lens) is down.

Prism induced is $d \times P$.
$d$ is 1 cm but what is $P$?

The power of this lens in the vertical meridian is gained from the sphere *and* the cylinder components: the spherical component is — 10 D: the cylindrical is + 2 D. (A + 2 × 180° has no power along its axis at 180° but + 2 D along the vertical meridian.) So the total $P$ in the vertical meridian is — 8 D and the prism induced is $8^{\Delta}$ Base Down.

*If* there were $8^{\Delta}$ of B.D. prism induced by *both* lenses, there would be no real problem for the patient since the prismatic effect would be symmetrical for the two eyes. The object *would* be displaced upward (toward the apex) from its normal position, but this causes no great tragedy. Though such a symmetrical "prism displacement" of the object *is* usually noticeable by patients with high powered plus or minus spectacle lenses, they tend to adjust rapidly to its presence and do not tend to be bothered by it.

Trouble can arise, however, in patients who have a *difference* of induced prism between the two eyes, as in our clinical example above; there is $4^{\Delta}$ *more* Base Down OS than OD. If a patient tries to read through the point 1 cm below the optical center of each lens, he encounters a vertical misalignment of the two retinal images of $4^{\Delta}$, which may be too great for him to compensate for by using his vertical motor fusion. The *non*-presbyope will usually have no problem. He avoids this area of the corrective lens like the plague by rotating his head to read so that he looks directly through the optical centers of both lenses where there is no prismatic effect; this neatly solves the problem for him. The *presbyope,* however, is the one who has difficulty. He will *not* be able to read through the optical centers of the distance correction since there is not enough plus power there to replace his lost accommodation. He requires a reading add which is usually built in as a bifocal segment positioned below the optical centers of the distance glass. This forces him to read there, through an area of the lens that has prism power induced. We will be looking at this "differential prism" problem and how to solve it when we look more closely at bifocals in a few moments.

Clinically, the alignment between the optical centers of the two corrective lenses and the corresponding visual axes is very important. Each of these variables must be measured, and *is* routinely by the optician as part of his job. The true separation of the *visual* axes is not easy to determine, so he uses its approximate equivalent, the interpupillary distance, which *is* simple to measure. If the spectacle corrective lenses are aligned so that each optical center is directly in front of the corresponding pupil, no prismatic deviation is induced. If both lenses are of equal corrective power, and the eyes rotate behind them so that the visual axes pierce them in comparable positions, the induced prism will affect the rotations of each eye similarly, and again there will be no *differential* prism induced. However, if the separation between the optical centers of the lenses is greater or less than the interpupillary distance (P.D.), or if these corrective lenses are of unequal powers, there *will* be a different amount of prism induced at comparable geometric positions of the lenses (upper right, lower left, etc.).

A typical problem is as follows (the figure is drawn from the *patient's* side):

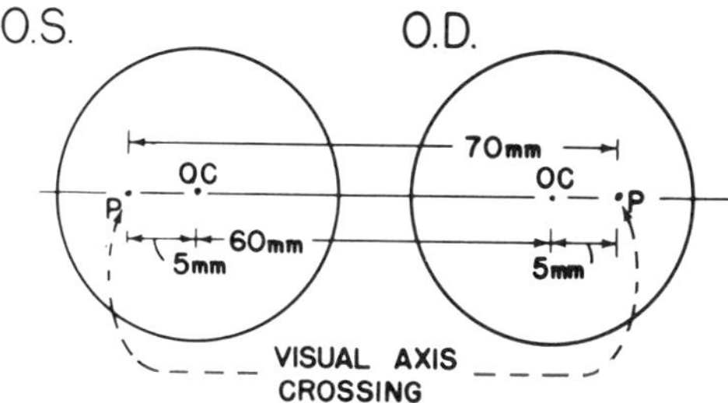


The distance between *optical* centers is 60 mm. The patient's P.D. is 70 mm. If the spectacle lenses are both — 6 D, at *each* point P (where the visual axis penetrates the lens) there will be (0.5 cm) $\times$ (— 6 D) or $3^{\Delta}$ Base Out prism induced — a total of $6^{\Delta}$ Base Out action for the two eyes combined. Base-out prism will deviate the images *inward* for each eye and, therefore, for sensory fusion of the two images to occur, the eyes must *over-converge* by $6^{\Delta}$ for objects at infinity. This convergence demand may or may not be desirable, that is, it may be either helpful or troublesome for this patient. In any case, you should be able to determine quickly the amount of prism induced at various points in any lens and to understand the effect it produces as far as the patient is concerned.

So much for prism displacement in the minus lens. What about that induced by a plus lens? As you might expect, looking through the comparable position of a plus corrective lens demands a *greater* rotation by the eye to see an off-axis object.

In the diagrams below, $\theta$ is the angular position of the object as subtended at the nodal point of the eye without any lens being present; $\theta'$ is its angular position (again subtended at the eye's nodal point) as the same object is viewed *through* the lens.

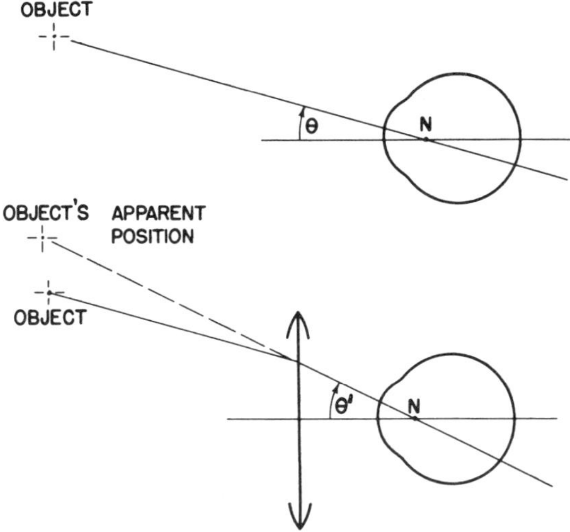


As you can see, the high plus corrective lens for aphakia is particularly adept at causing "object displacement". (The object shown above is displaced *away* from the lens axis.) This can discombobulate the hand-eye coordination for any aphakic patient; for example, when he eats (looking through the lower part of his specs) he will notice that objects not only appear magnified and closer (as you should by now recall), they will also be displaced *downward.* The combination of these three effects (plus other optical image distortions) make it particularly difficult to "get used to" aphakic spectacles. So be tolerant of your aphakic patients; they're not confabulating their symptoms.

## Rotational Magnification

Let us see how *much* ocular rotation is required to look up at an off-axis object through a spectacle lens and compare this rotation to that necessary without the lens while looking at the same object. This comparison, since it is in percentage terms, can also be looked at as a magnification — so, in addition to linear, angular, and axial, we have a *rotational* magnification. This type is unrelated to the others, and it is not an unimportant consideration; in fact it may be *as* or *more* significant for a patient (symptomogenically speaking) than is the *angular* magnification of a spectacle lens.

Look at the diagram given below:

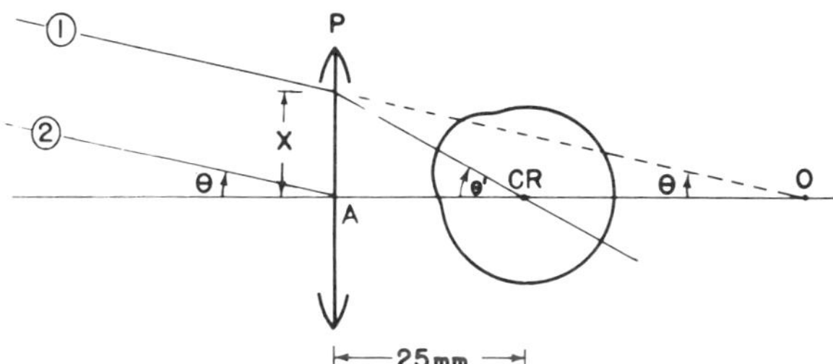


The eye is typically situated behind a corrective lens so that the cornea is about 15 mm away. The center of rotation CR of the eye is approximately at the eye's "center", perhaps 12 mm behind the cornea. The center of rotation then is *approximately* 25 mm behind the lens. (Don't quibble about 2 mm difference; the arithmetic becomes easier neglecting it!)

Ray 1 from an object at infinity strikes the lens (here, a plus lens). This ray was selected so that after refraction it would cross the axis at the center of rotation of the eye. The angle it makes with the axis is $\theta'$. This is the angle the eye would have to rotate through to see the off-axis object at infinity while looking through the lens. Without the lens, this eye would only have to rotate through an angle $\theta$ to see that same object. (Here $\theta$ is shown by ray 2 through the lens nodal point.)

Ray 1 is parallel to ray 2; so, if we continue ray 1 (past the lens but *without* refraction by it), it would also intersect the axis at an angle $\theta$. This straight continuation of ray 1 will cross the axis at point O.

For the moment, let's neglect the fact that both rays 1 and 2 stem from a single object point at infinity, and look at ray 1; it and the *axis* can represent two rays of light, both aiming toward O. Together then, they fix the location of point O as a virtual *object* point for the lens; you should then see that C.R. becomes the corresponding image point. Thus, axial points O and C.R. are *conjugate,* and the distance from the lens to O becomes distance *u;* that from the lens to C.R. (25 mm here) is *v.*

$$U + P = V; \text{ and } V = \frac{1}{.025} = +40 \text{ D}$$

Therefore, $$U = 40 - P$$

$$u = \frac{1}{40 - P}; \text{ and } v = \frac{1}{40}$$

For small angles, and with dimension x in the lens plane as shown,

$$\theta' = \frac{x}{v} = \frac{x}{\frac{1}{40}} = 40\,x$$

$$\theta = \frac{x}{u} = \frac{x}{\frac{1}{40 - P}} = (40 - P)\,x$$

The rotational magnification $= \dfrac{\theta'}{\theta}$

$$M_{\text{rotational}} = \frac{\theta'}{\theta} = \frac{40\,x}{(40 - P)\,x} = \frac{40}{40 - P}$$

This relationship holds true for any corrective lens (plus or minus) located in the spectacle lens plane. For an aphakic lens of + 12 D, the rotational magnification is

$$\frac{40}{40 - 12} = \frac{40}{28} = 1.43 \text{ or } 143\%$$

Thus an aphakic eye through its corrective lens must exert a rotation 43% greater than a non-aphakic eye viewing that same object. As stated, this stress on a patient's extraocular muscles may present more of a problem to him than the 25% angular *size* magnification of his retinal image!

## The Ring Scotoma

Another problem presented by aphakic (or high plus) lenses and caused by their prismatic effect is the "ring scotoma". The peripheral edge of the lens usually possesses the maximal prismatic power for that lens and will create the greatest deviation of rays.

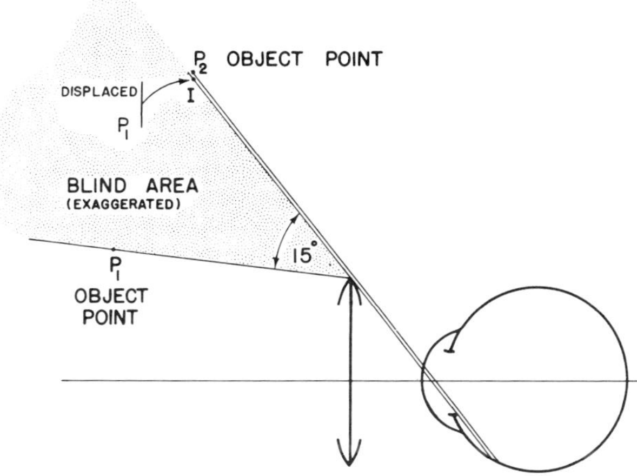


When the eye is directed straight ahead, some object rays from $P_1$ will hit the lens periphery and will be refracted so that they enter the eye through the pupil and will appear to arise from a more peripheral direction (at I). Another ray, drawn from $P_2$ — a point located just beyond I — does *not* go through the lens; (if it did, it would not be able to enter the eye — it would be refracted by the lens in too great a downward direction). This ray from $P_2$ is the most peripheral one that just misses the lens and is thus not affected by it, but that ray *does* get through the pupil. Rays from object point $P_2$ can be seen by the eye from "around" the lens. They will not be in sharp focus but still *can* be seen by the eye. Since $P_1$ is the most peripheral object that *can* be seen by this eye *through* the lens and $P_2$ denotes the next direction that is visible to this eye *around* the lens, the entire angular space between $P_1$ and $P_2$ is *invisible.* If some object is located within this blind area, the farther away that object is from this eye, the larger it can be and still remain completely hidden. The eye is as completely blind for this region as it is for the area subtended by the optic nerve head. It is a blind ring of space roughly 15° in angular extent, depending on the actual power and size of the corrective lens and its vertex distance.

You can demonstrate this field defect easily with a perimeter. The scotoma's overall shape will depend on the exact shape of the circumference of the lens; a square lens would give a square blind area.

The patient will not see a blank area; he will see space as continuous — filled in — *but* as if a strip of visual information has been snipped out.

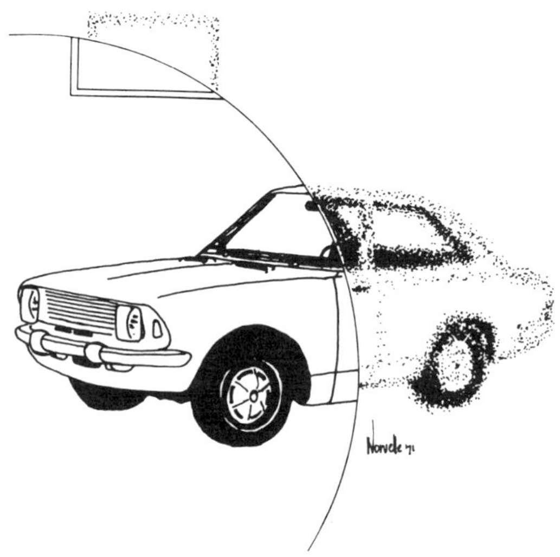

For example, a square window and a car might look like the sketch above, where "something" seems missing; the clearer parts of the image are those that are seen through the lens — these will be magnified.

A word now about "limiting rays" — I have drawn the limiting ray as that one just managing to spill through the lower pupillary edge; this is not quite conventional. In standard optics texts it is usually shown as the particular ray that passes through the *center* of the pupil. Actually it makes no practical difference which one is used, but the ray I chose makes for an "easier-to-visualize" explanation of the next phenomenon.

The exact extent of the ring scotoma depends on those particular limiting rays that can just skim by the corrective lens and still get through the pupil to fall onto the retina. The position of the pupil then, is key. The foregoing description of the "ring scotoma" was made with the eye directed straight ahead and motionless; so the pupil's position (and, in turn, the position of the ring scotoma) was stationary. However, we know that the eye does not normally stay rigidly in position behind the corrective lens but continually moves about, so the pupil will move. This will cause *new* rays to become the "limiting ones" and the scotoma will in turn be shifted.

This shifting scotoma accompanying eye movement creates a peculiar problem called the "roving ring" scotoma. This scotoma movement is always *towards* the lens axis as the eye rotates *away* from the axis to take up fixation of a peripheral off-axis object.

I would like to explain this in more detail so look at the accompanying figure A:

(See next figure.)

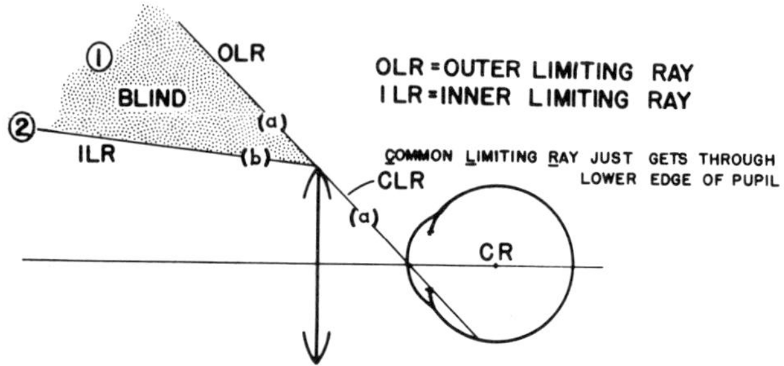


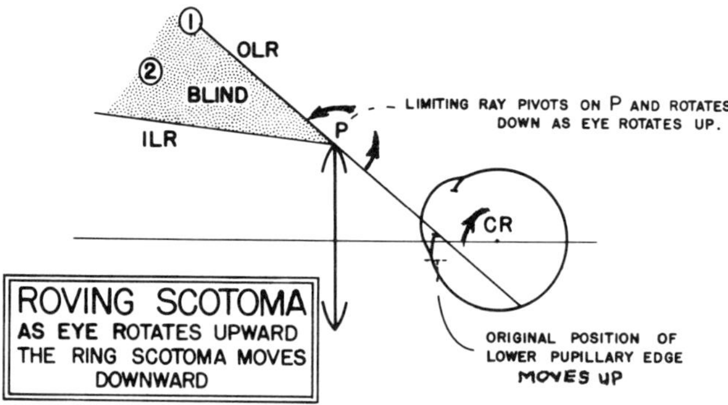


When the eye is in the straight ahead position, ray (a) is the limiting one since it just barely gets through the lower edge of the pupil. Ray (b) is refracted by the lens but also just barely gets through the pupil.

Any object in Area 1 (the ring scotoma) is invisible, but an object located at position 2 near the border of ray (b) *can* be seen (but not clearly, since its image is not on the macula). To obtain a clearer image of 2, the eye will rotate around its center of rotation CR to look towards it, and this rotation moves the pupil upward.

Look now to the accompanying figure B:

As the lower edge of the pupil moves upward, a new limiting ray is created. This new ray makes *less* of an angle with the axis than did the limiting ray with the eye in the straight-ahead position. (Imagine a piece of string connected between the upper edge of the corrective lens and the lower edge of the pupil. When the globe rotates upward, the angle that string makes with the axis *decreases.* That angle is maximal when the eye points straight ahead.) The fact that the limiting ray position has lessened its angle with the axis as the eye rotates upward (shown in the figure) indicates that the entire ring scotoma has shifted *downward* in this diagram, *towards* the lens axis!

Let's take it once more for clarity: (In figure A) the final limiting ray (a) between the lens and the eye is the final common path of the *outer* limiting ray (OLR, the one which just passes beyond the edge of the lens) and the *inner* limiting ray ILR (that one that is just barely refracted by the lens). (In Figure B) when ray (a) rotates upward ("pivoting" on the top edge of the lens), the entire ring scotoma must rotate toward the axis (downward here). Thus, with the lens fixed in position and the eye rotating upward, the ring scotoma rotates oppositely and swings into position downward.

Now look specifically at positions 1 and 2 in the figure: As this downward rotation of the "ring scotoma" occurs, the "blind area" will encompass the previously visible position 2 and gobble it up — it will disappear as the eye tries to see it! Also, objects previously in position 1 (invisible with the eye straight ahead) "pop" suddenly into view. This is the "Jack-in-the-box" phenomenon often spoken of in the literature.

These phenomena can occur with the lens perfectly still, with only the eye doing the moving; this yields the "roving ring" scotoma. However, this same effect is even more pronounced when the lens is moving too; with motion of the patient's head the ring scotoma will shift about in space, making objects disappear and then suddenly appear out of nowhere! This is extremely disconcerting to many patients but thankfully it can be gotten "used to" readily by most.

This effect is *most* noticeable in the mid-distances, say, between 3 and 12 feet. Up close, for reading, one's attention is on the reading material itself and most surrounding objects are visually large enough not to be completely hidden within the ring scotoma. At distances over 20 feet, as in driving, the central field of view is large enough to encompass most of the objects which might interfere with driving. So neither the near or far distances are usually too troublesome for the aphake. It is within the confines of a single room, however, that most of the difficulty is evident and the patient becomes most symptomatic.

Since the "Jack-in-the-box" has an optical basis, the only way to remove it completely is by not wearing spectacle corrective lenses which necessarily produce this prismatic effect. Contact lenses are free from this taint.

## Hand Neutralization of Lenses

This spot seems appropriate to introduce a practical application for the prismatic effect induced by spherical lenses. Since I have already written a short article on this subject,* I will reproduce it here.

## The Broken Lens Puzzle
## (or, Hand Neutralization Without Tears)

It is 4:50 p.m. E.D.T. You have just completed a day filled with the usual tribulations — demanding patients and complicated examinations, topped off with a disastrous glaucoma problem — a typical afternoon at the office. You eagerly anticipate some sun and fun with your foursome, which is scheduled to tee off at 5:15 at the Club. At this moment, into your suburban office dashes the bejeweled wife of the local bank president. It is obvious that you are not bursting forth with enthusiasm to see Mrs. W.O.B.P. (Wife of Bank President).

* Rubin, M. L.: "The Broken Lens Puzzle"
Survey of Ophthalmology, 15:105-108, 1970.

In her matronly hand she clutches some cracked lenses which were once a pair of vogue-ish spectacles. Naturally, she insists these be replaced (as with all other such demands) "yesterday." Well, no matter. Although the optician is over 40 miles away, a simple phone call of the prescription should do nicely to yield a newly-ground pair of spectacles in short order.

First to the files and the patient's previous record. Ugh; the W's are out being microfilmed this week. Aha, the lensometer in the corner. Horrors, a burnt-out bulb! Luckily, your next lane has another instrument. Double horrors and gnats!! You bulb-snatched that light last month!

Now what? The Geneva lens gauge at the back of your bottom desk drawer? No good. Mrs. W.O.B.P. has a high minus correction which has previously been ground in Thin-lite lenses (American Optical Company, Buffalo, New York) (index 1.70); the Geneva clock is calibrated for an index of refraction closer to 1.52. (Using this instrument would cause a great error in readings; so, it is certainly not appropriate for Mrs. W.O.B.P.!) Curses.

Wait . . . Could you possibly consider hand neutralization of the lens and do it effectively? It's been at least 10 years since you even held a loose lens in your hand!

Out comes a dusty box of trial lenses. In less than 1 minute you have the proper prescription. You race to the phone, call the optician and catch him at 5:00 p.m. just as he is closing. Joy. The lenses will be delivered tomorrow to a grateful patient, and punctually you meet your foursome, soundly trouncing each member in your state of euphoria.

In order to enable each reader to emulate our skillful ophthalmologist and to refresh his own memory as to the details of the useful maneuver of hand neutralization, let us now review the simple optics and rules involved.

## DETECTION OF MOTION

If you hold any minus lens in front of your eye (say, about 1 foot away) and view a distant object through the lens while moving it upward, the object will appear to move upward also. This apparent motion of the object in the same direction as the movement of the lens is called *with* motion (Fig. *A*).

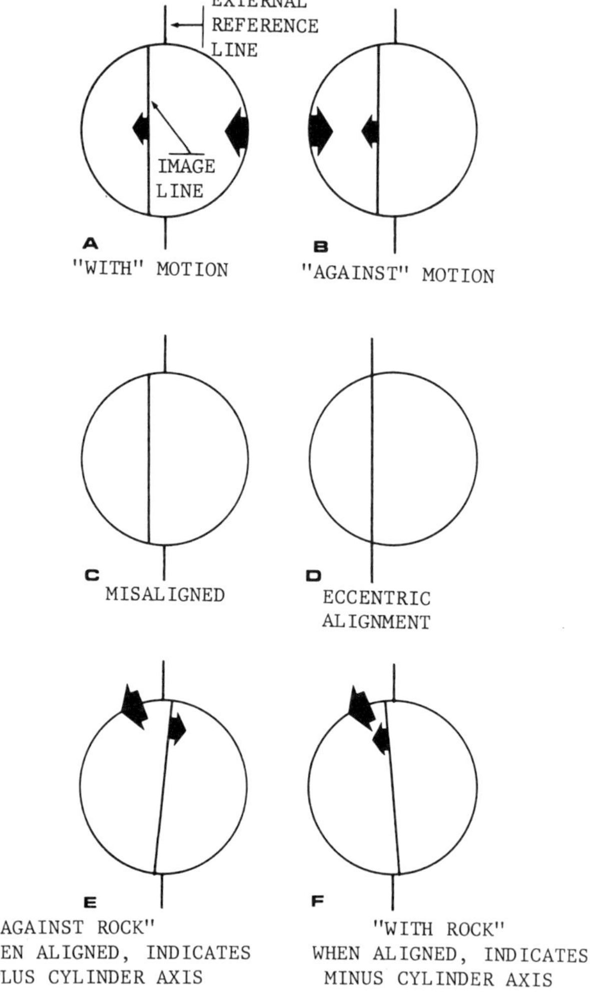

A plus lens similarly moved will provide an *against* motion (Fig. *B*). The detection of even minimal amounts of motion, either "with" or "against", is actually extremely simple — certainly one-quarter of a diopter should be patently obvious, and as little as one-eighth of a diopter is detectable with a bit of practice.

"Against" motion always serves as an indication that an unknown lens has plus power, and "with" motion usually labels a lens as minus. However, "with" motion can also be simulated by a *plus* lens, if it is held so that the eye-to-lens distance is greater than the focal length of the lens; in the latter instance, the image seen is inverted and real. In any case, there is rarely any difficulty in telling plus from minus.

## NEUTRALIZATION

Using the "with" or "against" motion as a cue, both the spherical and the cylindrical powers of any unknown lens can be determined with ease. The technique called *hand neutralization* involves the placing of lenses of known power by trial-and-error in juxtaposition to the unknown lens, in an attempt to eliminate (neutralize) any image motion.

As a practical point, it is easier to detect the minimal image motion produced by a weaker-powered lens (or any lens that is almost, but not quite, neutralized) if it is held further (a few feet) away from your viewing eye.

After some practice using a randomly selected spherical lens from your trial set, you should be able to detect the induced motion and neutralize it with facility, using sufficient plus-lens power to counteract "with" motion and minus for "against". Spheres present no particular problem.

## PRINCIPLES

Why do we see "against" motion with a plus lens?

It is somewhat surprising to find that Duke-Elder (*Practice of Refraction* (7th ed.), C. V. Mosby Company, St. Louis, 1963, p. 36) states that "The reversed motion (seen with a plus lens) is due to the fact that the image formed by such a lens is inverted." Perhaps I misunderstand his reasoning, but this explanation for the "against" motion does not seem to be correct. When your viewing eye is closer to the lens than its focal length, the image you see is virtual and upright, and *not* inverted; yet, "against" motion is visible. No, it seems to me that the motion itself results simply from "variable prismatic displacement" by the lens; the *direction* of that motion is dependent on the position of the "prism" apex (central in minus lenses, peripheral in plus).

A standard straight prism induces *constant* prismatic displacement. If such a prism is inserted into your line of sight, there will be an immediate displacement of the image *toward* the apex of the prism. However, moving the prism slightly back and forth, in the same manner described above for the lenses, will yield *no* movement of the image. A plus lens, on the other hand, is not a simple straight-sided prism. It can be considered to be composed of a number of different prisms of gradually increasing powers as one moves from the optical center toward the lens periphery. Therefore, viewing an image through progressively more distal portions of the lens forces you to utilize a variable, increasing prism power; this increases the apparent displacement of that object (toward the "prism" apex) through each "prismatic" segment. Motion is thus imparted to the image. As a plus lens is moved to the right, the image seen through it will move toward the left. (Diagram it out for yourself.)

With minus lenses, the apex of the "prism" is always at the optical center, so lens motion to the right will impart motion of the image to the right. Again here, as with the plus lens, the amount of prismatic displacement is variable and increases toward the lens periphery. The exact amount of prism displacement is given by Prentice's rule — the prism displacement induced by any point in a lens is simply the dioptric power of the lens at that point times its distance (in centimeters) from the optical center.

This variable prismatic displacement, then, is what I believe to be the explanation for the motion seen through a lens. (Is it really possible that Sir Stewart has erred?)

## NEUTRALIZING CYLINDRICAL CORRECTIONS

Initially, when you pick up an unknown lens for neutralization, you are faced with an immediate question. Does that lens have any cylindrical power? Detection of cylinder is your first obstacle, which can be surmounted more readily if you first partially neutralize some of the sphere, leaving no more than a diopter or so of spherical residual. (An unknown lens of high plus or minus spherical power tends to mask any coexistent cylinder.) To aid in this cylinder detection you will need some straight reference lines. Luckily, ubiquitous lines are usually available — a doorpost edge for the vertical, and a tabletop edge for the horizontal.

View your straight reference line through the "unknown" lens and carefully align the *image* line with the externally visible straight line. (That line will pass through the optical center of the lens.)* Rock the lens slowly about its optical axis with a wheel-like motion while holding the lens at arm's length — at this distance, it is easier to detect the minimal motion caused by low-powered *cylinders* as well as spheres; look for any rotating motion of the straight line in the image. If this rocking yields no rotating movement of straight lines, one can be fairly sure that there is no cylinder present. (To be absolutely sure, fully neutralize any sphere and recheck for cylinder.)

If you find that cylinder is indeed present, the exact positions of the two axes — the meridians of maximal and minimal power — must be determined. (The meridian of *maximal* plus power is the minus cylinder axis; that of *minimal* plus power is the plus cylinder axis). Using the lens-rocking maneuver described previously, mark the exact lens position which aligns (makes continuous) the externally viewed line and its image through the lens. This line position is one of the two cylinder axes; the other is 90° to it. If the image line seems to rock "against" the direction of the actual lens rock (Fig. *E*), the aligned meridian is the *plus* cylinder axis. It is the *minus* cylinder axis if the rock is "with" the lens rotation (Fig. *F*). Obviously, any cylindrical lens has *both* "with" and "against" rocks. (Do try it yourself to see what is meant!)

After locating these axes you will now neutralize each, using only spheres. This procedure is much easier than one which uses loose cylinders, since you can proceed to the neutralization of a meridian

* If the line seems somewhat eccentric in the lens (Figs. *C* and *D*), there may be some prescription-prism ground into the lens. At least, keep this possibility in mind.

without stopping to think whether you are dealing with a plus or minus cylinder axis. In other words, when you use loose spheres you can disregard the sign of a cylinder axis as a prerequisite for neutralization.

Align either axis with the externally visible line and, *without rocking the lens,* move it side to side, exactly perpendicular to the visible line. Note any "with" or "against" motion; neutralize whatever you find, using spheres and this side-to-side motion until no movement whatsoever is visible. This yields the spherical power of the meridian at right angles to the externally visible line.

After you complete full neutralization of this meridian, turn the lens 90° and again align the image line with the external line. Proceed to neutralize any additional motion present (in the side-to-side direction) with appropriate sphere. Then recheck the first meridian. (Using this technique with spheres, you obviously cannot neutralize all lens meridians simultaneously; however, that is not necessary.)

The maneuvers described isolate the spherical powers of the two principal meridians, and all that is left for you to do is to write the prescription from this information. (The determination of the axis will, of necessity, be through an educated guess via study of the lens fragment. Keep in mind that, for any lens prescription placed on the patient's face, the reference point — 0° axis — starts at the left ear lobe and proceeds counterclockwise — when viewed by the refractionist — for both the right and left eyes.)

For those individuals who might have difficulty in writing the final spherocylindrical prescription from the spherical data obtained above, one can learn to neutralize with spheres and loose cylinders. However, you *should* be able to transcribe the appropriate prescription.

Moral: Wise men are prepared for every exigency. Plan on enlisting in this elite corps by practicing hand neutralization for the day when your lensometer goes "on the blink". (By the way, have you checked yours lately?) *(Article finis)*

Back when we were discussing prisms, I promised to mention the Fresnel surface. So now, while we are elaborating on the many prismatic effects of lenses, I think it's time to pay up.

## Fresnel Surface Principle

Any prism introduces virtually the same amount of prismatic deviation over its entire surface — no matter what portion of the prism you may look through — the apex, mid-section, or base. So, any segment or strip of the prism, including the tiny wedge of prism at the apex, is *just* as effective as any other (e.g., the thicker base) in deviating the light.

What if you were to trim off the apex (say, a strip ½ mm wide) from each of a large number of prisms of identical powers and assemble the strips contiguously, laying them side by side as shown in the diagram below?

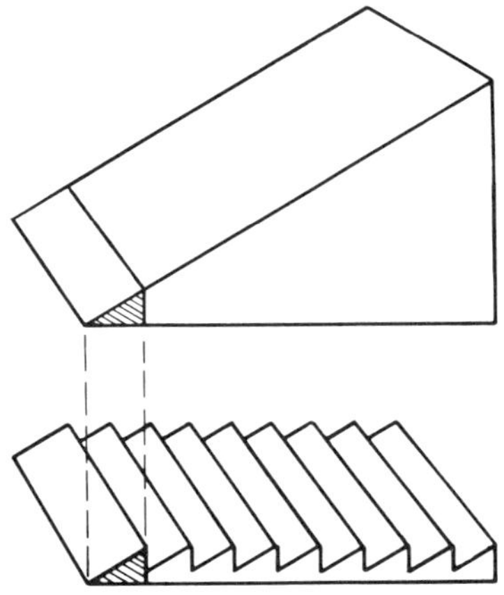

THE FRESNEL PRINCIPLE

You would then produce a ridged surface which would be able to introduce a fixed amount of prism power into any light path — just as did each large *single* prism from which each apical wedge was cut. What is gained by such a peculiar construction is a tremendous reduction in weight and bulk; what is lost is *some* optical clarity because of the number of ridges introduced by the strips — the amount of clarity reduction, though, depends on how carefully the strips are constructed.

Fresnel discovered this "stripping" principle back in the 19th century. It enabled him to duplicate the optical power of a large diameter and, therefore, heavy lighthouse lens with a much thinner and more manageable one.

For his *lens* construction, the strips were circular and of a gradually *increasing* prism power in each concentric "ring" as one moved away from the center. This gradual change of prism power across the surface, of course, optically mimics a plus lens whenever the prism apices of all the ring strips point away from the center — vice versa for a minus lens.

Now, Fresnel's principle joins hands with modern scientific technology to allow improvements in the method of production of the surfaces. So much so, that feasible and practical *ophthalmic* Fresnel prisms (and lenses too) of almost any desirable power can be manufactured. Various thin, flexible sheets of plastic incorporating a Fresnel surface of lightweight and of very good optical properties are presently commercially available.* These sheets can be cut to spectacle size and can be "pressed on" to the surface of any ordinary spectacle lens. They can create appropriate optical effects with very little of the distortion introduced by standard ophthalmic prisms and thick lenses, since much of that distortion is due to the *thickness* of the glass. Though not *yet* completely replacing these latter appliances, the "press-ons" are readily accepted by patients, especially aphakes and those requiring ophthalmic prisms. We are quite likely to hear more about other exciting applications of the Fresnel "press-ons" in the near future.

## CORRECTION OF PRESBYOPIA

We have already touched on accommodation and its progressive loss with age leading to presbyopia. The correction of presbyopia is done simply by the supplementation of the patient's waning accommodative power with plus lenses. We shall assume that any co-existing ametropia has been fully corrected.

---

* Optical Sciences Group, San Francisco, California

In order to prescribe the correct amount of reading add, you must determine the patient's near point of accommodation. This should be done with a good accommodative stimulus, one which will tax the patient's *full* power of accommodation, such as the smallest letters on a near acuity chart (that is, those corresponding in angular size to the maximally corrected distance acuity). These small letters will demand a continuous, conscious accommodative effort to be kept clear. A card with a graded series of print sizes (by Lebensohn, Sloan, etc.) is used to test each eye separately. Such a card should be supplemented with constant verbal exhortation by the examiner to "keep it clear" in order to help the patient reach his closest "Nearpoint". This tells us how much accommodation a given patient can muster up if he *had* to and will expose his *full* accommodative reserve.

The eye is a biological system with all the associated vagaries, so it has always been surprising to me that the accommodation reserve is so remarkably equal in the two eyes. Since we have always considered that the loss of accommodative power with age is a *local* one for each eye and probably not centrally controlled, why *shouldn't* one eye lose accommodation more rapidly than the other? (Cataracts and other bilateral eye problems are rarely so symmetrical.) In any case, the accommodative loss *is* typically symmetrical, and only rarely is it not. So in your refraction lane with a patient, if you happen to find a closer "Nearpoint" for one eye than the other, suspect the adequacy of your *distance* correction and recheck *that* before proceeding. You may have "under-plused" one of the eyes.

Once the near point is determined, you have an inkling of how your patient compares with others of the same age (Duane's accommodation table was given previously), but that is not of crucial import. What is definitely of major significance is how your patient's accommodative ability fits in with his *own* needs — what level of ability he requires for his own vocation and avocations. This varies tremendously and gives weight to my plea that you don't just read from a table the "proper" prescription for a reading add for a given age. The visual requirements of a machinist, a concert pianist, a stockbroker, and a shoe salesman are all so different that the management of their visual needs must be individualized.

It is said that the patient should be using about one-half his available accommodation to perform the task for which he requires glasses. As is true with rules, there are exceptions, but generally, this seems to work out in practice. Have the patient sit with his eyes closed and use his kinesthetic sense to place his reading matter or close work at a comfortable working distance for himself. Or, you can have him show you where he has to hold his head in relation to his work demand, as a dentist looking at a molar. You can then measure this distance, convert it to the dioptric equivalent and prescribe a glass which leaves one-half his available accommodation in reserve for that distance (after trying it first in a trial frame, of course!).

PROBLEM:

A nearpoint of accommodation (NPA) is found to be 33 cm, so the accommodative ability is 3 D. The patient would like to read at 40 cm. What add should I prescribe?

ANSWER:

Since we wish the patient to use $\frac{1}{2}$ of his 3 D amplitude for a task which stimulates an accommodative demand (at 40 cm) of 2.5 D, we arrive at the following: The demand is 2.5 D; the patient is to use 1.5 D of his own accommodation and this leaves 1.0 D to be supplemented. Thus, the appropriate lens to try is a + 1.00 D add over his distance correction. With this in place, he will have a good range of clear vision surrounding his reading "position" at 40 cm.

For this patient, the full extent of his near "range of accommodation" is 3 D; this is his *amplitude* of accommodation. With his reading glasses on, the maximal near position of clear vision is 4 D (as determined by his amplitude of 3 D plus the + 1.00 add). His clear vision range will extend to his furthest position of clarity, where no accommodation will be required; this is to a distance equivalent to 1 D, the amount of his add alone. (See figure A below).

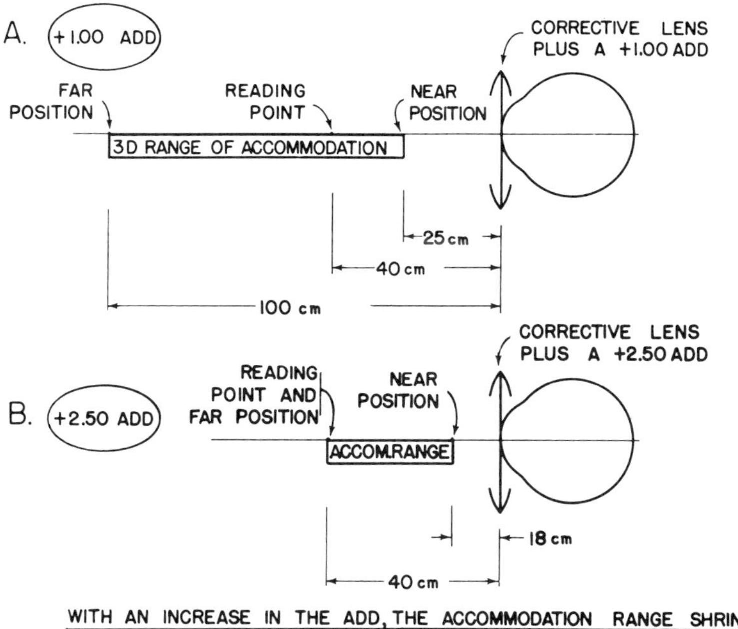


WITH AN INCREASE IN THE ADD, THE ACCOMMODATION RANGE SHRINKS

Always check the *range* of accommodation with the add you plan on prescribing so you can determine if it will be useful to *your* patient.

It is usually an advantage to give as *little* plus add as possible. Sure, you can bring the maximal near point closer by giving more add, but, if you do so, you will lose accommodation range on the far side of the reading position. If this same patient were given a + 2.50 add for his reading (see figure B above), he would not have to exert *any* of his own accommodation to read at 40 cm. However, he will be mighty unhappy since everything beyond 40 cm would appear blurry to him; he will not like this add even though his "new" near point is brought in to a (3 D + 2.5 D), that is, 5.5 D or 18 cm. So, while a 2.50 add *might* be of value to an embroidery seamstress, it will probably not be to a journalist.

Notice that with the 2.50 add, the range of accommodation (though still 3 D) has shrunk from 75 cm (100 — 25) to 22 cm (40 — 18). This shrinkage of zone is one of the main reasons not to prescribe the higher power adds to young presbyopes, that is, unless it is absolutely necessary. Personally, I rarely find it necessary to prescribe a patient's *first* bifocal add until he needs at least + 1.00 D or so. The *frequent* prescription of low power adds makes those who dispense glasses wealthy. (Take that as a hint or a warning, depending on your point of view!)

CLINICAL POINT:

It is amazing how often a 40 year old who has never worn glasses comes into your office with presbyopic symptoms. Do not be tempted to start right in with the near vision tests. You may find that he has some hyperopia which hitherto has been uncorrected.* It is very likely that if you simply correct the *distance* ametropia alone, he will require no near add whatsoever. You'd be surprised at how often this simple point is missed by refractionists.

Say you find that + 1.00 D distance spectacles are necessary for a patient; you prescribe them while telling him that he does not *have* to use them for distance viewing unless he wants to, but should use them whenever he reads and wants to see what he is reading. This is *not* a reading glass even if he uses it only for reading; it is a distance correction. However, it probably will not be too long before he uses it all the time. As he gradually loses more of his accommodative reserve, he'll find it more comfortable not to exert any accommodative effort at all for distance vision — a luxury bestowed by your prescription.

* * *

Of course, as patients get older and lose more of their accommodation, you will have to prescribe more add. However, in doing so, only very rarely will you have to exceed 2.5 D**, and when the patient has lost enough accommodation to require that much add, you will probably also have to add some sort of "in-between power" to take care of the patient's visual needs in the intermediate range

* Even in an adult (including the prepresbyope!), a *cycloplegic* refraction may well be required to *fully* expose the hyperopia — a fact not often appreciated.

** For an exception to this, read *"The Case of the Myopic Capitalist"*, which is reprinted in Appendix D.

(about 50 — 100 cm). In the standard *trifocal* lens, where the distance, intermediate power, and near lens are all put together in one spectacle lens, this "in-between" lens power is automatically about one-half the near add.

PROBLEM:

A 62 year old patient has an accommodative amplitude of 1 D and wears a + 2.50 D near add with a + 1.25 D intermediate add. Draw a sketch of his zones of clear vision.

ANSWER:

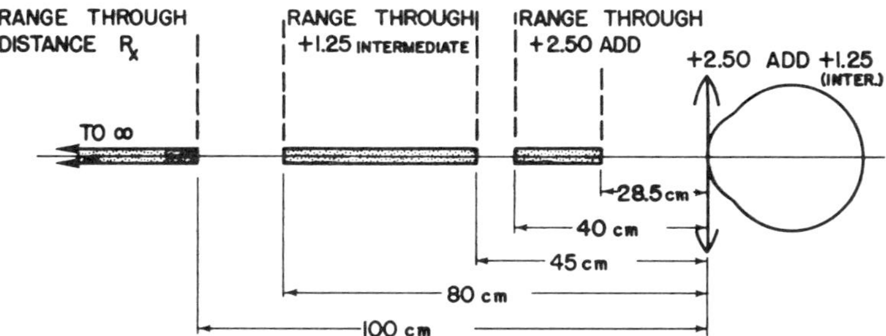


The small "gaps" between zones of clear vision are easily compensated for by shifts in head position. (Don't cheat yourself. Go back and make sure you check the ranges to find out if you understand how they were obtained. I didn't draw this out to demonstrate my prowess.)

CLINICAL POINT:

The last few diagrams have shown a corrective lens and the add placed *at* the cornea for schematic simplicity. However, in real life, the lenses are worn in the spectacle lens plane so the vertex distance becomes a factor. Remember, with high minus ametropic spectacle corrections, the accommodation demanded at a given distance is *less* than for an emmetrope; but moreover, the existence of a vertex separation between the lens and the cornea gives the moderate to high myope a bonus. By slipping his spectacles slightly (a few mm) down his nose while he is reading, he further lessens the demand on his own accommodation. This is a little trick all of us pre-presbyopic myopes have learned!

To explain this, let us use the same clinical data we did with the example given previously. This patient was an 8.7 D myope who required 3.55 D accommodation to see reading material held 20 cm from the spectacle lens, which was — 10 D (see figure A below).

If the lens now is allowed to slip down the patient's nose by 5 mm, the optical situation is as diagrammed in figure B below:

PROBLEM:

Calculate the accommodation required.

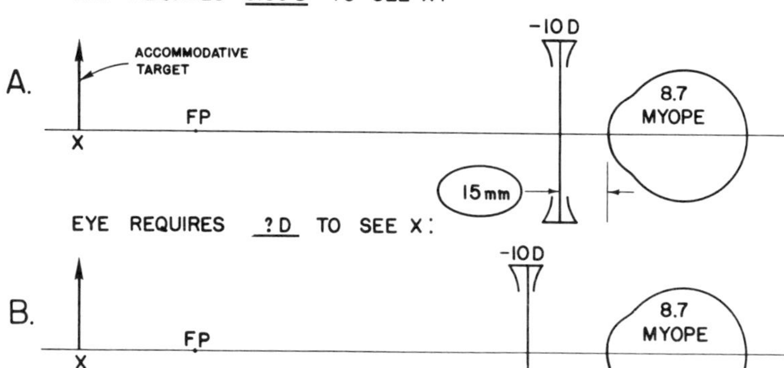


ANSWER:

$$U + P = V$$

$$-\frac{1}{.195} - 10 = V$$

$$-5.13 - 10 = V$$

$$-15.13 = V$$

$$v = -6.60 \text{ cm}$$

The image is located 6.6 cm in front of the *lens.* Since the lens is now 20 mm from the cornea (originally 15 mm + 5 mm "slip"), the image (which is the "object" seen by the eye), is 8.6 cm away from the *eye.* (That is, 11.62 D away). But the eye is 8.7 D myopic and so must exert 11.62 — 8.7 or only 2.92 D of accommodation to see it clearly.

We have thus shown that if the — 10 D spectacle lens slips 5 mm further down the patient's nose, his accommodation requirement to see material located 20 cm away has decreased from 3.55 D to 2.92 D, saving him 0.63 D of accommodation — a substantial quantity for a prepresbyope. But, please do not assume that this benefit is limited to *pre*presbyopic myopes. This same principle aids even the presbyopic myope who is wearing some bifocal add in his spectacles, as long as his *overall* lens power (minus correction plus bifocal add) is still minus. So, the presbyopic myope also gains "additional plus" help by letting his glasses slip down his nose; the higher the myopia, the more the "help".

The exact reverse is true for the high hyperope who has to keep pushing his glasses back up on his nose to keep the accommodative demand at a minimum*. (See also Appendix E.)

## Bifocals

The types, styles, shapes and forms of bifocal lenses commercially available are manifold. Each manufacturer is constantly plugging his own. I want only to introduce a few principles and not present a full treatise on this subject.

The precise methods of producing a bifocal are not germain here though you should have *some* idea about some of them.

FUSED TYPE:

Usually a plug of glass of high index of refraction is placed into a depression on the front surface of a standard glass lens and fused there at 700° C. The front surface of the add is then polished smooth, to the same curvature as that of the main lens. The bifocal segment area will have a greater dioptric power than the main lens since it is built of material with a greater $n'$, which makes *its* surface power $(\frac{n' - n}{r})$ the greater one.

* This last statement is true *only* when the object distance is less than two times the focal length of the plus corrective spectacle lens. The specific circumstances which govern the inconsistent effect on accommodation of a shift in the position of such a *plus* lens is the subject of another paper: Rubin, M. L., *The Sliding Lens Paradox*, Survey of Ophthalmology 17:180-195, 1972.

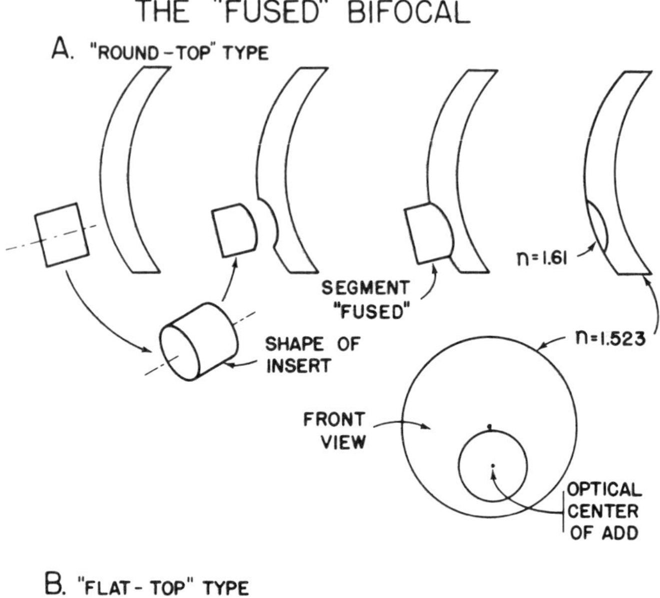


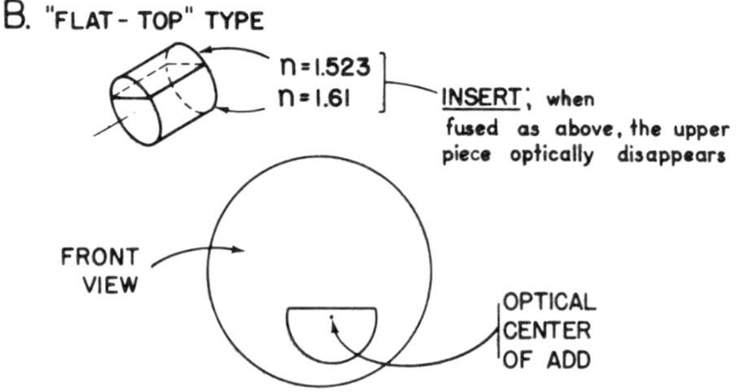


ONE PIECE TYPE:

Another type of bifocal (the "Ultex" type) is ground out of one piece of glass, but the reading area is cut with a different curvature. This area is usually on the back and provides the increase in power necessary to produce the "add".

The radius of the "add" circle (see figure below) may be varied, and each has a different name: the radius of an Ultex *A* is 19 mm, that of an Ultex *E* is 16 mm, that of Ultex *B* is 11 mm.

ULTEX TYPE BIFOCAL

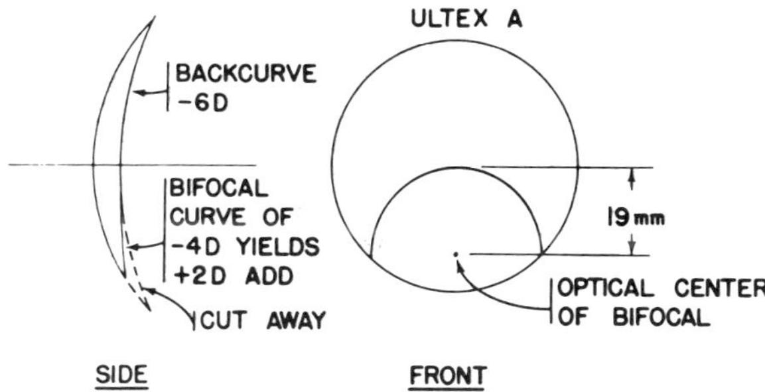


From here on, we will assume that whatever the manufacturing process, all bifocal segments will be equivalent, diopter for diopter, in providing a given amount of add.

Aside from cosmetic or vocational* reasons for choosing one type of bifocal over another, there are various optical considerations which would encourage you to use a specific type — we will consider only three: the first challenge is to decrease the "image jump" which occurs as the visual axis sweeps across the junction of the top edge of the bifocal segment and the distance lens correction; the second is to reduce the overall quantity of "object displacement" by reducing the total prismatic effect through the reading area of the lens; and the third is to reduce any induced prismatic *difference* in comparable reading areas of the two spectacle lenses, since this difference puts an unequal strain on the vertical eye movements of the two eyes.

## Image "Jump"

Prism "image jump" is the commonest and probably the most annoying aspect of wearing bifocals.

* Some types of occupational bifocals have the segments placed at the *top* of the primary lens, such as for inventory clerks, shoe salesmen, and librarians; that is, any presbyope who frequently does close work above normal eyelevel.

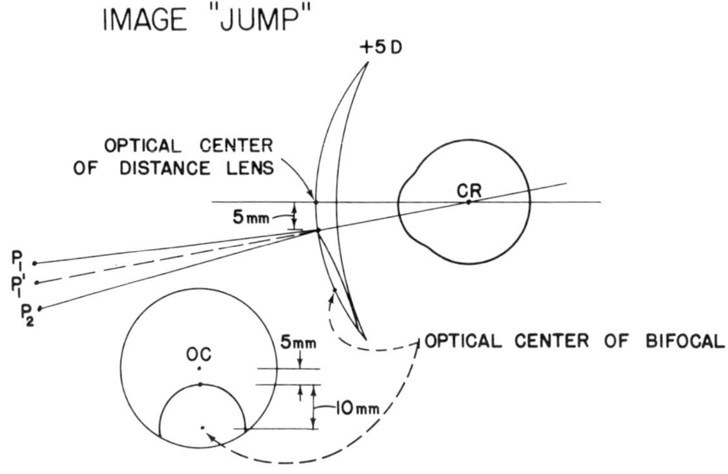


Say that to see object $P_1$ through a distance lens of + 5 D, the eye must rotate to look through a point 5 mm below the optical center of that lens. The prism induced at that point is (0.5 cm) × (+ 5 D) = 2.5$^{\Delta}$ Base-Up — Prentice's rule. So, object $P_1$ will appear to be *down* at $P_1'$.

As the visual axis moves down further, it crosses the top of the bifocal segment. If we consider that the bifocal segment is + 2.5 D add and has *its* optical center 10 mm below the segment top (see front view of lens in above diagram), then, as the visual axis crosses the top of the segment, it *suddenly* encounters (1 cm) × (2.5 D) or 2.5$^{\Delta}$ Base-*Down* prism. (The prism base of the plus lens segment is at the optical center of the *segment.)*

If there were *no* bifocal segment incorporated into a given plus distance lens, as one's visual axis swept downward, objects would appear to become more and more displaced downward — nothing dramatic, but a gradual shift. (A comparable movement with a *minus* lens would cause an *upward* object displacement.) However, when a bifocal segment *is* present, as the visual axis sweeps across its upper edge, there will be a *sudden* upward shift of any object being scrutinized. This shift is "image jump".

Now look again at the last diagram. As the visual axis crosses the top of the segment, the sudden introduction of base-down prism makes object $P_2$ appear in the same straight-on direction as that occupied by $P_1'$. Thus, all the *real* space between $P_1$ and $P_2$ is optically collapsed — no object point located therein will be visible. This blind area around the bifocal is identical to the "ring scotoma" around any plus lens; it is less noticeable here than in aphakia since the amount of plus is less.

Image jump occurs with any bifocal that does *not* have its optical center at the top of the segment; if this "center" *is* at the top, there will be no image jump. So, it is the location of the optical center of the bifocal *segment* that causes the jump — *not* the influence of the distance correction at all. A corollary is that the further away from the segment top the bifocal's optical center is located, the greater the jump induced for a given bifocal power. Thus, to reduce "image jump" for any patient, the clinician must choose a bifocal type that has its center at or near the top of the segment.

## Object Displacement

We studied this phenomenon when we examined how all lenses produce prismatic effects which result in a greater apparent displacement with lenses of high power. (I mentioned that the patient's sensory adaptation to the "new" position of objects caused by new spectacles is quite rapid, thank goodness.)

Bifocal segments can either exacerbate or reduce the apparent object displacement caused by the distance lens in the reading area of that lens. A plus corrective distance lens induces base-*up* prism in the lower fields. This can be counterbalanced somewhat by a base-down prism effect given by an appropriate bifocal segment with its optical center down low. Remember, base-down effect would be produced by the segment between its upper edge and its optical center, so the closer to the bottom of the segment the center is located (the center can even be beyond the segment entirely!), the greater the base-down effect created by this reading add.

For a plus distance lens, such an arrangement as just described can equalize the combined prismatic displacements, but *only* for one specific area.

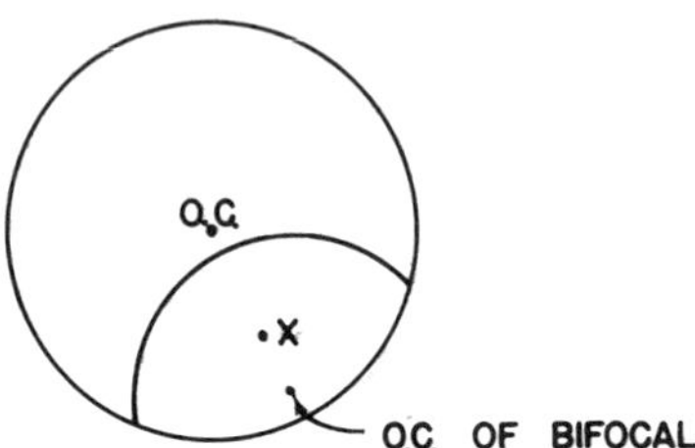


In the above diagram with the plus distance lens, let us arbitrarily say that point X (through which a patient may read) will have $3^{\Delta}$ of Base-Up prism induced by the distance lens and $3^{\Delta}$ of Base-Down prism induced by the segment. Above point X though, the B.U. effect of the distance lens is reduced, but the B.D. effect by the segment is increased, so overall there is a net *B.D.* displacement. Below X the reverse is true and there will be a net *B.U.* displacement. The exact position of this "null" point, of course, will vary as different lenses and different types of bifocals are considered.

CLINICAL POINT:

*Neutralizing* the prismatic effect near the chosen reading point of a bifocal lens may be of great value to someone who occupationally must spend much of his time at a desk (a lawyer, bookkeeper, or draftsman, etc.). To such an individual, reduction of image displacement may be a very important need, and he might gladly tolerate the increased "image jump" at the bifocal top which is obligatorily present in the above example. You should be able to see that attempting to correct "object displacement" for this patient (as for anyone who wears a high plus lens) will only make matters worse as regards "image jump". In contrast, to someone whose vocation requires a constant need to change from distance vision to near (with the visual axis continually criss-crossing the upper bifocal edge), "image jump" elimination would be the more important consideration. A steeplejack would be more likely to require help for "image jump" rather than for "object displacement", and so would a waitress.

For any hyperopic presbyope, then, you cannot have the best of both worlds. You'll have to choose which defect you want to reduce since correcting one will worsen the other. The myope, however, is golden in this regard. Since base-*down* prism is induced in the bifocal area by the distance lens, a bifocal segment with the center near the top would induce base-up prism to counter the base-down; in addition, it would obtain the fringe benefit of reduced "image jump" too!

Both "image jump" and "object displacement" are made manageable by manipulating the same bifocal variable — the distance from the top of the segment to its optical center, and so, depending on what you wish to influence, you can choose segments with the center near the top or near the bottom.

## Induced Prism Difference

The third defect caused by the prismatic effect of corrective lenses is the induced vertical prism *difference* at comparable reading points on two spectacle lenses which are anisometropic (of unequal refractive power). We saw such an example previously on p. 243. Since this optical problem may cause even more symptoms than the others discussed above, we should be concerned with ways of eliminating it, and fortunately there *are* a few ways to help reduce the induced prism difference.

1) Use two *different* types of bifocal segments — one for each eye: segments with different positions of their optical centers would vary their prism induction at the reading point desired; so, if you choose the two properly, you can sometimes eliminate the anisometropically-induced prism difference.

2) Actual compensating prisms *could* be ground into the segments, but this is not cosmetically acceptable to most patients. However, there *is* a feasible method available and it is occasionally used; it is not really a "compensating prism bifocal" but fits more nearly into this category. That method is called "slab-off" or "bi-centric

grinding'' in which a lens is ground so as to remove extra base-down prism, usually from that spectacle lens of the pair which has more minus power in its vertical meridian. In effect, then, ''slab-off'' grinding ''adds'' base-up prism to help balance out any induced prism difference in the reading position.

## *''Slab-Off''*

I am asked so often how this ''slab-off'' grinding is accomplished that, for those of you that are interested, this exposition is given below. This is *not* a very important detail, however.

First, a definition: A lens ''blank'' is a partly-finished lens having either the front or the back surface curvature completely ground and polished. (A lens ''blank'' which has a ''fused'' bifocal segment embedded on the front surface usually has *that* face already finished.) When an order comes to the optical shop to make a lens to a specific prescription, a lens ''blank'' is chosen which has the appropriate front surface power and the proper add already incorporated; the additional power may then be ground on the rear surface to provide the desired overall power. It is just such a lens ''blank'' with a fixed ''base curve'' on the front surface that we use in the ''slab-off'' method.

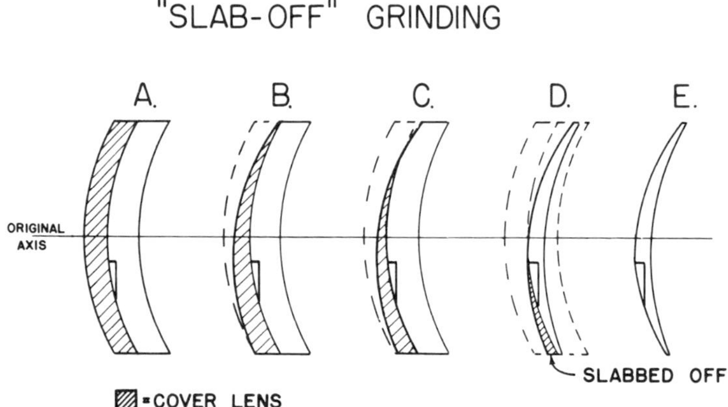

On top of this lens "blank" another "cover" lens (of immaterial refractive index) is cemented (A) and the entire combination is placed on a lens grinding machine; but, it is positioned *off* the original axis of the lens blank, so that the grinding wheel will cut the upper part of the lens first (B). The front surface of this combination is ground and polished with the same curvature as that of the *original* blank so the front surface power remains the same. This peculiar method of grinding creates a new, second optical center which, in the diagrams, is located *below* the old one.

After the lens is ground further (C and D) so as to reach the upper edge of the bifocal, the remaining piece of cemented glass below (covering the segment) is then knocked away, or "slabbed-off", leaving the original lens with two identical curves on the front providing the same surface power above and below. The line of intersection between these two curves (or between *any* two identical spherical curvatures) must necessarily be a straight line. Because of this, it is more cosmetically acceptable to use a flat-top bifocal segment in the "slab-off" method of grinding; the flat top is aligned with the straight intersection line on the front surface.

Now the lens is completed as any regular lens "blank" by grinding away the back surface to yield the proper total corrective lens prescription. The back surface is ground on the same "new" axis as that on which the front surface was just ground. This removes the back area of the lens as shown in diagram (D) and leaves the finished lens (E).

You should be able to see that the segment of glass "slabbed-off" has a base-down prism appearance and, therefore, this method removes base-down prism from the reading area, typically between $1\frac{1}{2}^{\Delta}$ to $6^{\Delta}$.

* * *

The need for the slab-off method of grinding (or, for that matter, any of these methods of compensating for the differential prism in the reading area) is only necessary for the presbyope, since the non-presbyope can always look through the optical centers of his distance lenses where there is no interference by any prismatic displacement whatsoever. This last comment leads us to the third method for avoiding a troublesome prismatic displacement.

3) You could forget completely about using a *bifocal* at all and revert to full, single-vision *reading* glasses (with the necessary plus add power incorporated into the distance prescription). Here you entirely bypass the problem of any anisometropic prism difference which may exist in the reading area of the lens.

## ABERRATIONS

When the term "first order optics" was introduced to describe the level of mathematics we would need in order to show the action of lenses on light rays, I elaborated on the necessary approximation, $\sin \theta = \theta$; I mentioned that for our purposes that was all the refinement necessary to present the material in this book. I lied. There *is* a bit more detail you should have, some of which would not exist were it not for "third order optics". (Have you forgotten this term already?) This material need only be touched on, but must be included to round out your basic understanding of some clinically useful information. It deals with the aberrations of optical systems — the term *abberration* signifying that an optical system fails to produce an image which is an accurate representation of an object.

Just as you might suspect, there is more than one type of aberration. Here we will discuss one which occurs because of the *multi-*wavelength nature of light and others which exist even when dealing with *monochromatic* light. Aberrations can either deform the overall image (such as distortion and curvature of the image plane), or destroy the sharpness of each image point (such as, spherical aberration, radial astigmatism and coma).

Though many special purpose optical systems are specifically designed to eliminate certain of these aberrations, *all* optical systems* possess these defects to a greater or lesser extent. For practical purposes, I will consider the aberrations in two groups — those mainly affecting the optics of the *eye,* and those most applicable to corrective *lenses.*

* Mirrors, however, do *not* introduce chromatic aberration since the angle of reflection is equal to the angle of incidence for *all* wavelengths.

# Aberrations — With the EYE as an Optical Instrument

## *Chromatic Aberration*

When the term "refractive index" was discussed, I stressed that the decrease in the velocity of light which occurred when light entered some medium was dependent not only on the medium itself but also on the *wavelength* of the light. The fact that *each* wavelength has its own, private refractive index number for each medium accounts for chromatic dispersion by a prism. In passing through a prism then, the blue (or short) wavelengths of light are always bent the most, the redder (longer) ones, the least. (This was diagrammed in the section on refractive index.)

As you should now know, lenses are "modified" prisms. In reality we *should* think of a "lens" as a *stack* of prisms of gradually changing prism power as we leave the axis. (In the figure below, such a "stack" comprises a plus lens.)

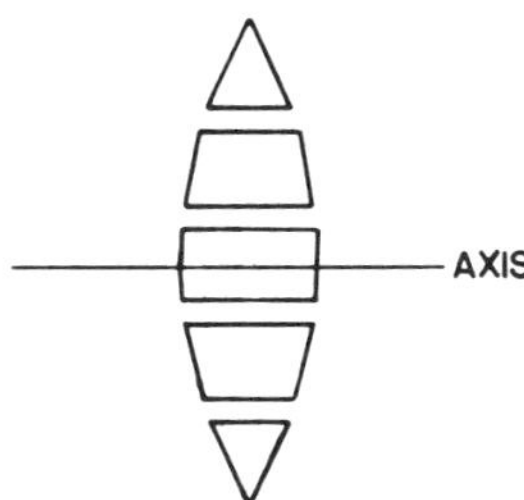


However, for schematic purposes, let's neglect the "changing prism" power aspect and depict a lens cross-section as consisting of only two straight prisms; a plus lens is represented by two prisms with their bases in contact — a minus lens, with the two apices touching.

# LENS ANALOGUES

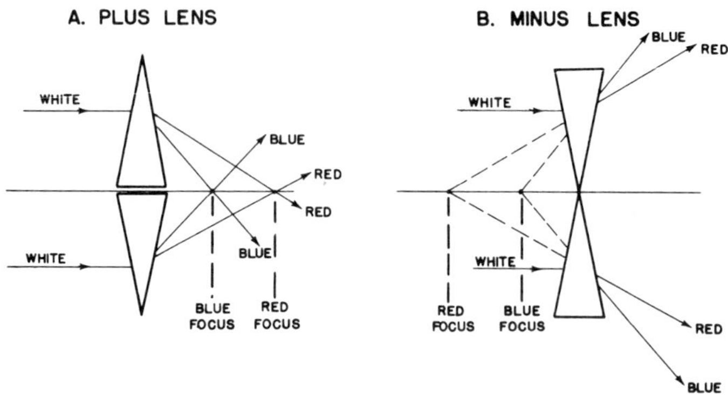


Our plus lens prism analog (figure A) would take bundles of white light (composed of many wavelengths) and disperse that light so that the shorter wavelengths crossed the axis in front of the longer ones.

So, when any plus lens deals with *multi*chromatic or white light, the focus for blue light will be found closer to the lens than that for red light. If *only* monochromatic blue light is emitted by an object, there can be *no* color dispersion in the image, and that "blue" image would come to a *sharp* focus in a single plane which is conjugate to the object plane. For a multichromatic object, however, every wavelength will have a *different* image plane. Since these image planes are consecutively placed axially and compose a "total image", each object *point* will be represented by a chromatic axial *interval* — this is *axial* chromatic aberration. (In figure A, if a screen is held at the blue axial focus, every blue image point will be surrounded by colored fringes and this is known as *"lateral"* chromatic aberration.)

The comparable situation for a minus lens is shown in figure B: the blue focus will always be closer to the lens than the red.

Chromatic aberration, then, is based on the finding that for all media denser than a vacuum, the index of refraction for blue wavelengths is greater than for red; thus, the focal length of *any* lens will always be shorter for blue, and this makes all simple lenses have a greater dioptric *power* for blue light.

Different transparent materials vary in their abilities to disperse white light (that is, they vary in their dispersion indices). So, it is possible to select a plus lens of one material and join it with a minus lens of another composition. The dioptric power of the combination lens can be whatever total one desires; but, since the minus lens' aberration is in the opposite direction to that of the plus, the overall chromatic aberration created by such a lens system can be effectively controlled. This is one method of reducing chromatic aberration in optical instruments.

The human eye is an optical system which *is* naturally subject to uncorrected chromatic aberration; the total refractive power is greater for visible blue than visible red light by about 1.25 D; that is, the chromatic interval in the image (for white objects) is about 1.25 D "thick". In an emmetropic system, the retina is situated in between the red and blue focal planes, probably very close to the focal area for *yellow* light (that portion of the spectrum having the greatest brightness-evoking capacity). The *am*etropic eye too has a chromatic interval, and use of this fact is made in the "Bichrome test" for clinical refraction.

## Bichrome Test

After concluding the clinical refraction of the patient in your examining lane and obtaining the best visual acuity by whatever means you wish, move to the Bichrome (or Duochrome) test to check the prescription monocularly. A standard Snellen chart is presented to a patient so that the one half (say, on the right) is illuminated with red light, the left half with green. (Here, green is substituted for blue as the short wavelength representative. It is used instead of blue because it affords an inherently brighter target than does blue, which would be interpreted as "quite dark" by any patient. Also, the center of the visible spectrum (yellow) is dioptrically closer to half-way between red and green, not between red and blue.)

The patient is asked to state on which colored half of the chart the letters appear clearer to him. If he says the "red" side, *you* can visualize that within his eye, it is the *rear* end of the chromatic interval that must be located nearer to his retina. (See next figure.)

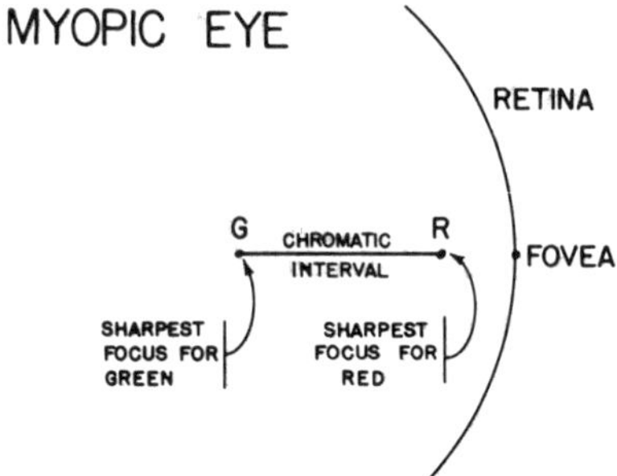


This means that your prescription must be too strong in plus (or too little in minus). To move the entire interval more posteriorly, you must add minus power. However, if the "green" side is clearer, the addition of more *plus* power to the prescription is proper. Our end-point comes when the patient tells you that the letter clarity is equal on the two colored sides. Repeat the test on the second eye; then quickly compare the two eyes for "prescription balance."

The Duochrome test provides a good, useful and simple check; but like all clinical tests, it must be interpreted intelligently. Its dictates should not be treated as gospel.

Another, similar manifestation of the chromatic aberration of the eye is displayed in a darkened projection room. When slide charts which are made up of brightly colored blue lines on a dark background are shown on a screen, the blue lines seem to be non-sharp or have a "defocused" appearance even though they really are in as sharp a focus as the projectionist can make them. The blue lines will appear slightly blurry even to emmetropes, but especially so to the under-corrected myope. Have you noticed this yourself? Maybe you need a little more minus in your spectacles?

## *Spherical Aberration*

So far, we have assumed that all rays which impinge anywhere on a lens surface arrive at a sharply focused image point somewhere. This is not strictly true. (You should recall that in "first order" optics we were really only dealing with *paraxial* rays.) If you study the object-image rays more carefully, you will find the following: Those light rays parallel to the axis which enter any optical system near its periphery are usually subjected to a *greater* refractive power than those closer the axis (see A below). If the entire lens surface is acting, you will find that for every axial object point there will be an axial image which is a "blur-circle" instead of a point. The *smallest* image "circles" will be found in a plane located somewhere within an axial *zone* (similar to Sturm's axial interval or to the "chromatic interval" just discussed), but nowhere will the image point be sharp.

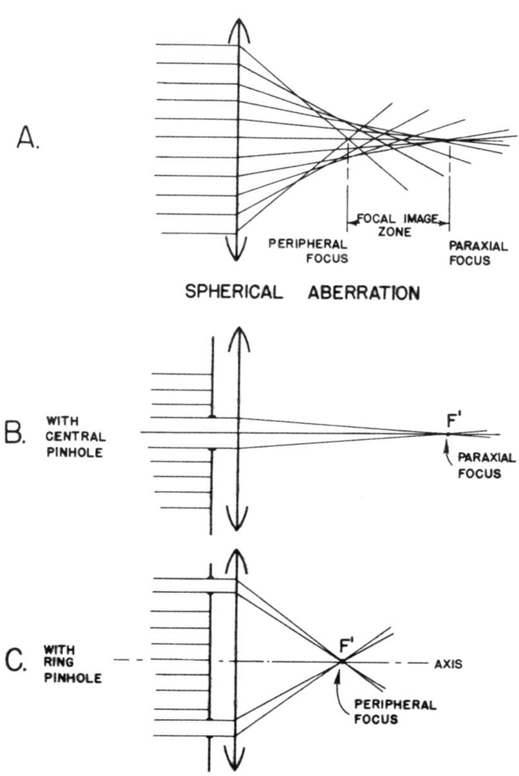

The overall image quality (sharpness) *can* be increased if you narrow the aperture (extent of the lens surface exposed) and thus limit the rays which will participate in the image formation (see B and C above). The aperture opening could be a small axial "pinhole" as described previously in the ametropia section, but it could also be a thin, ring-shaped opening which admits only the peripheral rays. In the latter instance, the lens power utilized would be *greater,* and the sharp, focal image plane would be closer to the lens.

CLINICAL POINT:

The fact that the pupil of the human eye dilates in dim illumination accounts for an interesting clinical finding: most individuals become somewhat myopic in low levels of illumination. The spherical aberration of the eye accounts for *some* (though *not* all) of this myopia ("night myopia"). When the pupil dilates, there is a greater contribution of peripheral rays to the makeup of the retinal image. This moves the clearest zone of that image (that position which provides the best acuity) somewhere forward into the vitreous, hence the myopia.

Since the myopia so produced can be as great as 1.00 to 1.25 diopters, some patients are quite disturbed by the prominent blurring and may even complain of seeing "rings around lights". (The causes of the "halos" here are the ametropic and lateral chromatic aberration blur circles — not the diffraction grating effect of an edematous cornea, as in acute glaucoma). Many such patients are markedly helped by your prescribing — 0.75 to — 1.25 D clip-on spectacles to be worn at night when driving. Keep this clinical problem in mind.

## Aberrations — With the CORRECTIVE LENS as an Optical Instrument

### *Distortion*

Distortion is the result of a differential magnification produced by an optical system. Say an object and its image are centered on the axis of a lens; if axial parts of that object are magnified *less* than parts farther off the axis, pincushion distortion is produced. If the reverse is true, barrel distortion ensues. Compare the following sketches:

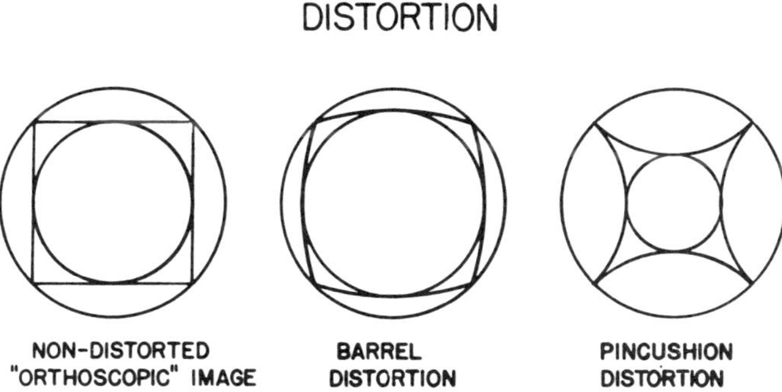


In other words, distortion exists when the overall image *shape* does not correspond well with the object's shape.

Distortion is influenced not only by the *type* of lens, but also by the position of the aperture "stop" used. (The "stop" is the true opening which limits the number and position of rays which will form the final image — like the iris diaphragm in a camera, or the pupil of the eye). For example, the very same plus lens can cause either "pincushion" or "barrel" distortion depending upon *where* the "stop" is positioned. When it is in front of the lens, "barrel" distortion ensues.

Typically, in *our* optics study, the final "stop" is the pupil of the eye. Since the pupil is always in front of the eye's crystalline lens, the eye itself actually exhibits "barrel" distortion, though this is not important clinically. However, the pupil "stop" is always located *behind* any *corrective* lens; so a plus corrective lens will generate "pincushion" distortion. This is very apparent to an aphakic patient. As he looks at a doorway through his specs, the side posts will bulge inward; in fact, so much so that he may fear he will not be able to squeeze through the passageway. (As he gets closer to the doorway, the passage "widens" optically to let him through!)

The finest camera lenses introduce very little image distortion not only because the lens curves are manipulated by computer to arrive at the best design, but also because the diaphragm opening which admits the light is positioned very carefully.

The following aberrations — radial astigmatism and coma — also disturb the retinal image detrimentally. These are created because the incident light does not strike the corrective lens perpendicularly, but at some angular inclination to the axis (glance at the next diagram). This situation does have clinical significance.

## *Radial Astigmatism*

By far, the most important of the off-axis aberrations is radial astigmatism. Look at a small bundle of off-axis rays which strike the lens obliquely:

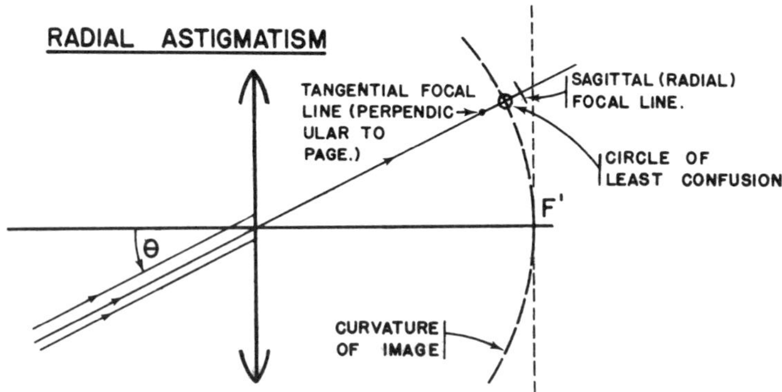


Instead of coming to a point focus in the secondary focal plane as predicted by "first order optics", we find there is an *astigmatic* image — a full, honest-to-goodness Sturm's interval, complete with circle of confusion. It is of interest that not only is astigmatic imagery induced by the obliquely incident rays, but the circle of confusion is moved forward, out of the secondary focal plane — that is, there is plus *sphere* induced also. The greater the obliquity, the greater the induced astigmatism *and* sphere.

In the above diagram, the increase in sphere is shown by the dotted line which is the locus of all the circles of confusion. In three dimensional aspects, these loci make up a curved surface. That curve represents the position of sharpest imagery — where film *should* be located if this were a camera, or where the retina should be if this were an eye. It is amazing how close to the true retinal curvature this image plane is!

(The *best* image plane created by any lens, even one fully corrected for radial astigmatism, is *curved* and is called a *Petzval* surface. This curvature is also considered to be one of the aberrations of an optical system and is called *CURVATURE OF THE IMAGE FIELD.* For obvious reasons, all fine camera lenses are designed to keep the image field as flat as possible.)

Back to the lens tilt — the quantity of astigmatism induced is proportional to the tangent² of the angle of tilt; the *sphere* is *approximately* 1/3 of the induced astigmatism. (*Calculated* — and forgive the formula but there is no other way to show you this — the actual amount of the sphere is $(1 + 1/3 \sin^2 \theta)$ for each diopter of power of the original lens for a tilt of $\theta°$.) What *is* important is that the *axis* of the induced cylinder is always the same as the meridian around which the lens is pivoted. When dealing with spectacles, you'll find that the increased tilt almost always occurs about the 180° meridian.

When a *plus* lens is rotated about the 180° meridian, *plus* cylinder axis 180 is induced. (In the above diagram, you can see that the induced plus cylinder × 180 has moved the *horizontal* line focus forward, closer to the lens) and the induced sphere is also *plus.* (The horizontal line here is called the "tangential" focus.) On the other hand, a *minus* spherical corrective lens rotated about its 180° meridian induces *minus* cylinder × 180° and *minus* sphere. This is easy to remember: A plus sphere tilted produces plus cylinder and plus sphere. A minus sphere tilted produces minus cylinder and minus sphere.

A brief look at a few specific examples will give you some insight into the *amounts* of radial astigmatism we are talking about when tilting about the 180° meridian:

*TABLE V*

*ASTIGMATIC and SPHERICAL ERROR INDUCED BY TILTING A CORRECTIVE LENS*

| Untilted, spherical lens | 10° tilt | | 20° tilt | | 30° tilt | |
|---|---|---|---|---|---|---|
| | Sphere | Cylinder × 180 | Sphere | Cylinder × 180 | Sphere | Cylinder × 180 |
| − 10.0 | − 10.10 | − 0.31 | − 10.41 | − 1.38 | − 10.95 | − 3.35 |
| − 6.0 | − 6.00 | − 0.22 | − 6.25 | − 0.83 | − 6.57 | − 2.00 |
| + 2.0 | + 2.00 | + 0.06 | + 2.08 | + 0.28 | + 2.19 | + 0.67 |
| + 8.0 | + 8.04 | + 0.20 | + 8.32 | + 1.10 | + 8.76 | + 2.68 |
| + 12.0 | + 12.12 | + 0.37 | + 12.50 | + 1.66 | + 13.14 | + 4.02 |

NOTE: If you keep in mind the *approximate* changes created by a *10 D* lens, the others can be easily figured out as they are all roughly proportional, diopter for diopter, to the inductions given for that one.

CLINICAL POINT:

If you look around a lecture hall the next time you are in one, you'll see many in the audience sitting with the temple pieces of their spectacles rotated upward. They will be peering through their lenses which have an exaggerated rotation about the 180° meridian. Most likely these are myopic individuals who are in need of more minus in their corrective lenses (and/or more minus cylinder × 180). They are tilting their lenses (as most myopes discover empirically) because this increases their acuity somewhat.

The principle is that the tilt of a minus glass in this direction increases minus cylinder × 180° *and minus sphere,* (that is, it provides more of a *minus* spherical equivalent). The axis-180 minus cylinder effect may disturb the retinal image quality slightly, but not much, since horizontal-axis cylinder is not too noticeable; *but* that minus spherical equivalent provides their lenses with some extra minus power. This helps to more "fully correct" an undercorrected situation. Additionally, the undercorrected myope will also benefit by having his minus corrective lenses *closer* to his eyes (enhancing the lens' minus effectivity, as previously described). The tilting procedure facilitates this proximity, and the newly created minus adds to that produced by the radial astigmatism.

So, now you know. The bizarre spectacle position is produced by all these undercorrected myopes in the audience and helps them read the blackboard or see some detail in a projected slide more clearly. (But there is another possibility; maybe the spectacle ear pieces are pinching their ears!)

*NOTE:* Although the "tilters" mentioned above are usually myopes (since the odds favor the presence of myopia in any audience), realize that some *could* be undercorrected hyperopes, who would profit by induced *plus* cylinder × 180° and *plus* sphere. However, in contrast to the myope, the hyperope will add to his tilt effect by sliding his spectacles further *away,* down his nose, to gain still more plus effect.

* * *

Another point you should understand: A corrective spectacle lens is *normally* kept somewhat tilted with respect to the *visual* axis of the eye. This tilt is a compromise position to allow the visual axis to strike the lens perpendicularly during most visual tasks.

The normal eye constantly varies the position of its visual axis, typically between straight-ahead-gaze and downgaze for reading; (upgaze is quite unusual). If the visual axis is to strike the corrective lens as perpendicularly as possible in the *reading* position (in an effort to minimize the aberration just discussed), the tilt of that corrective lens downward, away from strict verticality, would have to be about 15°. However, since those same lenses also have to serve for *distance* seeing, a tilt compromise is usually made at about 7½°. This is called "pantoscopic" tilt and is built into most spectacle frames so that the lenses are closer to the face at their bottom edges, about 7½° off verticality; but too much such pantoscopic tilt will engender significant radial astigmatism, especially with highly ametropic corrections as already detailed.

### "Corrected Curve" Lenses

Currently available spectacle lenses are designed to reduce as much as practical the radial astigmatism they generate. By properly choosing the front and back surface curvature for each distinct power desired (or, more practically, for each *range* of powers), this astigmatism can be effectively controlled. The "corrected curve" lenses manufactured by competing industrial giants do attempt to approach the proper shape. Punktal®, Orthogon®, Tillyer® — all are practical *compromises* to theoretical ideals to control radial (marginal) astigmatism and curvature of the image field.

## *Coma*

For us, this is the least important aberration. However, it also enters into retinal imagery considerations with off-axis objects and will tend to destroy the sharpness of details. It has a completely different optical basis than does radial astigmatism.

Coma is probably most easily defined as spherical aberration which occurs for off-axis object points. When you consider *axial* image "points" (or circles, in the presence of spherical aberration), the distribution of light in those blur circles is symmetrical and the "blur circles" are round. For off-axis single object points, on the other hand, coma creates an *asymmetrical image* (as shown), that spreads out to look like a comet, even though it still is supposed to represent a *single* object point.

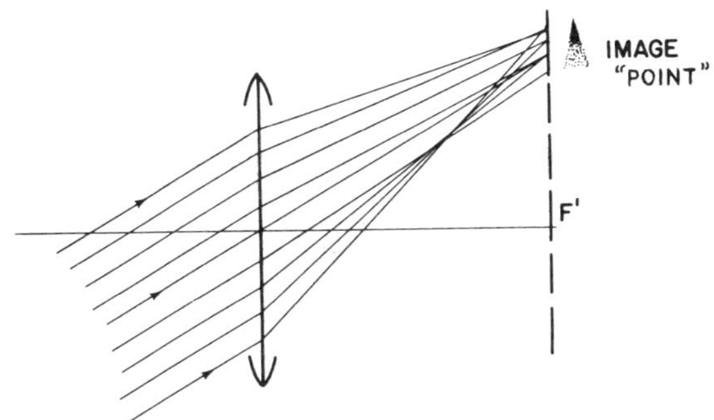


Coma is due to unequal magnification by different zones of the lens and can be reduced or eliminated in critical applications, such as fine camera and microscope lenses, by proper lens design and appropriate positioning of the aperture "stops".

## PRACTICAL INSTRUMENTS

We are well nigh bringing this opus to a close. Before doing so I feel compelled to elaborate on the optical principles behind the use of some instruments so useful that I cannot conceive of any study of *practical* optics leaving them out. They are the *indirect ophthalmoscope* (personal bias causes me to mention this first), the *retinoscope,* the *lensometer,* and the *keratometer.* Sure, I know I've neglected others. (Some have been mentioned earlier.) These four, however, occupy so important a role in your daily clinical lives that your intellectual curiosity should prompt you to discover how they work. Any others I must leave to you.

# The Indirect Ophthalmoscope*

You are familiar, I am sure, with the principles of a slide projector. An intense source of light illuminates a transparent slide from behind and makes that slide a luminous object on its own. That luminous object emits light which is picked up by the projector lens. This lens is situated so that the object is just beyond its primary focal plane; then the final image can be projected onto a screen. The key here is the *luminous* object.

An "opaque" projector — the kind you use to project images of sheets of paper or pages of a book onto a screen — also works by making the opaque object material "luminous", in this case by shining an intense light onto it; the object *reflects* that light energy instead of transmitting it, as does the slide. Because the illumination system here is not quite as efficient as that for the slide projector, and also because the material to be projected is usually bigger, the illumination chamber of this projector is large and the projection lens diameter is quite wide to allow it to gather as much of the reflected light as possible so that a bright image can be placed on the screen.

Now what has all this to do with the indirect ophthalmoscope? Everything. For an observer to see the fundus in *any* kind of ophthalmoscopy, it must be illuminated. With the direct ophthalmoscope, you are simply using "flashlight" illumination to examine the innards of a dark cave. In *indirect* ophthalmoscopy you also must make the fundus a luminous object. Here, however, the eye acts like a miniature opaque projector and projects that fundus image out into space. This is accomplished by shining an intense source of light through the pupil (the only way to get the illumination in); the fundus image is then projected out.

An emmetropic eye with an illuminated fundus would project the fundus image to infinity — the position conjugate to the retina. However, this distance is too impractical. (For one reason, by the time the image reached infinity, there would be so little light energy

---

* Condensed from, Rubin, M. L.: "Optics of Indirect Ophthalmoscopy" Survey of Ophthalmology, 9:449-464, October, 1964.

there that it would be too feeble to be seen.) If instead, we capture some of those parallel light bundles stemming from the eye with a hand-held condensing lens (say, one of + 20 D), a new luminous image will be created at *its* secondary focal plane. (Notice that this will *not* depend on how far the lens is from the eye!) While there is no screen stationed at the F′ plane to intercept the rays, a real fundus image *is* there nonetheless; it is called an *aerial* image, and is, of course, inverted when compared to the orientation of the fundus itself.

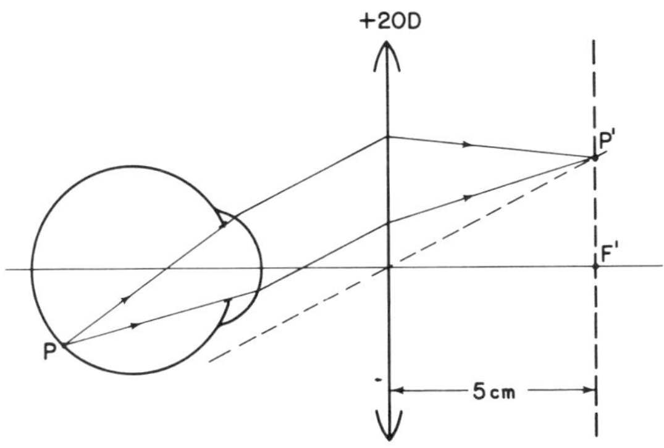


That aerial image has "thickness" representing any "thickness" present in the fundus itself; any excavation of the nerve head or elevation of a choroidal lesion will be manifested in the "depth" of the image. The *longer* the focal length of the condensing lens, the farther from the lens the image will be and, therefore, the more its *lateral* magnification. You should recall that the *axial* magnification of the image (that is, the "thickness" of the image compared to the "thickness" of the object) varies as the *square* of the lateral magnification afforded by the condensing lens. So, the axial image dimensions are *especially* thickened by the *lower* powered (longer focal length) condensing lenses, which provide a very "depthy" aerial image, and it is partially on the basis of this image depth that the observer's stereoscopic view of the fundus is determined.

The diagram below shows the retina imaged by a + 14 D lens and the magnifications attained:

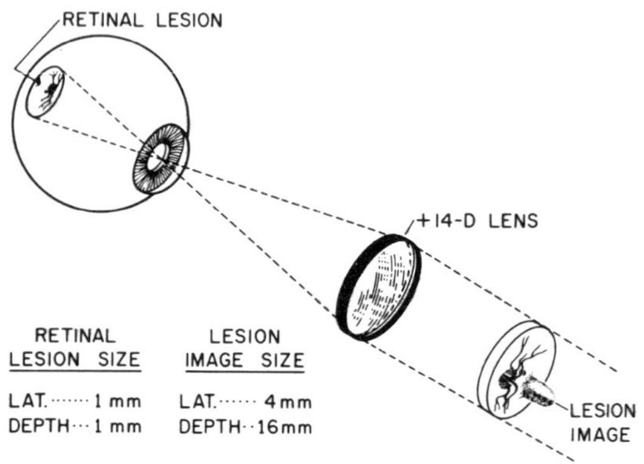


Fine. An image is there in space. But if someone is to see that image, light from it must reach the observer's eye and get through his pupil. But, only a small quantity of the light in that real aerial image can enter the observer's eye. (That tiny portion is shown in the shaded area of the next two diagrams.)

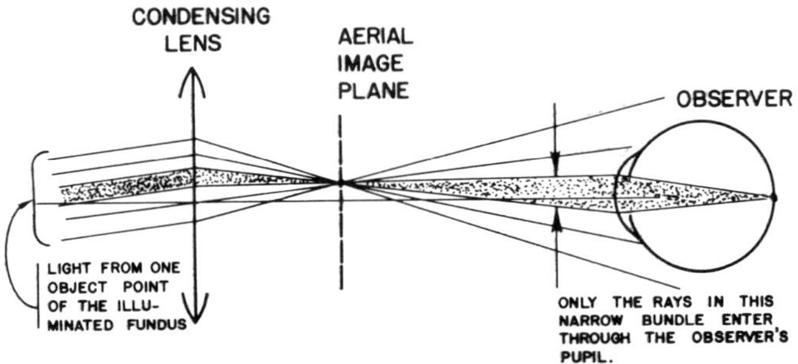

That light which does enter the observer's eye (if it is traced backwards), must originally have come through a comparably small area of the *patient's* pupil as shown below:

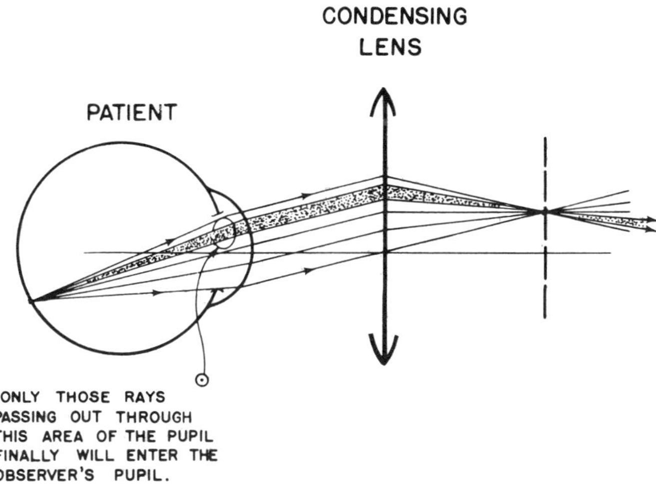


The further back from the lens the *observer's* eye is positioned, the smaller will be that area of the patient's pupil required to allow this visualization to occur. (So, if the observer moves backwards, he can more readily visualize a fundus through a small pupil.) It should take no great insight to see that the observer's pupil and that tiny area of the patient's pupil are conjugate; that is, the spot in the patient's pupil *is* the image of the observer's pupil by the condensing lens. (I call this object-image relationship a ''conjugacy of pupils''.)

So, although the condensing lens *could* be positioned *anywhere* in front of the patient's eye if its *only* job was to form an aerial image of the fundus (the image would always be at its F' plane if the patient's eye were emmetropic), the fact that that same lens must also image the observer's pupil into the patient's pupillary area *fixes* the position of that lens. If the observing eye approaches, in order to maintain this "conjugacy of pupils" the hand-held lens must be pulled toward the observer somewhat. An added benefit of moving closer is that the observer gets closer to the aerial image and obtains the reward of a larger image of it on his own retina.

As if that hand-held lens didn't have enough to do; don't forget we must somehow illuminate the patient's fundus. That same lens must also project *into* the patient's eye the image of the illumination source, *and* it must do so through an area of the patient's pupil which is not "occupied". As just shown, that image light which comes out of the pupil and heads for the observer's pupil has its own reserved area. This area must not be interfered with by the incoming illuminating beam. (This is another of Gullstrand's innovations — that is, he stressed the necessity of keeping the incoming and outgoing beams separated to keep glare at a minimum for the observer.)

So far, well and good. We have now shown the principles of monocular indirect ophthalmoscopy, where the observer uses only one eye. But one of the great advantages of examining a fundus by this indirect method is that it affords us the means of having *stereoscopic* visualization. For this to occur, *both* the observer's pupils must receive light from the "thick" aerial image, and the only way this can occur is if *both* his pupils are imaged into the patient's pupil. For all this to happen within the plane of the patient's pupil, it should be obvious that that pupil must be as dilated as possible.

However, we are presented with an optical problem. The observer's interpupillary distance is roughly 65 mm. This separation must be reduced if we are to squeeze the images of *both* observer's pupils into a patient's pupil which is only 6 or 7 mm in diameter. If the *observer* were far enough away (say 100 cm or more), the image of *both* his pupils *could* project into the patient's, but the observer's arms are not usually that long! So, we turn to another option and that is to optically reduce the observer's interpupillary distance with mirrors (actually, totally internal-reflecting *prisms*). These are placed in such a way so as to narrow the interpupillary separation from 65 to about 15 mm.

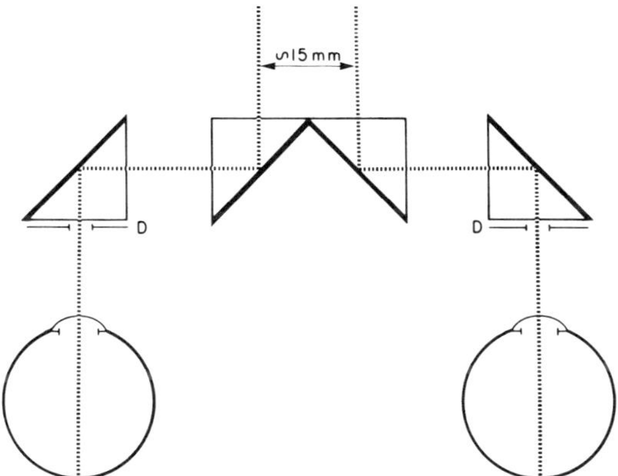


The beauty of this device (pictured above) is that the outer mirrors are made movable and can be adjusted to suit any individual observer's p.d. The actual, "optical" separation of the pupils, however, is *fixed* by the non-movable inner mirrors at about $\frac{1}{4}$ the *normal's* p.d.

There is also about 2 D of plus power built into the eyepieces to relieve any strain on the observer's accommodation (and concomitantly, on his accommodative-convergence) as he attempts to view the aerial image.

(Incidentally, the narrowed instrument p.d. reduces the image disparity available for the two eyes for *all* objects: So, as you look about through the eyepieces, space seems "flattened" — your stereopsis for everything is reduced; you may fumble and find it very difficult to pick up the traction sutures around the extraocular muscles during a retinal operation.)

So with the narrowed p.d., both pupils *can* be squeezed optically into the patient's pupil. The illuminating light can enter through a separate pupillary area.

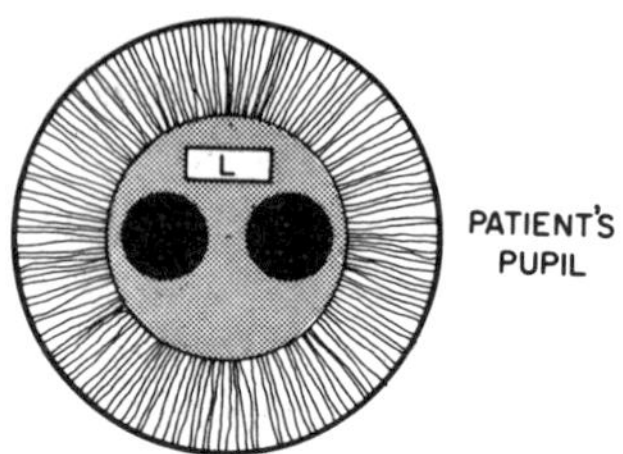


*Voila!* — a full, stereoscopic "Technicolor" view of the patient's fundus pops into view and seems to "fill" the condensing lens.

The extent (field of view) of the patient's fundus which is visible by this method is great — roughly 25° versus only 8° with the direct ophthalmoscope — almost 10 times as large an area becomes visible at one time. For a given power of condensing lens, the extent of field visible is directly proportional to the *diameter* of the lens. A 20 D lens of *large* diameter affords a greater field of view than one of smaller diameter. (Here, I'd like to "plug" the new Nikon 20 D aspheric lens which affords a beautiful, crisp aerial image with the same magnification (linear and axial) as that provided by a standard 20 D lens; but in addition, the diameter of that lens is large enough to provide the same wide field of view as that given by a standard *30 D* lens!)

## *Light Reflexes*

As we have pointed out, ideally, if a reflex-free field of view is to be obtained by an examiner, the illuminating and observation systems must be kept completely separated as the light rays course in and out — through the observing lens, the patient's cornea, pupil, and crystalline lens. If the image of the illuminating system is projected by the condensing lens into the patient's pupil through an area *away* from the position where the images of the observer's two pupils are imaged (last figure), the entering and outgoing light paths *are* kept separate and no corneal or crystalline lenticular reflexes would be visible. However, things are not always ideal. The light source utilized in the modern indirect ophthalmoscope headpiece is too broad to be focused easily in this manner, though it *is* possible — especially with the new MIRA* "small pupil" ophthalmoscope. The latter is a very flexible instrument. It is designed with enough adjustments to allow not only compensation for the observer's p.d., but also for the diameter of the *patient's* pupil; that is, the "instrument" p.d. can also be varied as can the position and size of the light source. All this allows examination of a fundus through a very small pupil with even some stereopsis possible. Furthermore, the "instrument" p.d. can be *widened* to take advantage of a fully dilated pupil to provide a greater stereoscopic effect than can the average indirect ophthalmoscope.

But, in the average clinical situation with a standard indirect ophthalmoscope, a completely reflex-free condition is unlikely. Careful observation by an examiner will reveal that there are three bright and annoying reflexes seen when he tries to visualize the fundus through the hand-held condensing lens; these are most troublesome to the beginning ophthalmoscopist. These three are the light reflections which occur from the anterior and posterior surfaces of the *condensing lens* plus the one from the anterior corneal surface. Since the condenser is a biconvex lens, the anterior convex surface forms a *virtual* reflected image of the illuminating headlamp, and the posterior surface, being concave to the illuminating source, forms a *real* image. These images therefore will move in opposite directions if the lens is tilted. The higher the dioptric power of the condensing lens, the more will be the tilting necessary to move the reflexes away from the center of the field of view. The small, bright *corneal* reflex moves only slightly with the lateral motion of the condensing lens, but it can be moved sufficiently so that the specific area being examined is not obscured. All this is usually discovered empirically by the clinician, who learns quickly to maneuver about even though these reflexes are present.

---

* Medical Instrument Research Associates, Boston

## *Image "Brilliancy"*

A widely dilated pupil facilitates the *admission* of light in any type of ophthalmoscopy. However, as you have seen, not all this light is used in the formation of the visible image. When one views an illuminated fundus with a direct ophthalmoscope, the diameter of the *emergent* beam through the dilated pupil is "stopped down" by the narrow aperture in the ophthalmoscope head (about 3 mm), but this is close to the patient's pupil, so a relatively high proportion of emergent light enters into the image, though, of course, *some* light is "wasted". In indirect ophthalmoscopy much more of the light is lost, since the only rays emerging from the patient's pupil which are *really* effective in the formation of the visible image are those which come through the tiny pupillary areas which contain the images of the observer's pupils. Thus, as long as the patient's pupil is sufficiently dilated to allow the unhindered entry of the illumination beam and the free emergence of the image-forming beams, having the pupil still more dilated can *not* further increase the brightness of the aerial image (except when you are looking at the fundus periphery, where the dilated pupil reduces vignetting). What the dilated pupil *does* accomplish, however, is it allows more leeway in the positioning of the observer's eyes, the viewing lens, and the illumination source, thus simplifying the entire viewing procedure.

So, if it's not the dilated pupil, what is it that makes the image seen in indirect ophthalmoscopy so bright? You should give most of the credit for the bright image and the good visibility to both the truly intense light source available in the modern instruments plus the large field of view which provides a "bird's eye view" of a large fundus area. Thus, an indirect ophthalmoscope seems to "cut through" the opacities in the optical media of the eye examined and affords the opportunity to see a fundus which may be hidden from the view of an observer who attempts to use a direct ophthalmoscope.

This is the story of indirect ophthalmoscopy and is complete enough for this time and place. You should now understand the basic

operations involved and should picture to yourself the following cartoon summary of the optics of your trusty instrument:

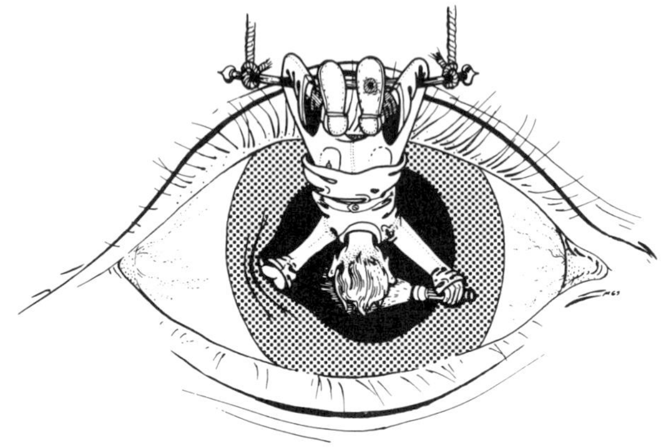

## The Retinoscope

### *Principles*

The retinoscope provides us with an opportunity to determine the refractive error of an eye objectively. For an anesthetized infant, for example, from whom we get no subjective response to "Which is better, one or two?", the retinoscope affords us the only practical way to measure the refractive state.

The retinoscope is a small projector which emits a *spot* or *streak*-like image of the lamp filament itself. (The "streak" scope is now the most popular design, and deservedly so since it makes the determination of astigmatism so much simpler). This "projector" is aimed at a patient's eye.

Most commercially available scopes provide for a variable focus of the filament image. The retinoscopist can project that image onto

a wall any distance away and focus it sharply, from infinity to any place behind or in front of the patient's eye; that is, he can vary the vergence of the focused image. In fact, the image can even be made virtual with *divergent* light; that is, it can optically be positioned so that it seems to arise from *behind* the observer.

If the projected light emitted from the retinoscope is made *either* divergent or parallel or convergent to a point somewhere *behind* the patient, the optical effect as far as the patient's eye is concerned is the same; and this is called the "plano-mirror" effect. If, on the other hand, the projected filament image is located somewhere between the observer's eye and the patient's eye (the only space left!), the retinoscope is said to have a "concave-mirror" effect.

This flexible focusing arrangement allows a swift change from "plano-mirror" to "concave-mirror" effect, but should only be considered an instrumental refinement; it is *not* absolutely necessary to the performance of clinical retinoscopy. Actually, the retinoscope's focus may be *pre-set* to yield either a "plano-mirror" *or* "concave-mirror" effect. In fact, even an ophthalmoscope can be used to function as a "plano-mirror" retinoscope!

The reason that actual focusing of the streak is not a critical requirement of retinoscopy is that the patient's eye is not meant to accommodate on that streak image anyway; the blurry image of the light streak which does fall on the patient's retina simply becomes a new, tiny *source* which can be thought of as "originating" there. The motion of that source on the retina is directly dependent on the retinoscope's movement. When the retinoscope is set for a "plano-mirror" effect, the direction of motion of the light source on the *retina* will always (*regardless* of the type or amount of refractive error of the patient's eye) be in the *same* direction as the motion of the retinoscope light as it sweeps across the patient's face; that is, any movement of retinoscope's image will impart a "like" motion to the retinal image (see figure A below). In contrast, the retinoscope with a "concave mirror" effect (which images the filament *between* the observer and the patient) will create a motion of the "new source" on the *patient's* retina which is always in the opposite direction to the motion of the retinoscopic streak. (See figure B below).

A. PLANO-MIRROR EFFECT

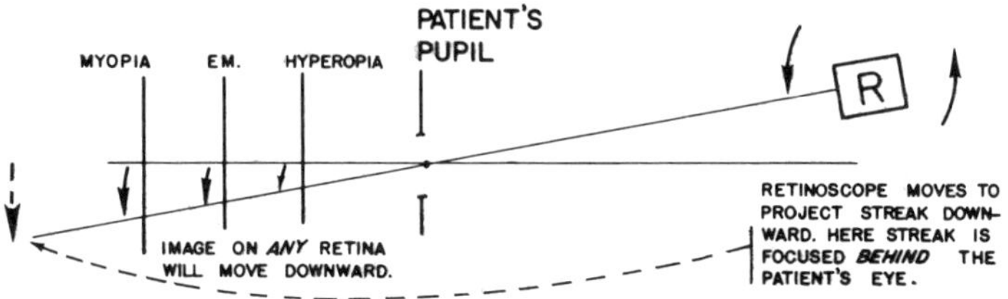


B. CONCAVE MIRROR EFFECT

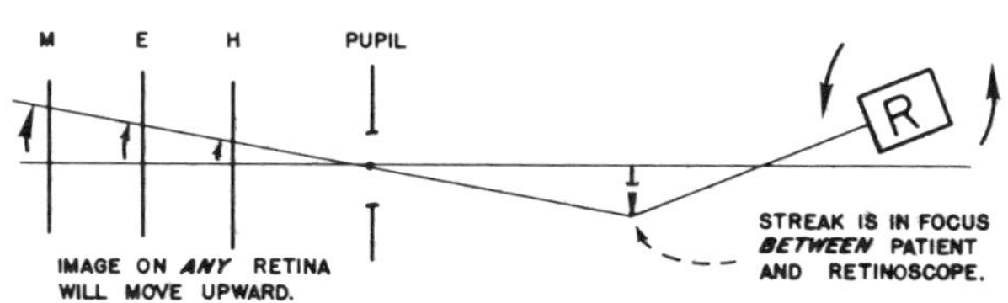


The relative motions we have just discussed are between the image of the retinoscope light and the patient's retinal image.

After the (streak) image is projected into the patient's eye, that eye will treat the bright but blurry image as a new source which now originates at the patient's retina. The dioptric power of his eye projects it back out toward the observer. With no spectacle lens present in front of the patient's eye, the projected image from the patient's retina *must* lie at the far point plane of that eye, since the retina and the far point plane are conjugate.

The observer watches this light through the peephole ("stop") of his retinoscope. The light he sees coming through the patient's pupil is actually coming from (or aiming towards) the far point of that eye. But it still *seems* to be located within the patient's pupil.

Depending on where the patient's far point plane is relative to the observer, the motion of that image projected from the patient's eye will either appear to the observer to move in the same ("with")

direction or in the opposite ("against") direction as the motion of his retinoscope light.

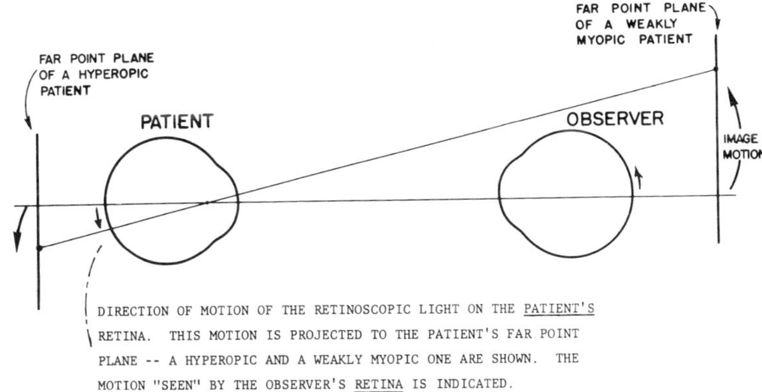

DIRECTION OF MOTION OF THE RETINOSCOPIC LIGHT ON THE PATIENT'S RETINA. THIS MOTION IS PROJECTED TO THE PATIENT'S FAR POINT PLANE -- A HYPEROPIC AND A WEAKLY MYOPIC ONE ARE SHOWN. THE MOTION "SEEN" BY THE OBSERVER'S RETINA IS INDICATED.

Study the diagram above:

If the patient's far point plane is located anywhere *except* between himself and the observer (where it *would* be in moderate to severe myopia), a downward motion of the image on the *patient's* retina will always produce a light motion in the *upward* direction on the *observer's* retina; and, naturally, any image movement which moves upward on the observer's retina is interpreted by his own consciousness as *downward* movement. Thus, with the retinoscope streak moving downward, he would perceive a "with" motion — a downward image movement coupled with a downward retinoscope movement.

But, if the far point plane (and, therefore, the specific location of the projected image from the patient's retina) is *between* the patient and the observer, then, as the light on the patient's retina moves down, the image *projection* (in the far point plane) moves upward (as shown below).

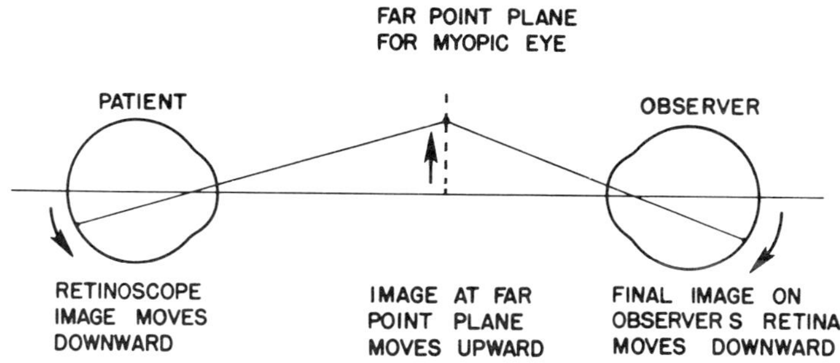


Since here, this image is in *front* of the observer's eye, its upward movement will cause a downward movement of the final image on the *observer's* retina. His brain will interpret this as *upward* and, therefore, "against" the downward motion of the retinoscope light.

The observer sees all this light-image motion as taking place within that tiny area on his own retina that contains his image of the patient's pupil. For you optical purists, look at a sketch of what *is* on the observer's retina:

## ON THE <u>OBSERVER'S</u> RETINA

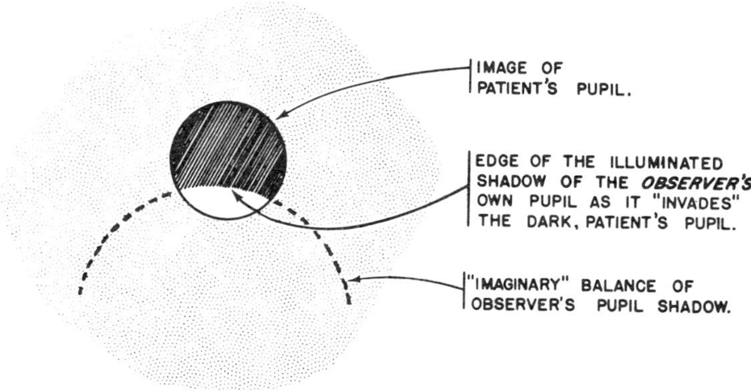

The image of the *un*illuminated patient's pupil is there as a dark spot. As the light emanating from the patient's far point plane is projected towards the observer's eye, it casts a shadow of the *observer's own pupillary margin* onto his own retina. The part of this latter shadow that corresponds to the *observer's* central pupillary *opening* (shown above) is obviously *illuminated* by the entering light and is not dark; however, the invading light is limited to that area shown because of the particular aperture "stops" of this optical system. What can be said is that the closer to the observer's eye the patient's luminous retinal image is located, the larger and brighter will be the illuminated shadow and the quicker it will move. In summary, the edge of the *light* reflex seems to invade the dark area which is the image of the patient's pupil; in reality, the *border* of this light represents the shadowy margin of the observer's own pupil (or, his *retinoscope's* aperture "stop", whichever is the smaller).

So, in retinoscopy, (actually *"skia*scopy", since one is viewing *shadow* movements), the observer's first task is to compare the motion of his retinoscope light across the patient's face to that of the light emanating from the patient's eye. The motion will be "with" or "against". His next step is to attempt to influence the motion he observes by placing appropriate lenses in front of the patient's eye — plus lenses are used to counter any observed "with" motion; and minus lenses, the "against". As these lenses are added, the observer continues to watch their effect on the image motion through the retinoscope "peep hole". His goal is to "neutralize" this observed motion with the lenses. At "neutrality", a small change in the added lens power will cause a prompt reversal of the direction of image motion.

## *Specific Optics*

As stated, when starting retinoscopy with *no* lenses in front of the patient, the image of the retina is projected to the far point plane of that eye. As lenses are systematically added, that retinal image is, in turn imaged elsewhere, plus and minus lenses moving that image as you'd expect. At "neutrality" the projected image from the patient's retina is at the "peep hole" aperture of the retinoscope (or the observer's pupil). When the image is immediately in front of it, a rapid "against" motion is seen; when just behind, a fleeting "with" motion is evident. But *at* "neutrality", the image of the patient's retina is imaged precisely at the retinoscope, *wherever* that instrument happens to be located. *This* is the key principle of retinoscopy. At the end-point (neutrality), the patient's retina is made conjugate to the retinoscope. Do not miss this forest for all the trees (the "trees" being the analyses of the various image motions), even though the specific trees and leaves are interesting.

## "NEUTRALITY" END POINT

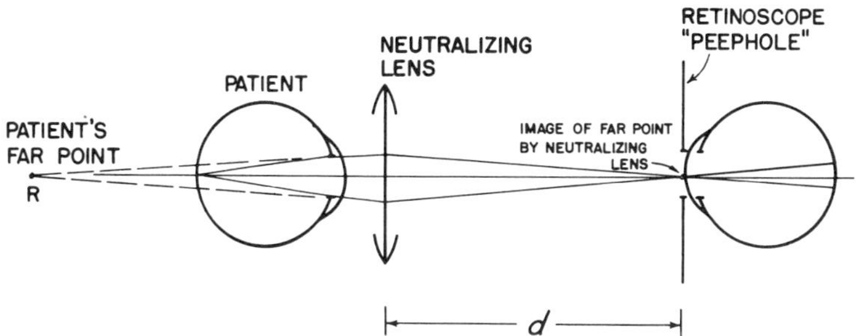


As shown above, then, at "neutrality" we will have found the neutralizing lens power necessary to image the retina at the retinoscope. But, that isn't enough; what we *really* want to know is the *"corrective"* lenses for this patient's eye, (that is, what the refractive *error* is). For this we must image the retina not at the retinoscope, but further back behind the observer, at *infinity.* This will always require some additional minus lens power, the exact amount of which would depend on how far away the retinoscope is from the patient's eye. If this latter distance is $d$ (called "the working distance"), the lens always has to be of power $\frac{1}{d}$ to accomplish its task. If $d = 50$ cm, a $-2$ D lens must be added to the lenses already in place at "neutrality" to yield the full "corrective" lens for the patient's eye.

For an astigmatic eye, each major meridian must be neutralized separately, and this may be done with spheres or cylinders. Though either plus or minus cylinders *can* be used for retinoscopy, *plus* cylinder neutralization (with the use of a *"plano-mirror"* retinoscope effect) is much to be preferred since it makes the observer's task easier. (For all practical purposes, it is much simpler to recognize and neutralize "with" motion than "against", and *plus* cylinders force you to work with "with" motion.)

The fine points of practical "streak" retinoscopy are presented by Jack Copeland ("Mr. Retinoscopy" himself) in his "how-to-do-it" manual.*

## *Monocular Diplopia*

This might seem to be an unusual spot for this subject, but you will soon see why it is placed here.

### THE WOMAN WHO SAW TOO MUCH**

A 50-year-old secretary appeared in my office with the following story: a few months ago she had begun to see that the lines she was typing were definitely "double." This symptom gradually worsened, so that now everything she looked at was tainted; that is, all subjects appeared as if they had a "will-o-the-wisp" quality, with a faint, offset, secondary image superimposed. Moreover, this montage was seen only with her left eye! I smugly asked how she was so sure that her right eye contributed nothing to the diplopia and she embarrassed me with the quick retort, "I lost that eye in a childhood accident." I tried to recover my shattered composure with an inane, "That sure is a beautiful prosthesis; just where did you have it made?"

Using a *single* Snellen target line, I found that her left eye showed a best-corrected visual acuity of 20/20 with her present glasses. During this test, she mentioned spontaneously, "There are *two* lines visible; the lower one is clear but the upper one is blurred." Moreover, I was amazed to find that the blurred line *could* be sharpened to 20/20 too, with a —3 +1.50 ×43° correction over her previous prescription. (This additional lens, of course, blurred the previously clear line.) Diagnostician that I am, I fully anticipated finding some slit-lamp evidence of an early cataract, but the examination with a slit revealed no sign — not even a hint that I could detect — of any lenticular (or corneal) bubbles or opacities!

On fundus examination, as I peered through her own eye's optical system with my direct ophthalmoscope, I had another surprise; I too saw double — a blurred *and* a clear image of each single retinal vessel! At least the existence of a bona fide monocular diplopia was confirmed. By using yet another objective test for a quick confirmation of the optical error, I convinced myself that she indeed did have an early cataract.

Following my examination, confidently, I sat down with the patient and analyzed her clinical problem. Since she is still engaged in her vocation without encountering too much difficulty, she could continue to do so as long as she can, especially since she has but one eye. The latter finding should cer-

* OPTEC Inc., 3543 N. Kenton Avenue, Chicago

** Reprinted from Rubin, M. L., Survey Ophth. 16:382-3, courtesy Williams and Wilkins, Baltimore.

tainly flash the caution signal and lead a surgeon to take a very conservative approach. However, when the diplopia does become too great a burden for her daily routine, one can try a simple trick which occasionally temporizes; prescribe a trial of pinhole lenses. *(See last sentence, page 190.)* If this doesn't work (or her visual acuity drops to a distinctly poor level for her everyday needs), only then will I recommend sharpening the keratome and extirpating the culprit of her difficulties!

Before I let *you* in on the objective test that I used to detect that there was indeed a real optical error present, let's discuss monocular diplopia for a few moments.

"Double-vision" is a most peculiar symptom, especially when it occurs with only one eye doing the seeing! This is monocular diplopia or polyopia. Sometimes patients describe a complaint which initially seems like polyopia; on careful questioning, however, the symptom really will be found to be a *distortion,* where "straight" lines appear kinked or crooked. Such patients may have a preretinal membrane with retinal wrinkles ("pucker"), submacular fluid, or choroidal elevations. They do not exhibit true diplopia. Real monocular diplopia has a number of causes but most have a purely *optical* rather than a *neural* basis. (Occasionally following squint surgery, during a period of spontaneous conversion from anomalous to normal retinal correspondence, there will be a few days of *"neural"* monocular diplopia. However, generally, the diplopia arises from a problem within the eye's refractive components.)

The commonest cause of monocular diplopia is probably *physiological.* Any "sharp" individual may notice this, wonder about it, and call it to your attention, though it usually will not bother him greatly. This type is due to a malalignment of the cornea and the lens so that their optical axes do not coincide — the lens being tilted somewhat. Remember that the human lens is not the simple structure that we usually describe, but a truly complex one composed of a number of layered, curved surfaces separated by different indices of refraction; a tilting on its axis may allow the edge of one of the refractive junction zones within the lens to straddle the pupil and create multiple retinal images. Typically, only 2 recognizable images are formed — the bright, primary one and a very faint and "ghostlike" secondary. This latter image is usually about 4—6$^{\Delta}$ offset superiorly from the main one. (Critical observation of a bright point or horizontal line target on a dark background will demonstrate this to you.)

Obviously, this same cause and symptom can be exaggerated if the lens is loose or partially dislocated, say, after trauma. In this instance, if the lens is also knocked slightly *backward* from its normal position, in addition to the possible diplopia, some *hyperopia* is induced too; If, instead, the lens falls somewhat *forward,* a myopic error will supervene. Furthermore, with either direction of lens shift, the more the lens is tilted, the greater the astigmatic error created (radial astigmatism).

Another extremely common cause of monocular diplopia is *early* cataract, (as in the case reported above). That diagnosis sounds as if it were a cinch to make, but apparently it is not since it is so often missed. Of course, when the lens changes are advanced, it is not difficult to associate the cause (cataract) and effect (diplopia). But very tiny cortical "bubbles" or a very slight increase in the lens' nuclear density can also cause a remarkable amount of diplopia and yet you might see nothing unusual during your slit-lamp examination. So, what *can* help you determine what's wrong? It is your trusty *retinoscopic reflex* which serves as a sleuthing beacon. It can provide the tell-tale evidence of an early cataract, or for that matter, *any* optical cause of the monocular diplopia. The reflex will show distortion much earlier and more predictably than will any other method of detection. So, use it wisely and learn to recognize small irregularities which even your "hotshot" colleagues might overlook.

Other possible etiologies for monocular diplopia are as follows: high astigmatic errors (with or without optical correction by lenses), keratoconus (even in very early stages), corneal scars, megalocornea, and spherophakia.

I trust you will now look less at the psyche and more critically at the retinoscopically created red reflex the very next time a patient complains of a superabundance of images! (*Article finis.*)

## The Lensometer

In clinical practice, somehow we have to be able to determine quickly the power, cylinder and axis of an "unknown" lens. I have already described the "hand neutralization" method and have alluded to a "computer" in the office which could take complex spherocylindrical lens combinations and produce a resultant without recourse to formulas or graphs — that "computer" is the lensometer. (This instrument of fantastic clinical utility can even measure the amount and direction of any prism which may be incorporated or induced by the lens).

The most clinically important information about any corrective lens is the distance between its back surface and its secondary focal point. This distance is the "back focal length" — its reciprocal is the *"vertex"* power (also called the "effective power").

The reason I stress the term *"back* focal length" is to contrast it with the "true" focal length, which is the distance between the secondary principal point and the secondary focal point. The position of the principal planes is determined by the "shape" of the lens (that is, the curvature of the front and back surfaces) — the more highly curved the surfaces (for *plus* lenses) the more anteriorly the planes are "shoved". (See next figure). In contrast, increasing the surface curves of a *minus* lens causes the principal planes to be pushed in the other direction — "out the back").

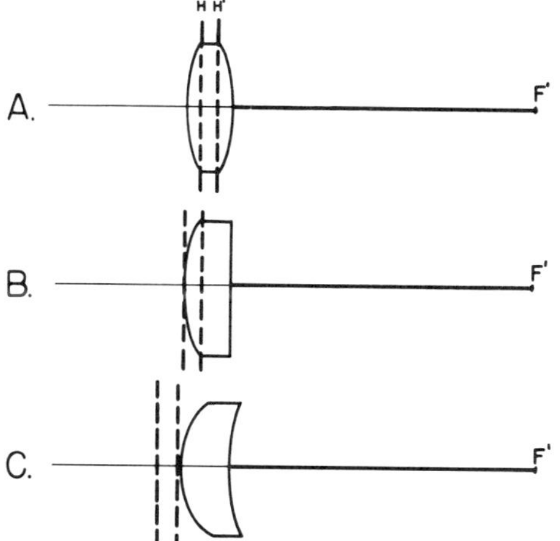


THE VERTEX POWERS ARE IDENTICAL; THE "TRUE" (EQUIVALENT) POWERS ARE *UNEQUAL*.

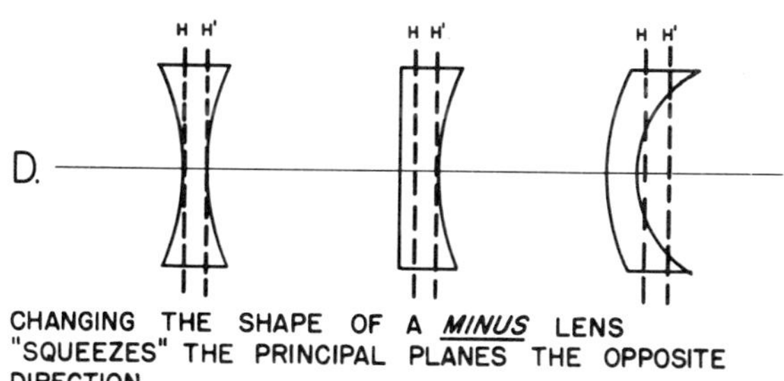


CHANGING THE SHAPE OF A *MINUS* LENS "SQUEEZES" THE PRINCIPAL PLANES THE OPPOSITE DIRECTION

Look at the three sample plus lenses above, A, B, and C. In each case it is the distance from H′ to F′ that is the *true* secondary focal length. This is longer in C than in A, and therefore, the *''true power''* would be greatest in A. Knowing the *true* power (also called the ''equivalent'' power) of a lens might be important in camera lens design, but for us, the important relationship is the *vertex* power. In the clinic, F′ must be capable of being pinpointed quickly, and relative to the readily accessible lens surface — *not* to the intangible principal plane. This is the task of the lensometer.

In each of the first three lens examples above, the distance from the *rear surface* of the lens to F′ is the same; thus, the "vertex" power of these lenses is identical — a lensometer would provide the same power reading for all.

Now, let's investigate how a lensometer *might* be constructed. The idea behind a simplified one is shown below:

## A SIMPLE LENSOMETER

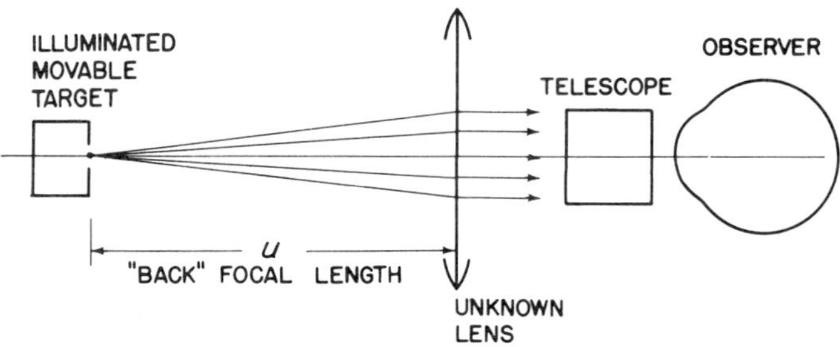


A grid-like, luminous target is moved back and forth behind an "unknown" lens until light rays leaving the lens are parallel (as detected by a small eyepiece telescope which is focused for infinity so that only parallel light bundles will appear to be clear). The target must then be located at F′ of the lens. Its distance *u* is simply measured and converted to diopters.

Now, what's so tough about that? Nothing, really, but this method is highly impractical — the instrument would have to be too long. (Think of trying to measure a lens of 0.25 D power — you'd have to have an instrument which was 4 meters long!) Another problem with this lensometer is that the dioptric scale would not be *linear;* a given, say 1 cm, movement of the target might correspond to 1 D for a short focal length lens (say a 10 D lens), but corresponds to only .01 D for a 1 D lens. Thus, our calibration scale (in diopters) would have to expand tremendously at the lower end of the power scale.

Both of these objections are taken care of by adding another fixed lens (field lens) into the system. It is carefully positioned (see diagram below) to take advantage of a neat optical trick called Badal's principle.

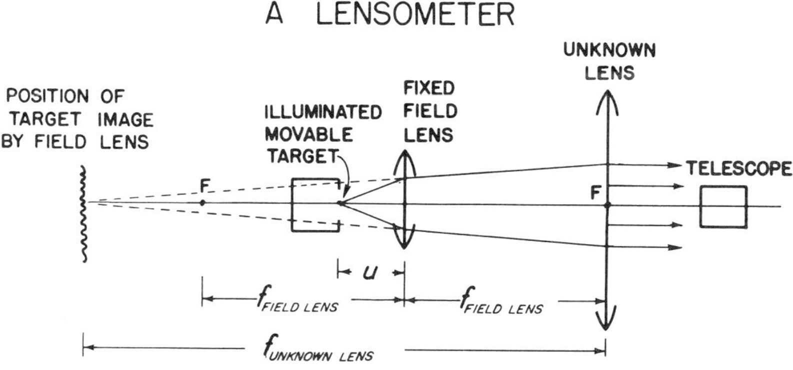


The fixed field lens is placed so that its focal point exactly coincides with the posterior vertex of the unknown lens. (The latter lens should be clamped in place to fix its position). The observer moves the illuminated target back and forth in relation to the fixed field lens until the reticle target appears sharply in focus through the eyepiece telescope. Then, parallel light bundles must be entering his eye.

Let's examine the diagram above to see what happens when the lens is so neutralized. The target is imaged by the field lens. Since the target is closer to this lens that is its own focal point F, a *virtual* image will be created over at the left as shown. That image position must also lie in the focal plane of the "unknown" lens since we know that the *final* image rays entering the telescope are parallel. Thus, adding a field lens to our lensometer allows us to condense instrument space; the field lens optically moves the target to a given distance instead of forcing *us* to move it there physically (as was required in our original, simplified instrument).

But the field lens does more than this: with the focal point of the field lens situated exactly at the posterior vertex of the "unknown" lens, the proper optical condition is set up so that distance $u$ (from the target to the field lens) is *directly* proportional to the power of the "unknown" lens. (This optical condition — Badal's principle — will be explained next). Because of this direct proportion, when we calibrate the lensometer scale to read directly in diopters, that scale will be perfectly linear — a given distance along it will always be equal to a fixed value of dioptric power, whether from 10 to 11 D or from 1 to 2 D — the scale will not shrink as higher powered lenses are tested, as it did in our first instrument example. The clinical lensometer is a beauty of simplistic design.

## *The Badal Principle*

First, a little background — when you are checking a patient's near point of accommodation, you usually ask him to read some fine print on a test card. You gradually bring the card closer, approaching the eye until he admits to a blurring of the print. Though this is practical for the clinical situation, it would not be satisfactory for accurate measurements in the laboratory, since, as the card approaches the eye, the angle subtended by the print at the nodal point of the eye grows larger. You are, therefore, making it easier for a patient to see that same print and less likely for him to recognize its initial blur.

There *is* a way to vary the accommodation stimulus yet still keep the (angular) image size on the retina constant. The optical principle which allows this to happen was discovered by Badal.

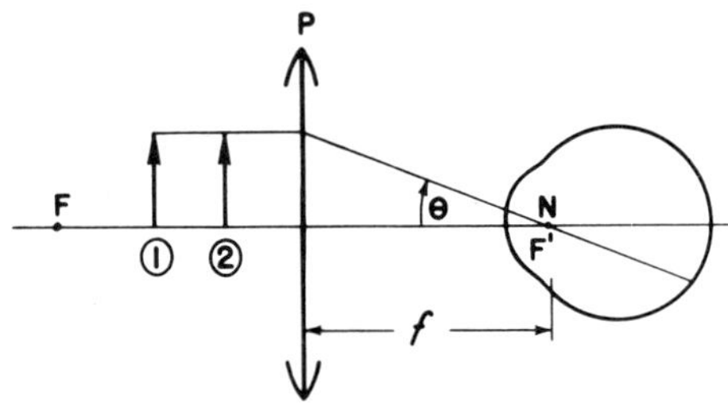

Place any plus lens in front of the eye so that the lens' F' coincides with the eye's nodal point. An object is put in front of that lens, but closer than its primary focal point F. The object tip (say, an arrow with its end on the axis) will emit *one* ray which is parallel to the axis. After refraction, that ray will pass through F'. Since F' was made to coincide with the nodal point of the eye, the ray will continue on undeviated. The angle at the nodal point sets the angular size of the image seen by that eye as $\theta$, no matter how far away that image actually is located. In the above diagram, object positions 1 and 2 both will create identical angular image sizes for the eye, although the image caused by the object at 2 will, naturally, be closer to the eye than the one formed by the object at 1. The image corresponding to the object at 2 will, therefore, stimulate accommodation to a greater extent. So you can see, with this optical system, the stimulation to accommodation varies while the angular image size remains the same.

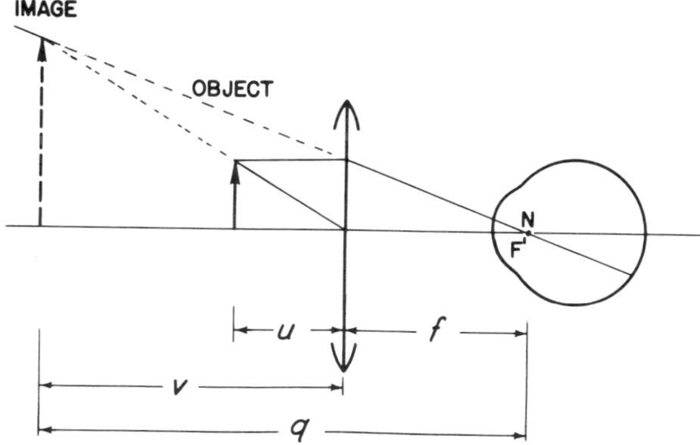


Pertaining to lens P above, the object is at distance $u$ and the image at distance $v$. Back again to $U + P = V$, but here, both *U and V* are divergent, and, therefore are negative.

$$-U + P = -V$$

$$-\frac{1}{u} + P = -\frac{1}{v}$$

When the two terms of the expression $(-\frac{1}{u} + P)$ are combined over one denominator $u$, we obtain

$$\frac{-1 + uP}{u};$$

So, $$-\frac{1}{v} = \frac{-1 + uP}{u}$$

Then, $$\frac{1}{v} = \frac{1 - uP}{u}$$

and $$v = \frac{u}{1 - uP}$$

Now, let's relate this information to the *eye.* The distance from the nodal point N to the image is $(v + f)$; let's rename this total distance $q$.

$$q = v + f$$

In this expression, substitute $\frac{u}{1 - uP}$ for $v$, and $\frac{1}{P}$ for $f$; we then arrive at

$$q = \frac{u}{1 - uP} + \frac{1}{P}$$

Combining, $$q = \frac{uP + (1 - uP)}{P(1 - uP)}.$$

In the numerator, $(uP)$ and $(-uP)$ cancel, leaving

$$q = \frac{1}{P(1 - uP)}$$

So, the *distance* $q$ from the eye's nodal point to the *image* is as follows:

$$q = \frac{1}{P - uP^2}$$

Then, the *divergence* $Q$ of this image at the eye is $-\frac{1}{q}$.

$$-\frac{1}{q} = -\frac{P - uP^2}{1}$$

Therefore $Q = uP^2 - P$

Let's take a specific instance with lens $P = +5$ D; when the object is 5 cm from lens $P$, $u = .05$ meters.

Then, $$Q = .05\ (5)^2 - 5$$
$$Q = .05\ (25) - 5$$
$$Q = 1.25 - 5$$
$$Q = -3.75 \text{ D}$$

Thus, when the object is 5 cm in front of the + 5 D lens, the image seen by the eye stimulates 3.75 D accommodation. This and some other values for *u* are plotted on the graph below, where *Q* and *u* are seen to have a straight-line relationship. Thus, we can say that *Q* is directly proportional to *u*.

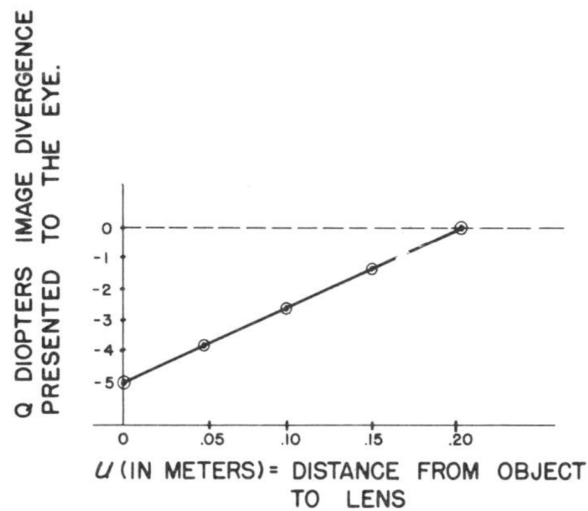


This relationship, the Badal principle, shows that the image divergence *Q* at the eye (that is, at F′) is directly proportional to object distance *u*, and, in the above example, F′ is located at N of the eye; but when we dealt with the *lensometer,* we positioned F′ of the field lens at the vertex of the "unknown" lens, and *u* was the distance from the field lens to the reticle target. It is clear that the two situations are exactly comparable, and now you can see what keeps the dioptric lensometer scale linear. Q.E.D.

CLINICAL POINT:

The exact method of using a particular lensometer should be ascertained from the instruction booklet accompanying the instrument. Usually, insufficient stress is given to the *proper* procedure for determining the power of a bifocal add; this is especially important when the add is incorporated into a high powered plus lens.

It is often stated that the correct method of reading the lens power of an add on the lensometer will depend on whether the add is on the front of the basic lens ("fused" type) or on its rear surface (Ultex type). This is only partly true. There is a more significant factor to be considered: any add, or for that matter, any *spectacle* lens — but *not* the theoretical *thin* lens with which we have been dealing — exerts a *different* dioptric power on object rays coming from a distant object compared to those arising up close! For example, a + 6 D spectacle lens has an effectivity of + 6 D on parallel object rays (with a vergence of zero) but only + 5.75 D on rays with a vergence of − 2.50 D from an object 40 cm away. This difference in power effectivity is related to the meniscus *shape* and the *thickness* of the spectacle lens. To demonstrate exactly why this is so, however, would require the use of "thick lens" formulas which I continue to ignore altogether.

As you already know, to obtain the basic *distance* power of any lens, you must use the lensometer to measure the *back* focal length; that is, an unknown lens must be turned so that its back surface faces *away* from the examiner. However, the lensometer can also be used to obtain a close approximation to the *near* effectivity (that is, near power) of this same lens if one measures its "*front* focal length"; this is because a spectacle lens' *front* focal length is a better indicator of the ability of a "thick lens" to deal with near objects. To determine this "front" focal length, you must turn the lens around in the lensometer lens clamp so that its back surface faces *towards* the examiner. The same procedure is necessary for measuring bifocal adds. (If the add on an aphakic lens is read with the lens in the "normal" instead of the "reversed" position, the indicated power will be too high in plus compared to its actual effectivity).

The correct way to find the power of any *bifocal* add on the lensometer is as follows: With the lens' back surface toward you, first read the total lens power through the add and then read the power *not* through the distant lens' axis but through a region comparably placed to the add, above the axis. (This allows you to compare powers as well as induced prismatic and distortional errors which *are* truly comparable between the two lens regions.) The two powers are then subtracted, the difference being the correct *add.*

What we have been speaking of is the add fused to the front lens surface; but even when an add is ground on the *rear* surface, its power should still be determined the same way as just described. However, for the rear add, it makes somewhat *less* of a difference.

Learn to read the bifocal add correctly.

## The Keratometer

This instrument for determining the corneal curvature is used mainly for the fitting of contact lenses. Its *modus operandi* is based on the ability of the anterior corneal surface to act like a highly-polished convex mirror and reflect light. The optical design of the keratometer allows a measure of the *size* of the reflected image and thereby, the radius of curvature of that anterior corneal surface can be ascertained. *That* is the key of keratometry.

As we learned when we studied reflection, the radius of curvature of a mirror will determine the *size* of the image it produces from a given sized object. So, if we illuminate an object of *known* size, place it a *known* distance from the cornea, and are able to *measure* the size of the reflected image, we can deduce the radius of curvature.

Since the cornea is such a high-powered mirror (— 250 D), an object does not have to be *too* far away to be effectively at the cornea's "optical infinity". (In the refraction section, I demonstrated this to you with an example). With the object at optical infinity, the reflected image will be situated practically in the corneal focal plane, positioned halfway to the center of curvature. This is shown in the diagram below where object O is at a relatively short object distance *u* from the cornea. Image I is located at just about the focal plane of the mirror.

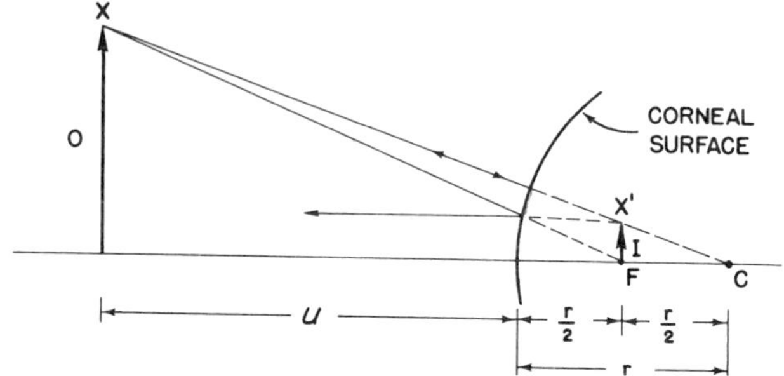


In the keratometer, the distance between the target object and its reflected image is fixed. This set distance is assured by the use of a small, fixed-focus, high-powered eyepiece telescope which will allow the corneal reflection to be in critical focus only when it is situated a certain, fixed distance away. The "object" is situated in the same mount as the telescope so when the image is seen in sharp focus, the distance between the object and image will always be the same. Thus, the distance from the object to the image $(u + \frac{r}{2})$ is constant (equal to some number K).

$$\frac{\text{Image size}}{\text{Object size}} = \frac{\text{Image distance from cornea}}{\text{Object distance from cornea}}$$

$$\frac{\text{I}}{\text{O}} = \frac{\frac{r}{2}}{\text{K} - \frac{r}{2}}$$

Since K is quite large compared to $\frac{r}{2}$ ($\frac{r}{2}$ is only about 3.8 mm),

$\text{K} - \frac{r}{2}$ is approximately equal to K

So, $$\frac{\text{I}}{\text{O}} = \frac{\frac{r}{2}}{\text{K}}$$

$$r = 2\text{K}\frac{\text{I}}{\text{O}} = \frac{2\text{K}}{\text{O}} \cdot \text{I}$$

Since distance 2K and object size O are known and fixed, the radius of curvature $r$ is exactly proportional to I, the size of the corneal image. Find I, multiply it by a known constant, and you have what you are looking for — *r*.

The trick is how to *measure* image size I. It is usually quite small, but that is not the problem since we could magnify it optically. If we were trying to find the *r* of a comparably-sized steel ball, we'd have no difficulty measuring the size of the reflected image, since the ball will remain nice and still. Not so with the eye which is constantly moving about; trying to get it to hold still so that we might measure the size of the corneal reflection would be one mean chore. We need some help. Aha! enter the principle of "doubling".

## *"Doubling"*

Let's demonstrate "doubling" using a biprism:

BIPRISM

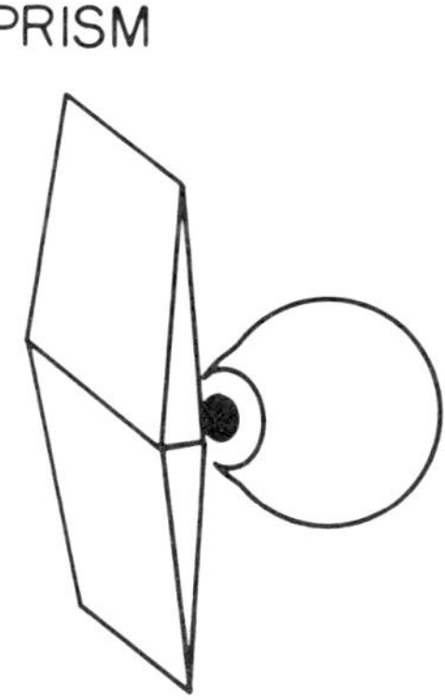

Look in your trial lens set and you'll find a "lens" which has two small $5^{\Delta}$ prisms mounted together in the same "mount" — their bases are in contact along a central line. Through such a biprism (while holding it so that the baseline is horizontal and "splits" the pupil) look at a distant, horizontal line. You should see *two* horizontal lines — one projected upward $5^{\Delta}$, the other downward $5^{\Delta}$.

CLINICAL POINT:

For the "squintologists" among you, if at the same time that the biprism is used in front of one eye the *other* eye looks at the same horizontal line, the latter will see that line located *between* the other two. Since binocular fusion has now been disrupted, if there is any tendency to a "cyclophoria", the central line (or the outer *pair* of horizontal lines) may appear tilted. Thus, any non-parallelity among the lines may indicate that a cyclophoria is present.

* * *

So a biprism "doubles" the view. Continue to hold the biprism baseline horizontal but shift your attention to a *vertical* line, say 10 cm long. You should see two vertical lines superimposed to a varying degree which depends on how far away you are from them; they could be quite separated if you are far enough away. As you approach the line, the two image lines will come closer together. When they just touch, you have "arrived"; you have created an optical condition that allows you to obtain an exact measure of the length of that line.

Let's reason this through:

Here each prism is $5^{\Delta}$ and will cause a displacement of 5 cm at one meter. So, at one meter distance, one prism displaces the line image 5 cm up, the other, 5 cm down. If the images touch, the *total* displacement must be exactly $10^{\Delta}$ or 10 cm at 1 meter, so the size of the object *must* be 10 cm. This is because "at touch" each prism must have displaced the line image a distance of one-half its actual length. So, if you know the distance from the prism to the object and you know the amount of prism causing the doubling, when the images touch, you can always learn the size of the original object.

The rule is as follows: the dimension (in cm) to be measured by the doubling device (there *are* others besides the biprism) is the product of the total deviation in prism diopters and the distance (in meters) between the device and the object to be measured. With the same biprism used above, if the "touch" of the inner extremity of the line took place when you were 40 cm away, the line would have to be $(10^{\Delta})$ (0.4 m) or 4 cm long.

The great advantage of "doubling", and this *is* the mechanism employed in the keratometer to measure the size of the image reflected from the cornea, is that any oscillation of the cornea during measurement will affect both the doubled images equally; that is, motion of the eye will not cause the *separation* between the "doubled" images to change. This allows you to twiddle the knobs to arrive at the "contact" position accurately and without difficulty.*

So now you know how the image size I is measured and how the corneal radius *r* is attained.

The keratometer knob of most instruments reads in diopters of *refractive* power. This requires an assumption (not a *measurement)* of some refractive index $n'$ for the cornea. Then, knowing $n'$ together with the radius of curvature, you can find the refractive power *P* through the relationship $(\frac{n'-n}{r})$. This is the power indicated on the calibrated dial.

This is how the keratometer works. Be gracious and toss a little credit to the fantastic genius Helmholtz who discovered all this for himself!

## *The Keratoscope (Placido) Disc*

This disc is usually a flat card or lighted source which presents a series of luminous, concentric circles. This disc is placed in front of the cornea at a 5 to 10 cm distance and the examiner views the reflection of these circular rings by the cornea through a small peephole cut in the center of the disc. There is usually (or should be) a

* By the way, the "doubling" principle is also used in the Goldmann applanation tonometer biprism, which optically creates the two "half-circles" you see from the "whole" fluorescein circle formed at the contact edge between the flat prism face and the corneal surface. If you increase the pressure of this prism against the cornea, you will increase the size of contact zone and thus the "half-circles" will enlarge in size. The biprism is built so that when the proper "contact" end-point is reached with this instrument, you will know you have flattened the cornea to a fixed diameter of 3.05 mm. *That* is the fixed size of the circle you are measuring with the biprism. The tonometer itself is calibrated to measure the force required to flatten the cornea to this fixed amount. It is this *force* (times area) that is the equivalent of the intraocular pressure you are measuring.

high plus viewing lens built into the peephole so that the viewer can get a very close look at the reflected rings without straining his accommodation too violently!

Corneal astigmatism, if present to more than a minimal degree, will show up as an elliptical, though regular flattening of the reflected circular object rings. Via this test, moderate to marked "with-the-rule" astigmatism would look to the observer as a series of horizontally elongated ovals. Irregular astigmatism, as that given by a conical cornea, will cause an asymmetry and distort the ring shapes out of perfect smoothness.

Keep in mind that to provide an *obviously* distorted image, the cornea must be quite distorted itself. If the distortion error is only a small one, while it may be terribly effective at reducing a patient's visual acuity, it is *not* likely to be visible by this crude test. *Much* more sensitive as an indicator of irregularity is the *retinoscopic* light reflex or even the "red reflex" given by an ophthalmoscope. (These were mentioned previously). Close observation of *these* reflexes (in contrast to those given by the Placido disc) will indicate even tiny surface or refractive disturbances of the cornea or lens. Remember, a distortion of the reflexes as seen through these instruments can be even a more sensitive indicator of trouble than a careful slit lamp examination of the refractive media.

You don't need a Placido disc to perform "keratoscopy". Every time you watch the reflection from the corneal surface of a moving hand-held flashlight — as when you are searching for a minor surface dent from a foreign body or a shallow ulcer — you are using keratoscopy. This flashlight method is just about as capable of indicating a corneal irregularity as is the Placido disc itself.

## *The Purkinje-Sanson Images*

The catoptric (formed by reflection) images from the major eye surfaces are known by the eponym Purkinje-Sanson (P-S). The four major ones are created by the anterior and posterior surfaces of the cornea and lens. Only the posterior lens surface is a concave mirror; the other three act as convex reflectors.

In an experimental setup, if one reflects a standardized object source from an eye which is changing its accommodative status, there will be a relative shift in the size and position of the P-S images. This experiment will show that it is mainly the change in the anterior surface of the lens that is responsible for the functional physiology of accommodation.

The P-S images also allow one to *measure* the curvature of all four of the eye's curved surfaces. You have already seen how the first, very bright P-S image — that formed by the anterior surface of the cornea — permits the measure of the corneal radius of curvature via the keratometer. But one can also measure the others. However, when doing so, you must take something "extra" into account; the sizes of the other three P-S images are dependent not only on the reflection occurring at the corresponding surface, but also on the *refraction* of that image by each of the surfaces anterior to it *and* on the re-reflection possibilities of each of the images!

The clinical importance of the P-S images is that the anterior one forms the basis for keratoscopy and keratometry, plus the flashlight reflection tests (Hirschberg and Krimsky) which can grossly indicate the alignment status of the visual axes of the two eyes.

# LIGHT: Physical Aspects

# LIGHT: PHYSICAL ASPECTS*

A few fundamental properties of light, usually described under a heading of *physical optics,* must be understood as a basis for comprehending many physiological phenomena of vision.

"Light" which effectively stimulates the visual cell is a form of electromagnetic radiation, a type of energy with a very wide spectrum. It is identifiable to us because of the specific effects it can produce. One end of this energy spectrum encompasses the portion which functions as a carrier of information — light, radar, television, broadcast, and power transmission; the other end, the very high energy gamma rays, x-rays, and cosmic rays.

Two separate ways of describing this electromagnetic (EM) spectrum have evolved:

(1) The *quantum* explanation treats the radiations as *particles of energy* with a characteristic frequency of vibration. When we are concerned with energy effects (absorption, heat, etc.), we almost always are concerned with *frequency* of vibration ($\nu$) since the *quantity* of energy is always directly proportional to it. (2) The *wave* explanation describes the radiations as *waves* of varying length. This analysis seems better to describe some observed phenomena — diffraction, interference, and polarization.

Thus, some phenomena can be interpreted more conveniently if light is considered in one category than in the other. The quantum theory provides a description of light emanating from an emitter or absorbed in a visual cell photoreceptor, while the wave theory beautifully describes light moving through the cornea and lens, or squeezing through an opening in a slit or pupil.

These two ways of looking at the spectrum are *not* mutually exclusive. Neither theory is wrong since both are necessary to explain the observed facts.

In a vacuum, the velocity of either the waves or particles of the EM spectrum is constant. The waves, frequency of vibration, and velocity

---

* Reprinted here through the courtesy of C. C. Thomas Publishers from Rubin, M. L. and Walls, G. L.: *Fundamentals of Visual Science,* C.C. Thomas, Springfield, 1969.

of propagation are all linked by the formula $c = \nu\lambda$, where

$$c = \text{velocity of light in a vacuum (cm/sec)}$$
$$\nu = \text{vibrations per second}$$
$$\lambda = \text{wavelength in cm}$$

This expression is closely allied to that tying together a more familiar relationship, the formula for determining how *far* a car travels in a given amount of time if it moves at a fixed rate:

$$\text{distance} = \text{rate} \times \text{time}$$
$$d = r \times t$$
$$r = d \times \frac{1}{t}$$

Since $\lambda$ (wavelength of the waves) is equivalent to d,

$\nu$ (frequency of vibration in cycles/sec) is equivalent to $\frac{1}{t}$,

and c (velocity or rate of travel) is r,

$r = d \times \frac{1}{t}$ becomes $c = \lambda\nu$.

This holds true for electromagnetic *waves* or *particles* traveling in a vacuum. The speed of light in a vacuum, denoted by c, is one of the most characteristic constants in nature and is approximately $3 \times 10^{10}$ cm per second.

When light enters a medium which is denser than a vacuum, say glass, the light slows down (i.e. its velocity ($V_m$) decreases) and the wave front direction is refracted (or bent). The exact amount of refraction is stated by Snell's law ($n \sin i = n' \sin i'$), the fundamental relationship of geometrical optics. (Recall pages 45-47.)

As stated, the velocity of light in a vacuum is $3 \times 10^{10}$ cm per second. The ratio of its velocity in a vacuum (c) compared to its speed in another medium ($V_m$) is called the refractive index (n) of the medium and is constant for any given wavelength.

$$n = \frac{c}{V_m}$$

Since $V_m$ is always $< c$, n is always $> 1.0$.

As an example, if a diamond has a refractive index of 2.42 for light of $\lambda$ 555 nanometers (nm), what is the speed of the light within the diamond for that wavelength?

$$2.42 = \frac{3 \times 10^{10}}{V_m}$$

$$V_m = \frac{3 \times 10^{10}}{2.42} = 1.22 \times 10^{10} \text{ cm/sec}$$

For every different frequency (or wavelength), there is a different index of refraction. Thus, if white light enters a prism, each $\lambda$ contained therein is refracted a *different* amount and a color spectrum becomes visible. The prism *disperses* the white light. The dispersive power for prisms varies with the type of glass used. (Remember that dispersion is the factor responsible for the chromatic aberration of lenses and of the eye.) However, the n associated with each $\lambda$ is not *linearly* related to the actual wavelength; usually, there is relatively more spread of blue wavelengths than the red ones. (More about this in the following discussion of diffraction gratings.)

It is clear now that when light passes into a denser medium, it slows down — $V_m$ has decreased. Since $V_m = \lambda\nu$, one or both of the variables ($\lambda$ or $\nu$) must also decrease. Which? Does the $\lambda$ shrink and become shorter or is there a decrease in its vibratory rate? Or both?

After careful experimentation, it was determined that light (and indeed all electromagnetic radiations) is best characterized by *frequency* of vibration; *this* unit is independent of the medium through which the radiation travels. Thus, the answer to our query — it is the *frequency* that stays constant, unchanging, and is the standard way to describe the electromagnetic radiations since *it* is independent of actual velocity in any medium; on the other hand, the *wavelength shortens* somewhat as light (of a given $\lambda$ in a vacuum) moves from the vacuum into glass, and lengthens again as it moves out from the denser medium.

We have already mentioned that the quantum particles vary in energy proportional to the $\nu$; this relationship is expressed by the formula $E = h\nu$, where h is called "Planck's constant." It is interesting to see how the energy of certain "rays" compare — say that of a gamma ray quantum and that of a red light quantum.

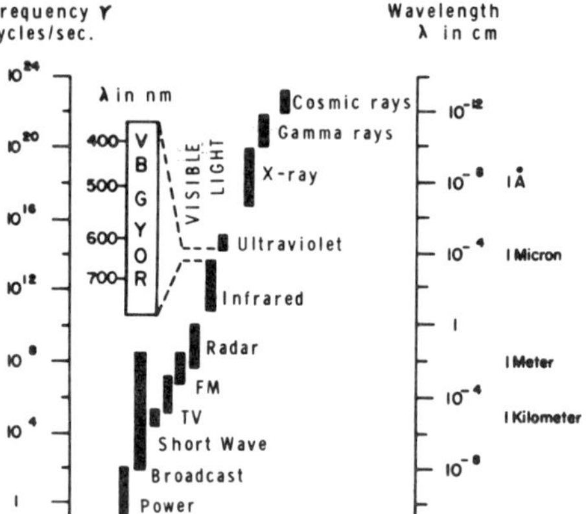


FIGURE 1. The electromagnetic energy spectrum. (Modified from Riggs*)

From Figure 1 (the EM spectrum), the $\nu$ of gamma is about $10^{22}$; that of visible red light, $10^{14}$: thus a gamma ray quantum has $\frac{10^{22}}{10^{14}}$ or $10^{8}$ (100 million) times more energy than a quantum of red light.

Quanta of *visible* light are called "photons"; here also, each particular photon (with its own frequency of vibration) carries with it a given amount of energy as noted by $E = h\nu$. From this formula, it is obvious that the shorter wavelengths of light (e.g., violet) have higher energy photons. Since violet light has a wavelength about one-half that of red light, violet photons carry about two times as much energy as red.

Photons can be produced whenever some outside source of energy knocks an orbital electron (of some element) from a shell of one energy level to another. The energy difference is released as a photon. A few practical illustrations are given below.

*"Luminescence"* is the production by some material of visible light of certain fixed characteristics when energy (or light of an "exciting" wavelength) falls on it. The color and frequency pattern of the luminescent light is "characteristic" of each particular material which possesses this property.

* Riggs, L. Chapter 1 in *Vision and Visual Perception,* pp. 1-38 ed. Graham, New York, Wiley, 1966.

We all know of gases which are used commercially to give a typical, colored luminous emission when an electric current is passed through the gas. Two common ones are *neon* and *mercury.* Here, the *current* electrons knock *orbital* electrons off the atoms into higher energy levels. The new position of these misplaced electrons is unstable; thus, the electrons jump back into their normal ''shells'' and release photons of characteristic *frequency* and, therefore, $\lambda$. Gases, thus excited, emit a *lot* of energy output at certain fixed wavelengths and little or none in between; this is called a *discontinuous* emission spectrum, in contrast to the *continuous* spectrum given off by heated solids which emit some energy at most visible wavelengths.

Other examples of luminescence include bioluminescence (as by fireflies), — fluorescent lights, — cathode ray tubes for television, and — electroluminescent panels.

''Luminescence'' is a *general* term signifying the absorption of *any* type of energy (chemical, electrical, etc.) by a substance which then gives up some of this energy as luminous energy. ''Fluorescence'' is a specific type of luminescence in which the ''charging'' (absorbed) energy is also light energy — usually ultra-violet, but not necessarily.

The material *fluorescein* exemplifies fluorescence. Clinically, fluorescein is used to demonstrate retinal and choroidal blood flow patterns in ocular disease diagnosis and for fundus photography. The ''excitation-energy spectrum'' of fluorescein dye in human serum centers about 490 nm; this means that light in the vicinity of this wavelength must fall on the fundus to cause the dye to ''light up,'' and it is *the* most effective wavelength to cause this fluorescence. In the physical process of fluorescing, the dye absorbs the energy contained by the 490 nm $\lambda$ and *converts* it to a characteristic spectrum which is centered about 530 nm. This yellow-green glow emitted by the irradiated fluorescein is what is meant by ''characteristic'' of that specific dye.

In contrast to the *electronic* activity yielding luminescence, *molecular* agitation (as by heat) emits luminous energy by ''incandescence.'' The total energy yielded varies with the temperature to *the fourth power.* The temperature is expressed in degrees Kelvin ($273^\circ$ + number of degrees Centigrade). Spectra emitted by hot solids or liquids are usually *continuous.* The hotter the solid (or more energy supplied), the more light of short wavelength is emitted.

Different substances vary in ability to radiate by incandescence. The "standard" is the theoretically perfect radiator which emits all energy imparted to it — the "black body." A curve of the emission of the black body at various temperatures is given in Figure 2. It can be seen that the hotter the black body, the more energy of short wavelength is added to the long wave emission already present, thus "whitening" it.

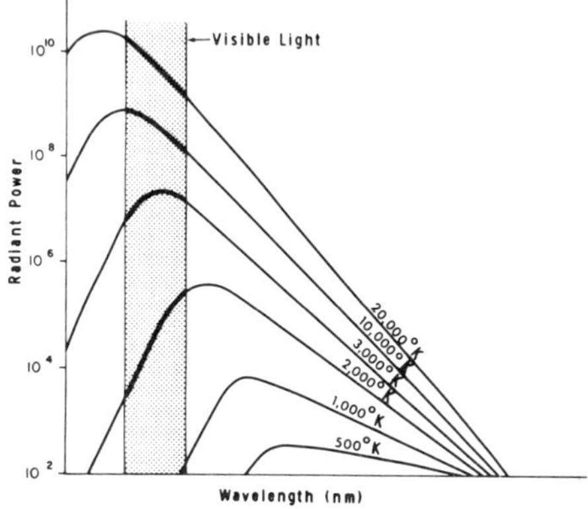


FIGURE 2. Radiated power by an incandescent source; graphs show the increase of short wavelength emission accompanying an increase in the temperature of the source. (Modified from Riggs.)

We are able to classify the spectral composition of a light source by stating only the temperature a black body would have to be heated to yield a given color of "white" light — we call that the "color temperature." A heated tungsten filament has a color temperature of 2700°K. This identifies the coloredness of the source as a yellowish white. If we place a blue filter in front of the source, we raise its color temperature toward a more blue-white *without* changing the *actual* temperature of the filament itself. The control of the color temperature of sources and illuminators is extremely important in color photography and color-vision testing.

We have just discussed a few of the phenomena concerned with the quantum nature of light. The phenomena of the EM spectrum which have to do with its *wave* nature will now be detailed. We will limit our discussion, for convenience, to that portion of the spectrum we call light — somewhere between 400 and 800 nm — but realize that this holds true for all electromagnetic radiations.

## Polarization

One property of the *wave* nature of light is polarization.

When a stone is dropped into a pool of water, waves seem to move outward in ever-widening circles in a direction along any radius. The wave motion itself, however, is up and down, so that a cork in the path of the wave will only bob up and down in a single plane. The motion of waves of a *light beam,* on the other hand, is not in a single plane. The wave motion can be considered to be taking place in all planes which include the line of direction of travel (propagation) (Fig. 3) and is said to be unpolarized. If wave motion, for one reason or another exists in only one plane, the light is then linearly (or plane) polarized.

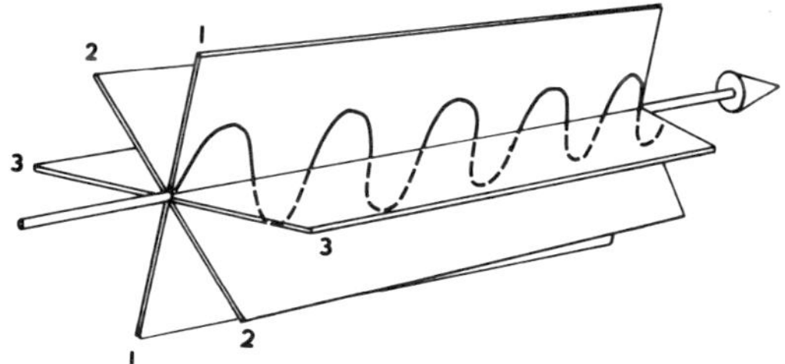


FIGURE 3. Nonpolarized light: The wave motion shown above in one plane is actually present in *all* planes encompassing the direction of propagation.

If polarized light is given the ability to cut paper, *plane* polarized light would cut a thin, linear slit on intersecting the paper; still other types of polarized light could cause an ellipitical or circular hole cut in the paper. Though these latter two types of polarization will not be discussed here, they are very intriguing; actually, circularly polarized light does have commercial significance as does plane polarized light.

In this next section, we will deal only with linearly polarized light and a number of ways it can be produced.

1. *Double refraction (birefringence).* Certain naturally occurring crystals such as quartz and calcite have the property of dividing a beam of unpolarized light into two distinct beams, each of which is equal in intensity and is plane polarized, the polarization of the two being 90 degrees apart. For most practical purposes, one of these two polarized beams is eliminated, such as in the Nicol prism; a very efficient linear polarizer remains.

2. *Reflection.* When unpolarized light strikes a flat, smooth surface at an angle, the reflected light will be either partially or completely linearly polarized. The degree of polarization depends upon the angle of incidence and the index of refraction of the reflecting surface. That incident angle at which the degree of polarization is complete is defined as the polarizing (or Brewster's) angle. You should be familiar with the fact that many glare-producing reflections from various surfaces are indeed polarized and can be eliminated by properly introducing a Polaroid* filter which will cancel and effectively remove the annoying polarized reflections.

3. *Scattering.* Any particulate matter, dust particles, or even molecules of a gas or atoms of a crystal, can produce polarization by means of light scattering. The best example of the production of polarized light by scattering is the blue sky light. A more thorough discussion of scattering follows shortly.

4. *Dichroism.* The large majority of commercially available synthetic polarizers are dichroic, such as Polaroid sheet. Dichroism is the property of a material to absorb light selectively to an extent which depends upon the polarization form of the incident beam. These polarizers have an axis along which maximum transmission of incident light occurs — the polarizing axis. If incident unpolarized light falls on such a dichroic polarizer, the light vibrations in planes which are oblique to the polarizing (transmission) axis are selectively absorbed, and only that light vibrating along the transmission axis is transmitted. In the production of polarized light by Polaroid dichroic sheet, about 40 per

---

* Trademark of the Polaroid Corp.

cent of the incident light is absorbed and results in a rather large light loss compared with "reflection" as a method of production.

If a second sheet of Polaroid sheet (called "the analyzer") is positioned with its transmission axis aligned perpendicular to the first ("the polarizer"), the plane polarized light transmitted through the first will be absorbed and will not be allowed to pass — in effect, it will be filtered out. This polarization principle is used to eliminate unwanted reflections as described above and also is used commercially in analyzing physical strains in glass and plastics. Polarization is useful in our study of vision to demonstrate the "Haidinger brush" entoptic phenomenon.

## Interference

The waves of two beams of light may interfere with, or add to, each other. If the trough of one wave train coincides with the peak of another, the result may be a cancellation of the intensity of light. Historical importance is made of this principle: Michaelson and Morley proved the absence of an "ether" carrier by their interferometer experiments and also helped to determine the speed of light with this instrument.

Practical use of interference is made in the optical coating of lenses. Troublesome surface reflections are greatly decreased by coating the surfaces of a lens with a very thin layer of material $\frac{1}{4}\lambda$ thick. The two reflections — one from the front and one from the rear surface of the coating layer — are bounced forward $\frac{1}{2}\lambda$ out of phase and thereby cancel each other (Fig. 4). Another practical use is the inter-

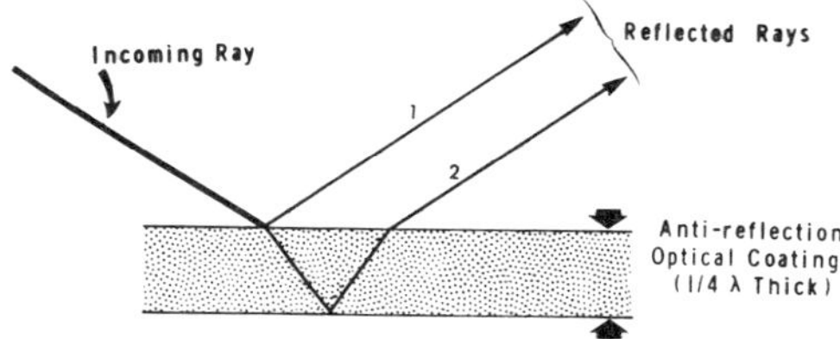


FIGURE 4. "Anti-reflection" coating utilizes the principle of interference of waves reflected from the front and rear surfaces of the coating one-fourth of a wavelength thick; thus, reflected ray 2 is one-half of a wavelength out of phase behind ray 1 and tends to cancel it.

ference filter, which is structured in such a way as to allow a very narrow band of wavelengths (even on the order of 1 nm) to pass. These are used in very exacting work with color stimuii.

## Diffraction

Diffraction is the ability of light to apparently "bend around corners" or to appear in areas of an image or shadow where one using strictly geometrical optics reasoning would not expect to see it (Fig. 5).

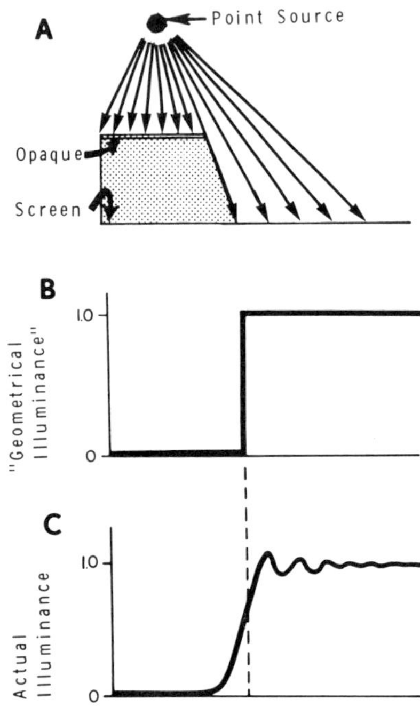


FIGURE 5. Diffraction of light by a "knife edge." *A.* Light from a point source is impeded by an opaque edge, the shadow falling on a screen. *B.* Graph of the theoretical, "geometrical optics" image illuminance on the screen. *C.* Graph of the actual diffraction illuminance on the screen.

The effect of diffraction is negligible if you consider *large* openings and distances relative to the minute wavelength of light; thus, if you are considering the *macroscopic* effects of light — the light-focusing effects of a lens system or the optical components of the eye, object-image relationships — these can be quite adequately described in terms of strict geometrical optics and the *refraction* of light based on Snell's law. However, when one considers the "microscopic" aspects of light (e.g., the finer details of retinal image illumination), account must be taken of *diffraction.* Even when light passes through as large an opening as the pupil of an eye, the retinal image is, in reality, the diffracted image of that source — light has to "squeeze" through the pupil before it falls on the retina.

The diffraction image of a small source of light has the appearance of a central "hot spot" surrounded by a series of bright, regularly spaced rings, each decreasing in intensity (Fig. 6). If the source is not a point, but an extended object, there is a diffraction image corresponding to *every* point making up the object.

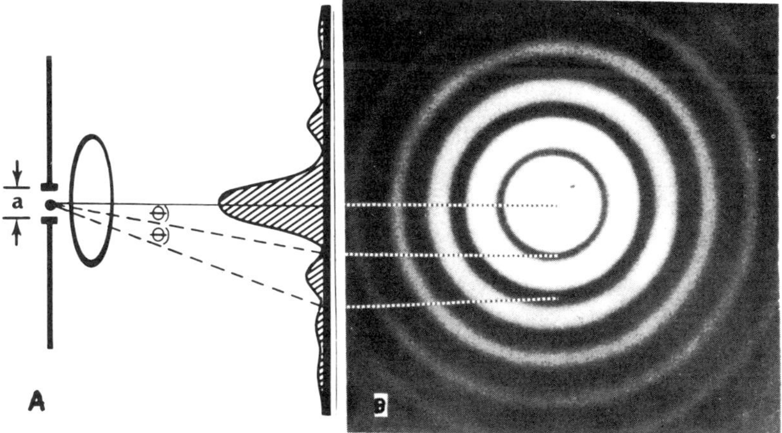


FIGURE 6. Diffraction by a circular opening. *A*. Graphical, intensity of diffraction pattern. B. Photograph of actual diffraction pattern. *a*, the width of aperture. $\theta$, the angular separation between the center of the "hot spot" and the first dark ring — also the angular separation of each of the succeeding dark rings. The total width of the hot spot is $2\theta$.

Diffraction can occur only when a light beam passes by an opaque object. Thus, we are constantly aware of its effects in any optical instrument, where light is almost always impeded by apertures. The diffraction pattern produced by them will determine the limiting sharpness of the image. With circular apertures such as the pupil, the angular separation $\theta$ (in radians) between the center of the "bright spot" and the first *dark* ring is equal to the angular separation of each of the succeeding *dark* rings (Fig. 6A).

$$\theta = 1.22 \frac{\lambda}{a}$$

Where $\theta$ = angular separation (in radians) at the center of the aperture
$\lambda$ = the wavelength of light
and a = diameter of the aperture (units same as $\lambda$).

The diffraction pattern is obviously larger with smaller pupil diameters and has visual significance with pupils less than about 2.5 mm. (Visual acuity is actually decreased somewhat when one looks through a 1-mm pinhole because of the diffraction effect caused by the pinhole. (See pages 187-8.)

Even with large openings or slits, if the edge of the light pattern in the image (or shadow) is closely examined under high magnification, the effects of diffraction can be noticed (See Fig. 5).

In contrast to a *circular* aperture, if one *slit* diffracts light to form an image on the screen, the angular separation of the lines visible will be $\frac{\lambda}{a}$ (where a is the slit width); here, the 1.22 is missing since it is a constant necessary only when dealing with *round* apertures. One can see, then, that for a given width of slit with incident *white* light (as occurs with a *prism*), the various contained wavelengths will be spread out, each a different amount; the shortest will be bent least, the longest, most. Note that this spread is in the *opposite* direction from prismatic dispersion; and here, the spread is *linearly* related to the actual $\lambda$. Individual wavelengths can be more efficiently spread out and intensified if *multiple* slits are utilized instead of a single slit. This is the so-called diffraction grating, where very fine and closely ruled lines are made in glass (10,000 to 20,000 in each inch). Here the effects of both interference and diffraction are made constructive, to work together in a useful instrument. Gratings are used for various optical purposes, such as in quantitative spectroscopy, where different metals are analyzed for chemical content.

## Scattering

Another interesting phenomenon which is related to both diffraction and reflection is called "scattering." You are familiar with the fact that light beams (when viewed perpendicular to the direction of travel) are invisible — light from a projector will illuminate a screen, but the space between the projector and screen will be dark unless there is some particulate matter present in the air to scatter the light away from its projected direction. The particles serve to make the light bundles "visible," (for example, in the familiar anterior chamber reaction of "flare and cells" in a slit lamp view of an anterior uveitis). Here the particles act like tiny mirrors. Thus, when the air is filled with particulate matter such as smoke particles or water droplets (which are *large* relative to the size of a wavelength of light), the particles act as diffuse reflectors and reflect all wavelengths about equally. On the other hand, if the particles are minute and compare in size with the wavelengths of light (such as molecules of gas in the earth's atmosphere), most of the incident light is transmitted and only a small portion is reflected; however, the quantity that *is* reflected is not the same for each wavelength along the spectrum. The intensity of the scattered light is inversely proportional to $\lambda^4$; thus, the blue (or shorter) wavelengths are scattered much more than the red, and light from a clear sky appears blue because it consists primarily of light which is scattered by the gas molecules of the air. With no such air outside our atmosphere, the "sky" will appear black to the astronaut since there is nothing out there to scatter the light and reflect it back towards his eyes.

To an observer back on earth, the sun appears to get more and more red as it approaches the horizon for the reason just described. At the horizon the sun's rays must traverse a much longer course through the earth's atmosphere to reach an observer's eye. Therefore, the shorter wavelength rays are more likely to encounter atmospheric molecules and thereby be scattered. As the sun's color loses more and more light of short wavelength, it first becomes yellow, then orange, then red.*

## Summary

In summary, for our purposes then, it does not make too much difference whether we consider light as particulate (in energy packets), or as a wave motion, or even as a change in an electromagnetic field — as long as we are familiar with the effects it is capable of producing.

* Many wonderful explanations of assorted visual phenomena which have their bases in the physical nature of light can be found in Minnaert's fine book, *The Nature of Light and Color*, Dover, New York, 1954.

# APPENDICES

# *APPENDIX A*

(from page 75)

## LINEAR MAGNIFICATION BY A SURFACE

On page 38, we saw that the linear magnification M of a thin *lens* in air was equal to $\frac{I}{O}$, to $\frac{U}{V}$, and only incidentally, to $\frac{v}{u}$. With a refractive *surface*, you should recall that the object and image vergences $U$ and $V$ must take into account the refractive indices of their respective spaces; that is, $U = \frac{n}{u}$ and $V = \frac{n'}{v}$. (These are called "reduced" vergences.) Thus, $M = \frac{U}{V}$ necessarily means $M = \frac{\frac{n}{u}}{\frac{n'}{v}}$, which is the same as $\frac{v}{u}$ times $\frac{n}{n'}$. So, M will equal $\frac{v}{u}$ for a thin *lens*, but for a refractive *surface*, $M = \frac{v}{u} \times \frac{n}{n'}$.

For completeness, there is another way of looking at the linear magnification of a surface:

For both the lens and the surface, $v$ and $u$ are measured from the axial vertex, (actually, from the principal planes which here are located *at* the vertex). But, if one is to determine the magnification of a surface geometrically, by using similar triangles, the *proper* distance measurements should *not* be to the object (and image) from the surface's axial *vertex* but from the surface's *nodal* point, i.e., the center of curvature of that surface. (It is only the ray through the center of curvature which is undeviated and allows you to construct similar triangles and thereby compare the object and image sizes "by proportion" — draw it out for yourself and see.) The center of curvature, of course, is located a distance $r$ from the surface's axial vertex; so, for a surface, M will not equal $\frac{v}{u}$, but instead, M will equal $\frac{v \pm r}{u \mp r}$.

To reiterate, when one deals with surfaces, to calculate the magnification using the *distance* formula, one must know how far *the center of curvature* is from the object and image. Since this reference position is an unorthodox one in our standardized way of denoting object and image distances, and since the results will be identical anyway, it is probably safer for you to remember the particular relationship mentioned in the first paragraph which always holds true for *both* lenses and refractive surfaces; that is, the linear magnification $M = \frac{U}{V}$. But, don't forget that each vergence term *must* include the appropriate index of refraction!

## *APPENDIX B*

(from page 111)

### LIGHT CONVENTIONS

By now, you should have acquired skill in handling the sign conventions for object and image distances and the radii of curvature for both refractive and reflective surfaces; therefore, I now feel comfortable in exposing you to a single, unifying concept dealing with the light convention which ties all these together! The basic key lies with the actual direction that light travels; all distance measurements — *u, v,* and *r* — are compared to this direction. In other words, these distances are not just absolute quantities; they are *vector* quantities — they not only show distance they simultaneously indicate direction. The vector (directional measure) — and this you must remember — always *begins* at the axial vertex of the refractive (or reflective) surface and *ends* at the object, image, or center of curvature. If that direction happens to point the same way that light actually travels (in the particular problem being considered), then the vector distance is *plus.* If it is opposite to the direction of light flux ("against the light"), it is considered a *minus* vector.

Let's give a few examples:

As discussed in the text previously, *any* reflective surface (mirror) will reverse the direction of object rays arriving *from* the left, so image rays will physically be reflected back *towards* the left. In reflection situations, we learned that the position of the image (given by distance *v*) is *plus* only when measured to the left of the surface. This was our dogmatically-stated convention and should be "old hat." Now you should also see that this fits in perfectly with our present unifying concept of *vector* (directional) distances. Since *v* must be measured *from* the mirror vertex *towards* the final image, *v* is plus since this is in the same direction that the light actually travels after reflection by the mirror.

On the other hand, if you consider a *refractive* surface, the actual light travel across that surface will always be from left to right and will never be reversed by the surface. Therefore, if *v* turns out to be *minus*, this would indicate the distance *to* the apparent (virtual) image — remember, it is to be measured *from* the surface's axial vertex — must be measured in the direction opposite to the light propagation; thus, *v* will here be measured to the left of the refractive surface.

The radius of curvature vector *r* fits neatly also into this same convention. Since light which is in the object space moves from left to right, (and *r* is always considered to be in the object space,) *r* is *minus* when representing a *concave* surface (reflective or refractive) as it is measured *from* the surface vertex *back* (opposite to the direction of light travel) to the center of curvature of that surface. So, any *concave* surface will always have a minus radius vector (and convex, a plus), as we have already pointed out (footnote page 71).

But try this: using our unifying vector concept, examine the sign of the primary focal length *f* of a *plus* lens. Remember (by definition) that *f* is an object distance (the image of F is at infinity), and being an object distance, it has to be measured *from* the refractive surface vertex *towards* F, the primary focal point. With a plus lens, this direction is "against the light flow," and so *f* of a plus lens must be *negative!* Surprised? Using similar reasoning for the vector distances, you will find that *f* of a *minus* lens is plus. Although these may seem like "flies in the ointment" — exceptions to our "unifying convention" — they are not. Let's probe a little deeper:

Remember? $U + P = V$

When $U$ is the object vergence of the primary focal point, the image vergence $V$ will be 0; so, $U + P = 0$.

Then, $U = -P$ and since $U = \frac{n}{u}$, $\frac{n}{u} = -P$

Thus, $u = -\frac{n}{P}$

Since here (by definition) the object distance $u = f$, $f = -\frac{n}{P}$; and when $n = 1$ (as it usually does), $f = -\frac{1}{P}$.

So, for a lens of *plus* power, $f$ will be a minus vector; and $f$ of a minus lens will be a plus vector. This is correct even though, until now, you *may* have assumed that $f$ of a plus lens was plus. Actually, when we discussed focal length in the early pages of the text, we tended to treat both $f$ and $f'$ similarly as regards their signs — the reciprocal of either was the lens power $P$ in air. For such a determination, we *should* have limited ourselves to working with $f'$, since *its* reciprocal is *the* proper lens power. So, now, if I tell you "the focal length" of some lens is —50 cm, you should automatically assume that I'm referring to the *secondary* focal length; then, the lens' power will be correctly stated as — 2 diopters. Had I introduced the fact that $f$ of a plus lens is a *minus* vector early in this course, you might have been confused. And even now, I only mention it for completeness. Remember, for all clinical purposes, treat $f'$, the *secondary* focal length, as *the* reciprocal of a lens' power.

# *APPENDIX C*

(from page 195)

## THE LITTLE POINT THAT ISN'T THERE*
## (or, THE CLINICAL INSIGNIFICANCE OF THE NODAL POINT)

A healthy 45 year-old prison guard complains of a gradual decrease in his visual acuity over the past year, but interestingly, only when working in the bright sunlight, as when guarding the prisoners chipping away at the local rock pile. Recently, his loss of acuity has been so great as to allow two convicts to escape under the cover of his dim vision! Needless to say, the warden is as unhappy as the patient, and both come pleading to you for help.

Surprise! A check of the guard's acuity in your refraction lane reveals 20/20 OU with his present corrective lens. "Malingering?" you ask yourself. No; a more careful check finds a substantiating acuity of 20/400 OU in bright light levels, and your slit lamp examination confirms your suspicions of small posterior subcapsular cataracts OU.

The question this case prompts is as follows: "Why are such small posterior cataracts so adept at knocking the acuity measurement for a loop?" I've asked this question of many groups at many teaching sessions, and nine out of ten individuals respond that the answer has something to do with the fact that the "nodal point" of the eye is located there. When I let out a little more rope and ask *why* they feel this might be true, I'm usually briefed on the obvious "fact" that all light rays forming the retinal image must pass through the eye's nodal point, so that even a tiny opacity situated there is particularly apt to cause a devastating loss in acuity.

Do you agree with this explanation? Do you also believe Lucy when she tells her little brother Linus that snow "grows" from the ground up, like grass, or, do you recognize these as mere coincidences "explained" by the embellishments of an over-active imagination? Really now, you're not as gullible and as easily swayed as Linus, . . . or are you?

---

* Reprinted from Rubin, M. L., *Survey Ophthalmology* 17:52-53, 1972 through courtesy of Williams and Wilkins, Baltimore.

I do not know where most refractionists have gained the impression that the nodal point's position is what makes posterior cataracts so potent. On top of that, where they "learned" the completely erroneous dictim that "all light rays cross the axis of any optical system at its nodal point" is completely beyond me. Both "facts," of course, are pure bunk. But if you think I am exaggerating how widespread (even sacrosanct) these beliefs are, you are greatly mistaken. (Ask your own buddies and you will find that I am *not* overstating my case.) So let's try to understand the straight scoop about nodal points and dispel their presumed relationship to cataracts.

First, you must understand that the nodal point is not an actual, physical spot within an optical system but only a very convenient optical reference position. The nodal point concept allows one to make geometrical constructions of object and image rays and makes it easy to calculate the angular sizes of objects and images. Think of the nodal point as a technical, schematic device, not a physically real location.

Actually, any optical system has two nodal points, a primary and secondary one. These two points are superimposed in very simple optical systems — such as a single "thin" lens but, they are separated in more complex systems (though only by 0.3 mm in the schematic model of the human eye). However, for most of our purposes in simplified optics, these two nodal points will be considered as one single point.

The nodal point is defined as that axial point at which the object and its corresponding image subtend identical angles. Thus, if one object ray happens to be drawn through the "nodal point" of the optical system, that ray will be continuous, *as if* it were undeviated by any of the optical elements of that system. This is the way one *draws* optical diagrams to help visualize, for example, the size comparison between an object and its image via the "similar triangles" route. But this is only a schematic representation, since any ray which enters an optical system *must* actually be deviated (refracted) by each and every element of it; that is, every ray which enters the eye is bent by each surface that the ray encounters, and so there is no *real* ray which is completely undeviated.

Even though the nodal point's calculated position can turn out to be located anywhere along the axis of an optical system, the location itself has absolutely no importance as a physical spot within that system. Consider a concrete example: I can take any simple trial case lens — say, one which is 5 cm in diameter — and drill out the center 2 cm. I am left with a ring-shaped lens having a hole in the middle. The nodal point of that lens is still optically located at its previous vertex, even though there is no lens material present there. In fact, it makes absolutely no difference to the image formed by that lens (as regards its *sharpness* or its *position)* whether there is a hole in the lens, actual lens material there, or even a 2 cm black paint-spot or cross covering the axial area. So, you must think of the nodal point as an optical simplification and not a physical reality; and wherever the nodal point diagrammatically appears to be, it makes no difference to the overall functioning of that optical system. Savor and digest that morsel!

Also, to dispel once and for all the second incorrect notion (that all image rays must cross the axis at the nodal point), you should look closely at any optical diagram. You will then see that the nodal point position is *not* a fulcrum for all image rays. In fact, only coincidentally will *any* of the infinite number of possible image rays cross the axis in that location.

Back to our patient, the prison guard. If it is not the position of the nodal point that is the basic cause of his problem, what is? Why does a cataract located in the axial position of the ocular lens, especially posteriorly, have such a profound effect on the retinal image quality? We will now learn that sometimes it does and sometimes it doesn't (just as in *this* case report), and that the key element here is the size of the patient's pupil.

Study the following sketch; please note that the pathways of the light rays as they course through this eye have been somewhat simplified (straightened and dotted) to demonstrate some basic optical principles.

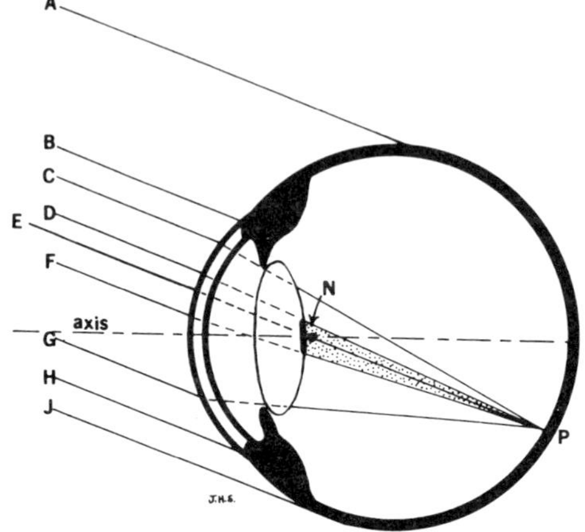


1. All the rays between A and J have been emitted by *one* single object point which is located at infinity; therefore, all the rays that reach the eye are essentially parallel. The object point is situated above the axis, and so these rays are shown to be inclined (arbitrarily at 20°) to the axis. Schematically again, the *one* object ray (E) which happens to be directed straight for the nodal point's position (N) is also inclined at 20°, and we say that the original object point "subtends 20° at the nodal point."

2. All rays located between A and B and between H and J certainly do not enter into the formation of the retinal image point since they are absorbed by the sclera. Those rays between B and C and between G and H, while they *do* get through the cornea, are absorbed by the iris and so also do not contribute to the final retinal image point. Only those rays passing between C and G enter through the pupil and thus may possibly help in the formation of the retinal image.

3. Prior to impinging on the cornea, one object ray (E), headed (at 20° to the axis) directly toward nodal point N, will, after final refraction by the posterior lens surface, have an identical image direction (20° to the axis) within the vitreous. This ray can be considered to be "undeviated" schematically; however, as we now know, in-between the anterior corneal surface and posterior lens surface, that same ray must

have been deviated by each of the refractive surfaces of the eye that it encountered. It is only for our simplified optical description that this fictitious dotted ray segment crosses the axis at the nodal point. (Of course, that dotted segment will connect the *real* incoming object ray and *real* outgoing image ray.) So we may *draw* the complete ray as if it were in a continuous straight, undeviated direction, and doing so enables us to locate the final retinal image point at P.

4. The image rays which pass into the vitreous will arrive at a *sharp* focus at point P since the eye is emmetropic, (and we are neglecting the effect of diffraction). Even if *some* of these rays happen to be intercepted first by a corneal, lenticular, or vitreal opacity, it will make no difference; there should still be a sufficient quantity of properly focused rays remaining to leave the sharp image point at P. (However, because some of the image-forming rays will have been absorbed by such an opacity, there probably will be a reduced retinal illumination at P, but this does not preclude sharp imagery.) And, even if a two millimeter dab of black paint were placed on the posterior vertex of the lens, as long as that dab was not large enough to block *all* the light flow through the lens, some light rays will continue to pass around "the opacity" and these will be sharply focused at P, exactly as if the opacity were not there.

It should not take too much insight to see that the situation is no different if, instead of a paint dab, a small lenticular opacity is situated near the posterior lens pole. It also can cause a marked loss of intensity of the retinal image point if the pupil is small, (that is, of comparable size to the opacity). If, on the other hand, the pupil is well dilated, some light rays will then be able to pass around the opacity; these might form a surprisingly sharp image. (Remember that our patient had 20/20 acuity with a dilated pupil!)

If the opacity is irregular and does not have a neat, sharp outline, and/or if it is not truly opaque but translucent, it may serve to "scatter" (in addition to "absorb") some of the intercepted rays. This *may* degrade the sharpness somewhat of each of the retinal image points which compose the overall retinal image thus blurring it. This scattering of light is another reason that whitish opacities located anywhere within the ocular light path can degrade a retinal image, especially in bright ambient luminance levels. But the point I wish to emphasize here deals not with the possible scattering by the opacity, but with the size of the pupillary opening itself. Here the pupil is *the* critical and limiting factor which either allows the opacity to effectively block the formation of an image point or renders the cataract relatively innocuous.

Consideration of this influence of pupillary size generates another clinically important point: if any patient with a cataract (like our prison guard) happens to maintain relatively good vision when the pupil is dilated, you might lay away your knife for awhile and delay operating on the cataracts. Keep the eyes chronically dilated with regular instillations of homatropine or neosynephrine eye drops, which work as a satisfactory temporizing tactic amazingly often. Try it; you'll like it — this simple expedient!

Now, you should have a firm handle on the simple optics of this situation and know that even though a lens opacity happens to lie close to where the "nodal point" of the eye is, this does not explain "why" the vision is so adversely affected. *The* important factor is nothing more than the size of the pupillary opening and its relation to the size of the opacity. So, keep up your guard and don't be teased by plausible "conventional wisdoms" which are being palmed off as "explanations." Fallacious refraction dictums can usually be refuted by careful probing with the tools of sound, basic optics and the application of cold, dispassionate logic.

# *APPENDIX D*

(from page 266)

## THE CASE OF THE MYOPIC CAPITALIST*

There is a gentleman in our city who has always felt that he has had good vision. He knew that his visual acuity was good since distant road signs were always easy for him to read, even when his wife was having difficulty. But now, at age 55, he noted a gradual loss of this ability. Lately he found he could not even see the signposts that his wife could read. It was time to seek ophthalmic help; Dr. Elsewhere was chosen to do the honors.

There was no history of distortion or excessive dimness of vision in either eye. He was last examined 3 years previously and the following lenses were prescribed:

OD —2.50 +0.50 ×90 +1.75 add
OS —2.75 +0.25 ×90 +1.75 add

with a resultant 20/20 visual acuity in each eye. At present a check of his visual acuity with his old lenses revealed a distance vision of only 20/80 OU. However, this vision could be corrected to 20/20 in each eye by OD —3.75 +0.75 ×85
OS —3.75 +0.50 ×105

Slit lamp examination with a dilated pupil showed some very minimal nuclear changes, but essentially a normal appearing lens, ocular media and retina. On the other hand, with the retinoscope, a moderate "scissors type" reflex was visible; the central portion of the light band reflection clearly moved at a different rate than the outer portion. (The central reflex seemed much more myopic in character than the peripheral portion.) A diagnosis was made of bilateral nuclear sclerosis with correspondingly induced myopia. (However rare it might be in this age group, keratoconus was looked for, but absolutely no corneal sign of it was evident.)

---

* Rubin, M.L. Survey of Ophthalmology, 13:122-23 (1968). Reprinted through the courtesy of Williams and Wilkins, Baltimore.

The lenses giving the best visual acuity were prescribed along with a +2.25 add. This near add gave him clear reading vision of Jaeger 2 type at 33 cm. (Dr. Elsewhere was very proud of the fact that he never overcorrected the near range of his presbyopes by giving too much add!)

One week later our friend excitedly obtained his lenses from the optician and drove home with them, teasing his wife about his newly regained visual acuity.

Following a wonderful celebration dinner, he began reading his evening newspaper and became mighty disturbed. While attempting to research his stock market accomplishments for the day, he found the eye strain intolerable. He just barely managed to read the quotations and, then, only with the help of a bright reading light. What had happened? The previous day, with his "old glasses," he certainly was able to see the quotes with absolutely no difficulty. Today, with his brand new glasses, he could barely wrestle with those same figures. Annoyance tore at his disposition. Those glasses set him back nearly three-quarters of a point, on top of the one-quarter which went to Dr. Elsewhere!

As a sleuth of world renown, I was approached by our capitalist friend, who strongly desired an explanation for his troubles. A check of his distance correction revealed that it was perfectly adequate; the base curve of the lenses were identical with his previous lenses. His previous add, which was only +1.75, was increased to +2.25. Why should he now have more presbyopic symptoms than before? With a "Sherlock Holmes" pipe clinched between my teeth, I smiled wisely. You see, the old distance correction was undercorrected by about —1.25 diopters in each eye; thus, a residual, uncorrected refractive error of that amount of myopia remained. Through his old bifocal of +1.75, our friend utilized his myopic error to his benefit and obtained an additional 1.25 D of "add"; this yielded him a full +3.00 D to read the fine print. Now, with his new glasses, he has only a +2.25 add available, and he certainly misses the additional 0.75 D help for close work.

Elementary, my dear Watson!

## MORAL

When you correct induced lenticular myopia with more minus power in the distance correction, try to ascertain the total amount of add that the patient requires to maintain his near point at about the same position as it was previously, especially if he was not unhappy with his near range. You may even wind up with an add of +4.00 diopters, or more! If you worry that giving such a high add will move his usable near range too close, remember that the patient has already adapted to it. At any rate, at least do *check* for his range of comfort with the add that you plan to prescribe, and find out how it compares to what he is used to.

# *APPENDIX E*

(from page 269)

## CALCULATION OF ACCOMMODATION

Although you have been shown how to determine the accommodation required of a corrected *myope* performing near vision tasks, you should know the *general* method for *any* object distance with *any* ametropic eye and *any* spectacle lens. This general, "common-sense" approach will now be covered.

We know that if any object is imaged by a lens at the far point of any eye, that eye will not have to accommodate to see the image. It really doesn't matter whether the original object is at infinity or closer. If it happens to be at infinity, the lens which places the image at the far point is, by definition, a *corrective* lens and, naturally, no accommodation is demanded; if the object is closer than infinity, an appropriate spectacle or contact lens (or any lens placed at *any* other position) may still create an image at the eye's far point plane, and when it does, no accommodation will be necessary to see it. Using this fact, it is very easy to determine the exact amount of accommodation required to see any object, whether the eye is ametropic or not.

METHOD: With any lens ("corrective" or not) worn somewhere in front of the eye, locate the exact image position that lens creates of any object. (As long as you know the object distance and the power of that lens, you can find the distance between the lens and the image.) Now, as far as the eye is concerned, this lens-created image is indeed an object and to find how much accommodation is required of any eye to see any object, you must first locate the position of that object relative to the eye; that is, find the object distance between this lens-created image and the anterior corneal surface. If this "object" lies anterior to the cornea, the object rays presented to the eye will be diverging, and the object vergence will of course be *minus.* On the other hand, should this "object" lie posterior to the eye, the object rays at the corneal reference plane will be converging, and thus a *plus* object vergence is presented to the cornea.

The accommodation required of this eye to clearly visualize this new "object" will be that amount of refractive power which must be added (at the corneal surface*) to the existing dioptric power of the eye. This *additional* power must create an image of that "object" precisely at the far point plane. (As long as the amount of ametropia or the position of any "corrective lens" is given, you will know the exact location of the eye's far point plane.) Again, the amount of *additional* dioptic power (*at* the cornea) *is* the required accommodation and can be optically represented by a fictitious "lens" situated at the corneal surface.

An example will help explain this method:

PROBLEM: A + 6 D "corrective" lens is worn 20 mm anterior to the corneal surface. A target letter object is located 25 cm in front of this lens. How much accommodation must the eye exert to see the object clearly?

ANSWER: First, find the object and image relationship relative to the corrective lens:

---

* In a schematic, "reduced" eye, we assume that all accommodative action takes place at the anterior corneal surface.

The object distance $u = -25$ cm

$$U = -4\text{ D}$$
$$P = +6\text{ D}$$
$$U + P = V$$
$$-4 + 6 = V = +2\text{ D}$$
$$v = +50\text{ cm}$$

So the image distance is + 50 cm (behind the lens).

Now, relate this to the eye:

Since the lens is 20 mm (2 cm) anterior to the cornea, the image formed by the lens is 48 cm (50 — 2) *behind* the anterior corneal surface, and this distance is now the object distance for the eye.

It is given that the + 6 diopter lens is a "corrective" lens, so we know its secondary focal point must be coincident with the eye's far point. Since F′ of this lens is 16.67 cm ($\frac{1}{6\text{ D}}$) behind it, and this lens is 2 cm anterior to the cornea, the far point of this eye must be 14.67 cm behind the cornea.

The accommodation required (since it is here considered to take place at the corneal surface) is simulated by a fictitious lens (P) at the cornea. This lens will move the object for the eye (that object being 48 cm behind the cornea) to the far point plane (14.67 cm behind the cornea). Thus $u = +48$ cm and $v = +14.67$ cm.

Since $u = +48$ cm, $U = +2.08$ D,

and since $v = +14.67$ cm, $V = +6.82$ D;

$P$ is the power of the fictitious lens representing the accommodation required.

$$U + P = V$$
$$P = V - U$$
$$P = +6.82 - 2.08$$
$$P = +4.74$$

Thus this eye will require 4.74 diopters of accommodation to see the object letters clearly.

So when *any* eye has *any* lens positioned in front of it, and that eye is asked to scrutinize an object at *any* distance away, you should now easily be able to determine the amount of accommodation demanded.

## POST SCRIPT

Well, here we are at the end of my exposé on optics. In it, I attempted to direct this subject to a very broad audience — to individuals of very diverse backgrounds and interests — the pre-clinical resident as well as the practicing clinician. For all, a strong foundation has been laid for the understanding of practical optics. But for some, especially for the ophthalmic "initiate" with an insufficient clinical exposure, this book may seem to be uneven; certain of the text's more clinical sections (especially those dealing with the clinical tests and clinical instruments) may appear too detailed. Yet, these same sections *will* be perfectly appropriate for one *with* such clinical exposure. So, if you are a pre-clinical resident and want to extract the maximum benefit from this manual, return to it again later after gaining at least *some* personal experience with refraction techniques and instrumentation. The second time through, you will likely only need to skim the basic areas for review; but the applied clinical optics will become much more vivid when viewed in the light of your own clinical experiences.

Now, all readers, please glance back at the "Table of Contents" and see what you've actually covered. However much that seems to you, please realize that there's more — much more than I could ever present here. I had no problem in deciding what subject material I had to *include;* my difficulty was in deciding what to *exclude* to keep this opus relatively short. I trust, then, you will understand that, in spite of the quantity of subject material presented, it is still but a drop in a vast ocean. No matter. If you only occasionally re-read the material given herein to maintain and strengthen your basic foothold, you will be well rewarded; you will have gained as full and enriched a background in practical optics as will ever be required of you as a clinician. Should you aspire to reach even greater mastery by consulting more profound books on the subject, I shall consider my effort as having been very successful and, "my heart will soar like a hawk".*

Shalom and good luck.

---

* Paraphrase — Indian Chief from "Little Big Man" by Thomas Berger.

## BIBLIOGRAPHY

Borish, I.M.: *Clinical Refraction,* 3rd Edition, Professional Press, Chicago, pp. 1381, 1970.

Davson, H.: *Physiology of the Eye: Visual Optics,* 2nd Ed., Little Brown & Co., Boston, pp. 378-477, 1963.

Duke-Elder, S.: *System of Ophthalmology,* Vol. V, *Ophthalmic Optics and Refraction,* C.V. Mosby, St. Louis, pp. 879, 1970.

Emsley, H.H.: *Visual Optics,* Vol. I, Hatton Press, London, pp. 454, 1955.

Epting, J.B. and Morgret, F.C.: *Ophthalmic Mechanics and Dispensing,* Chilton, Philadelphia, pp. 339, 1964.

Fonda, G.: *Management of a patient with subnormal vision,* 2nd Ed., C.V. Mosby, St. Louis, pp. 167, 1970.

Fry, G.A.: *Geometrical Optics,* Chilton, Philadelphia, pp. 290, 1969.

Jenkins, F.A. and White, H.E.: *Fundamentals of Optics,* 2nd Ed., McGraw-Hill, New York, pp. 647, 1950.

Johnson, B.K.: *Optics and Optical Instruments,* Dover Pub., New York, pp. 224, 1947.

Linksz, A.: *Physiology of the Eye; Optics,* Vol. I, Grune and Stratton, New York, pp. 334, 1950.

Pascal, J.I.: *Studies in Visual Optics,* C.V. Mosby, St. Louis, pp. 800, 1952.

Ogle, K.N.: *Optics — An Introduction for Ophthalmologists,* 2nd Ed., C.C. Thomas, Springfield, pp. 264, 1968.

Southall, J.P.C.: *Mirrors, Prisms, and Lenses,* 3rd Ed., Dover Publications, New York, pp. 806, 1933.

Stimson, R.L.: *Ophthalmic Dispensing,* 2nd Ed., C. C. Thomas, Springfield, pp 604, 1971.

While the following two books are not strictly "optics" references, they both represent information in closely allied areas and are aimed for the student who wishes to embellish his knowledge therein. Modesty keeps me from further elaboration.

Rubin, M.L. and Walls, G.L.: *Fundamentals of Visual Science,* C.C. Thomas, Springfield, pp. 435, 1969.

Rubin, M.L. and Walls, G.L.: *Studies in Physiological Optics,* C.C. Thomas, Springfield, pp. 125, 1965.

# INDEX